Advances in Neurobiology

Volume 27

Series Editor

Arne Schousboe, University of Copenhagen, Copenhagen, Denmark

Advances in Neurobiology covers basic research in neurobiology and neurochemistry. It provides in-depth, book-length treatment of some of the most important topics in neuroscience including molecular and pharmacological aspects. The main audiences of the series are basic science researchers and graduate students as well as clinicians including neuroscientists (neurobiologists and neurochemists) and neurologists. *Advances in Neurobiology* is indexed in PubMed, Google Scholar, and the Thompson Reuters Book Citation Index.

Gabriela González-Mariscal
Editor

Patterns of Parental Behavior

From Animal Science to Comparative Ethology and Neuroscience

Springer

Editor
Gabriela González-Mariscal
Center for Research in Animal Reproduction
CINVESTAV-Universidad Autonoma de Tlaxcala
Tlaxcala, Mexico

ISSN 2190-5215 ISSN 2190-5223 (electronic)
Advances in Neurobiology
ISBN 978-3-030-97761-0 ISBN 978-3-030-97762-7 (eBook)
https://doi.org/10.1007/978-3-030-97762-7

This Springer imprint is published by the registered company Springer Nature Switzerland AG
The registered company address is: Gewerbestrasse 11, 6330 Cham, Switzerland

Contents

Parental Behavior in Rodents

Mariana Pereira, Kristina O. Smiley, and Joseph S. Lonstein

Abstract Members of the order *Rodentia* are among the best-studied mammals for understanding the patterns, outcomes, and biological determinants of maternal and paternal caregiving. This research has provided a wealth of information but has historically focused on just a few rodents, mostly members of the two *Myomorpha* families that easily breed and can be studied within a laboratory setting (including laboratory rats, mice, hamsters, voles, gerbils). It is unclear how well this small collection of animals represents the over 2000 species of extant rodents. This chapter provides an overview of the hormonal and neurobiological systems involved in parental care in rodents, with a purposeful eye on providing information known or could be gleaned about parenting in various less-traditional members of *Rodentia.* We conclude from this analysis that the few commonly studied rodents are not necessarily even representative of the highly diverse members of *Myomorpha*, let alone other rodent suborders, and that additional laboratory and field studies of members of this order more broadly would surely provide invaluable information toward revealing a more representative picture of the rich diversity in rodent parenting.

Keywords Brain · Hormones · Lactation · Maternal behavior · Paternal behavior · Rodentia

---

M. Pereira
Department of Psychological and Brain Sciences, University of Massachusetts, Amherst, MA, USA
e-mail: pereiram@umass.edu

K. O. Smiley
Centre for Neuroendocrinology & Department of Anatomy, School of Biomedical Sciences, University of Otago, Dunedin, New Zealand
e-mail: kos24@cornell.edu

J. S. Lonstein (✉)
Department of Psychology & Neuroscience Program, Michigan State University, East Lansing, MI, USA
e-mail: lonstein@msu.edu

G. González-Mariscal (ed.), *Patterns of Parental Behavior*, Advances in Neurobiology 27, https://doi.org/10.1007/978-3-030-97762-7_1

1 Introduction

Rodents (order *Rodentia*) are the single largest group of mammals, constituting over 40% of extant mammalian species. This large order involves 29 families containing over 2052 species that utilize many different ecological strategies (Carleton and Musser 2005; Fig. 1). Thus, rodents have been successful in almost every niche, from the Arctic tundra to forests to deserts, and live on every continent except Antarctica. Many rodents live mostly in subterranean burrows (e.g., tuco-tucos, gophers, lemmings), but others are arboreal (e.g., porcupines, spiny rats, some squirrels) and live above ground on bare earth or under low shrubs (e.g., maras, pacas), and some are even semiaquatic (e.g., capybaras, beavers, nutrias). Rodents are predominantly omnivorous – foraging on nuts, seeds, fruits, tubers, corms and bulbs, leaves and grasses, bark, insects, snails, slugs, spiders, and fish – but there are herbivorous, insectivorous, and carnivorous species. A number of species do not drink water but instead obtain all their water from their food (e.g., mole rats, tuco-tucos). Most are nocturnal although some are diurnal in both the field and in the laboratory (e.g., Nile grass rats, degus). Rodent reproduction is often seasonal although some are capable of breeding all year round such as rats, mice, and lemmings (Gromov 2010; Hasler and Banks 1975; Potapov et al. 2012; Shilton and Brooks 1989). Other species, such as spiny mice that live in arid deserts, breed opportunistically in response to rainfall (Breed and Leigh 2011; Dewsbury and Hodges 1987; Laurie and Goodrich 1946; Phifer-Rixey and Nachman 2015; Sarli et al. 2016).

Rodents are also highly diverse in their social organizations. Some live naturally in complete solitude with males and females only coming together for mating (e.g., porcupines, some tuco-tucos, dormouse, pocket gophers, pocket mice, Syrian hamsters), whereas others live in family groups (e.g., beavers, capybara, spiny mice, some voles), in extensive colony "towns" (e.g., prairie dog), or in highly complex multigenerational families with individual members performing specialized tasks (e.g., eusocial naked mole rats and Damaraland mole rats). Rodents exhibit highly diversified reproductive strategies, from monogamy (e.g., beavers, California mice) to polygyny (e.g., Norway rats, house mice, squirrels) to promiscuity (e.g., white-footed mouse, meadow voles). Some monogamous species also form a lifelong social and/or sexual bond with their mates (e.g., prairie voles, agoutis). Lastly, several rodents have flexible mating systems that can vary between monogamy, polygyny, and promiscuity depending on environmental pressures (e.g., muskrats).

Rodents are thus incredibly diverse in many ecological, social, and life history factors. They are also undoubtedly the best-studied mammals for understanding the biological mechanisms underlying parental behavior. Despite their history as premier study organisms that have provided a tremendously rich and informative scientific literature on parenting, the far majority of studies have been conducted on just a few rodent species, most in the suborder *Myomorpha*. This is due to numerous practical reasons, including their ease of maintenance and breeding in captivity, short gestation times, large number of offspring per litter, widespread availability,

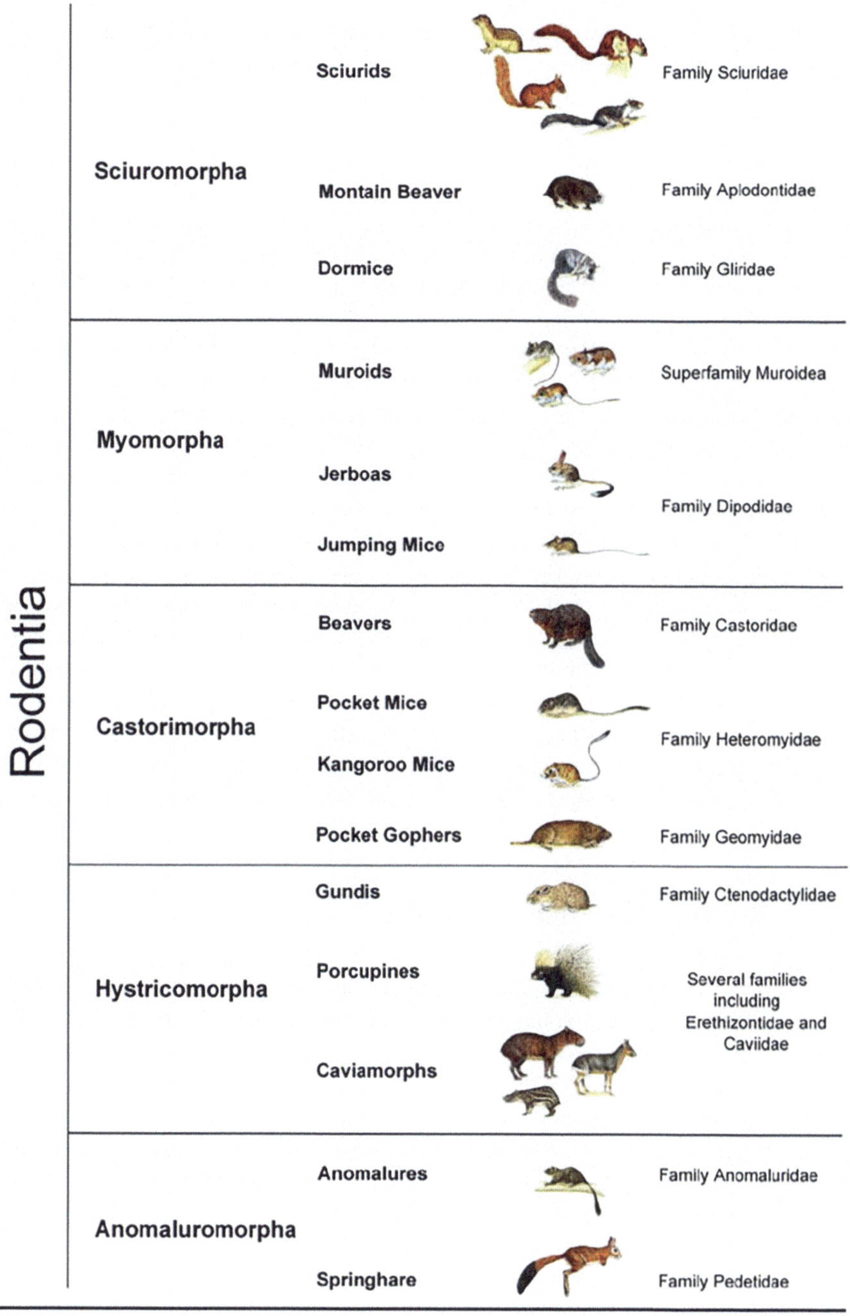

Fig. 1 The five *Rodentia* suborders with some examples of species in each. (Modified from Paco 2014)

and high social acceptance of their use in research. The result is that parenting in common rats (*Rattus norvegicus*) and mice (*Mus musculus*) has been studied in the laboratory for more than 50 years and comprises most of the existing literature. Of course, reliance on rats and mice as primary research models is not unique to the study of parenting, but it begs the question of whether rats and mice accurately represent most or even much of *Rodentia*, let alone parenting in other mammalian orders.

In most rodent species, parenting is sexually dimorphic with mothers carrying most, if not all, of the burden alone. Hence, most of what we know about *Rodentia* parenting is known from mothers. Information about paternal care is, of course, particularly interesting to many of us because human caregiving of infants often includes fathers. Although paternal care by males is displayed by a number of rodent suborders, including those containing California mice, porcupines, beavers, voles, and common degus, fathering has also been studied in only a handful of species. The goals of this chapter are to provide an overview of the patterns of maternal and paternal caregiving behaviors in some well-studied, as well as some not very well-studied, rodents. We further discuss the endocrine and neural bases of these behaviors, mostly known from laboratory rats and mice. We do not exhaustively review these large literatures because this has been done numerous times in great detail (Numan 2020; Lonstein et al. 2015; Gonzalez-Mariscal and Melo 2013). Instead, we wish to highlight some of the major advances and concepts from the scientific literature on common laboratory rodents thus far and discuss how this information is sometimes universal and sometimes exceptional compared to what little is known about less-commonly studied rodent species. This endeavor will provide insight into the divergent and convergent evolution used by *Rodentia* for this complex, motivated, and essential mammalian social behavior.

2 Patterns of Maternal Caregiving

In any animal species, maternal care provided to each member of a litter is not a one-way street (mother to young) but is instead regulated by the dyadic, reciprocal relationship between the participants. That is, the amount and type of mothering provided strongly depends on the immediate and near-future needs and capabilities of the offspring, which quickly change over their early life development. Maternal behavior first arises around the peripartum period and is maintained until the offspring reach independence. Many species such as squirrels, rats, and mice give birth to relatively underdeveloped or *altricial* young, while others like guinea pigs, porcupines, spiny mice, and pacas have *precocial* young that are relatively mature and mobile from the moment of birth (Fig. 2).

Altricial young are typically born without the ability to see or hear, have limited motoric capabilities, are small, have no fur or thermoregulatory capabilities, and are toothless so they rely on suckling for mother's milk as their only food source. In addition, rodent mothers often give birth to many altricial young simultaneously, usually

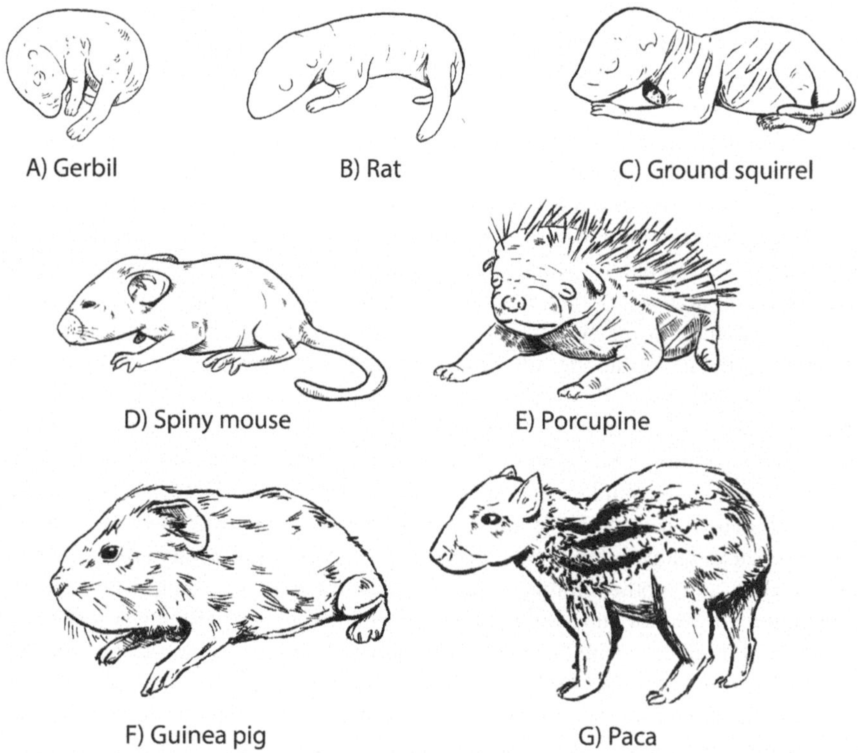

Fig. 2 Altricial rodents like the gerbil, rat, and ground squirrel newborns shown (**a–c**) are relatively undeveloped and are born without fur, teeth, thermoregulation, or much mobility. They can be compared to more precocial young, such as spiny mouse, porcupine, guinea pig, and paca neonates (**d–f**), that are variably more developed at birth. Images not to scale. (Figure created by Stephen R. Thomas)

around 8–12 or more (e.g., laboratory rats, hamsters, mice), but can produce as many as 28 in a single litter (e.g., naked mole rats; Sherman et al. 1999). Early postpartum maternal caregiving in these rodents is extremely time consuming, with mothers (and sometimes fathers and/or alloparents) spending the 70–90% far majority of their time in contact with their litters (Grota and Ader 1969; Jarvis 1981; Pereira et al. 2008). In contrast, rodent species that give birth to precocial young often have longer gestation periods (e.g., ~68 days in guinea pig; ~90–120 days in porcupines, pacas, agouties and maras; ~150 days in capybaras; and up to 283 days in pacaranas) compared with altricial species and give birth to smaller litters, often one to three pups at a time. Precocial young are often born with fully functional eyes and ears, are mostly furred, and possess advanced locomotor abilities. Young from several species of the infraorder *hystricognathi*, including gundies and pacas, are born extremely precocial ("superprecocial") with the ability to roam, eat solid food, climb trees, and/or swim within a day of being born. But regardless of the stage of development at birth, the young of almost all rodents still require substantial parental care.

Caregiving displayed by most rodent mothers can be categorized into numerous "active maternal behaviors" that involve high motoric activity and "inactive maternal behaviors" that do not. Active behaviors include building a nest where the mother and young spend almost all of their time together, particularly during the early postpartum period. Nests are built days before parturition in anticipation of the young's arrival and are usually located away from most conspecifics and within the safety of an unnoticeable and inaccessible niche (often underground but also in trees, rock crevices, and within dense vegetation on ground). Parents use their mouths, snouts, and paws to move dry vegetation and similar items into the nest for soft insulation and will sometimes bury the young under nesting material and/or plug the entrance to the burrow when they leave to forage (Butterworth 1964). Maintaining the nest continues after the pups are born, when other active behaviors have the opportunity to be displayed. These include moving the young from one place to another, which involves the mother gently picking up her offspring with her mouth and repositioning them within the nest, or transporting them back to the nest if they are displaced. While retrieval of young is very easily and thus often studied in the laboratory, it is probably quite rare in nature during the early postpartum period because relatively immobile, young pups cannot wander much from the nest on their own, and mothers rarely abandon a valuable nest site and move the entire litter in the absence of substantial environmental disturbance (Brewster and Leon 1980). After mothers ensure that any stray pups are returned to the nest, she joins the litter and hovers over them while licking and grooming them as well as herself (i.e., self-grooming). During this time of high maternal activity, the pups can be stimulated by her licking to excrete feces and urine (Friedman et al. 1981) and are behaviorally activated to search for a nipple and begin suckling. Lastly, heightened aggressive behaviors toward potentially threatening conspecifics or allospecifics emerge during lactation and serve to protect the tremendously valuable nest site and offspring within it (Bosch 2013; Lonstein and Gammie 2002).

Interspersed among periods of these active behaviors are prolonged bouts of inactive nursing. In fact, when tested under undisturbed conditions, nursing is the most frequently displayed maternal caregiving behavior in common laboratory rodents (Champagne et al. 2003 and 2007; Grieb et al. 2017). Nursing by the dam is induced by pup suckling, which causes mothers to transition from a highly active state to a state of relative quiescence. This quiescence maintains her presence over the pups in the nest and avoids their dislodgement from the teats. In some rodents and other animals, including humans, this suckling-induced quiescence involves periods of slow-wave sleep and is the only time when milk letdown occurs (Voloschin and Tramezzani 1979; Benedetto et al. 2017). Nursing in small laboratory rodents is displayed in distinct postures that can be measured through careful observation: (1) Kyphosis/upright crouching/arched back nursing – this nursing posture is characterized by complete immobility by the dam and standing over the litter with stiff, splayed limbs and high dorsiflexion of the spinal column. Kyphosis is thought to provide room for the litter to easily find a nipple and breathe while suckling and is the most common nursing posture during the early postpartum period in laboratory rats and probably many other small rodents (Stern 1996; Fig. 3); (2) lying prone on

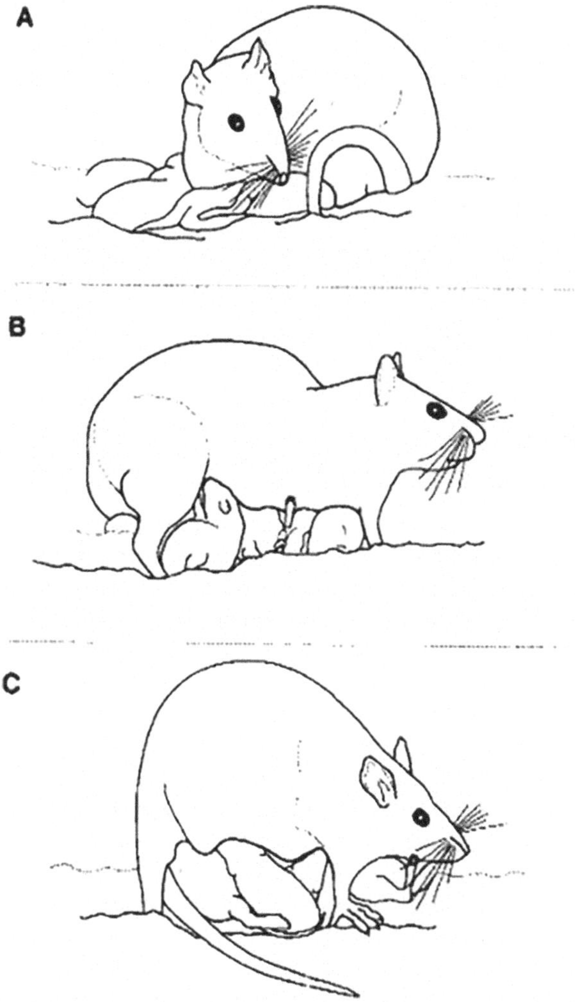

Fig. 3 Laboratory rat hovering over the litter while motorically active (**a**) or displaying a low (**b**) or high (**c**) kyphotic nursing posture. (Modified with permission from Stern and Johnson 1990)

top of the litter in a "blanket" position with little or no limb support; and (3) nursing while passively lying on the side in a supine position with the pups attached to the exposed teats. The latter two positions are more common during very long episodes of nursing probably because the rigid kyphotic posture is fatiguing to mothers. The latter two postures are also more common later in lactation when older offspring no longer easily fit underneath their standing dams and are motorically capable of suckling without her postural assistance anyway (Stern and Johnson 1990).

The particular behaviors detailed above have been best described for small laboratory rodents including rats, mice, hamsters, and voles. One could assume that other rodents that give birth to relatively altricial pups show the same or at least very similar caregiving behaviors. But other parental behaviors and patterns exist and may be species-specific even within *Myomorpha*. Most obvious may be that caregiving is likely to be more time consuming for *Myomorpha* mothers that produce altricial vs. precocial young. For instance, while rat mothers spend >70% of their time in the nest with their altricial litters during the first few days postpartum (Grota and Ader 1969) and lick them very frequently, guinea pig mothers show a more rapid decline in their time spent in contact with their pups and lick them less often (Hennessy and Jenkins 1994; Schiml and Hennesy 1990). Even within altricial species, there are differences in some details of maternal care. For instance, there is a seemingly unusual and stereotyped posture observed when spiny mice (*Acomys cahirinus*) mothers interact with their pups, involving the dam's mouth kept slightly ajar, which allows the pups to lick around her partially exposed teeth and gums. This activity is important for development of food preferences by the offspring (McFadyen-Ketchum and Porter 1989). And unlike laboratory rats or mice, prairie vole mothers rarely need to or can be induced to retrieve pups in the laboratory (Lonstein and DeVries 1999), presumably the consequence of their fairly precocial young being born with teeth that securely latch onto nipples even when the mother is highly agitated (McGuire et al. 2011). Also dissimilar to almost all postpartum laboratory rats and most postpartum laboratory mice, parturient hamsters are well known to kill and ingest much of their newborn litter, which helps regulates their energy balance while they lactate in their arid desert environment (Day and Galef 1977; Schneider and Wade 1989).

While *Myomorpha* is indeed the largest suborder of rodents and contains the well-studied common laboratory animals discussed above, most animals within the remaining four *Rodentia* suborders have been largely overlooked as models in which to study parenting behaviors. These other *Rodentia* suborders are the *Sciuromorpha* (squirrels, chipmunks, marmots, flying squirrels, dormice, among others), *Hystricomorpha* (porcupines, agoutis, pacas, guinea pigs, capybaras, tuco-tucos, chinchillas, among others), *Castorimorpha* (beavers, pocket gophers, kangaroo rats, and pocket mice), and *Anomaluromorpha* (African scaly-tailed squirrels, spring hares, and Cameroon scaly-tails). As discussed below, many of these animals have evolved unique adaptations to their environments and therefore would be of interest to study for their parental strategies. It is imperative to have a greater understanding of the behavior of these less- or rarely-studied species as the similarities and differences across rodents within and across suborders can inform us about divergent and convergent evolution of behavior and the underlying biological determinants.

Sciuromorpha

Suborder *Sciuromorpha* contains three families (*Aplodontiidae, Sciuridae*, and *Gliridae*) containing 307 extant species. One of these families, *Aplodontiidae*, has just one member – the mountain beaver (*Aplodontia rufa*) or "mountain rat." These are misnomers, though, as *Aplodontia rufa* are not beavers or rats. They are thought to be the most primitive rodent, having an extremely long fossil record (Feldhamer et al. 1999). Their natural habitat most commonly encompasses the densely vegetated, humid, and temperate Pacific Northwest of North America (Novak 1991; Feldhamer et al. 2003), where these nongregarious animals live alone, although their home ranges can overlap (Martin 1971; Nolte et al. 1993). These seasonal breeders synchronize reproduction with other members of their colony (Pfeiffer 1958) and give birth to two to six altricial pups (mean litter size = 2.8; Pfeiffer 1958) after a 28- to 30-day pregnancy (Pfeiffer 1958; Lovejoy and Black 1974). Dams begin nursing their pups within 20 min after parturition (Cramblet and Ridenhour 1956), and the young suckle for a relatively long 6- to 8-week preweaning period (Lovejoy and Black 1974). Mountain beavers have limited vision and ability to hear, which is unsurprising given their fossorial nature (they spend >90% of their time in underground burrows; Naughton 2012), but their tactile and olfactory senses are very well developed (Camp 1918), so presumably they are the most relevant for their maternal caregiving. This can be contrasted to laboratory rats and mice that readily use their altricial pups' vocalizations to guide their behavior (Wohr et al. 2010). Nothing appears to be known about the endocrinology of pregnancy or any details of postpartum caregiving behavior in *A. rufa*.

Family *Sciuridae* involves 279 species broken into five subfamilies that include tree squirrels, ground squirrels, chipmunks, prairie dogs, and marmots (including woodchucks/groundhogs). Many members of these polygynandrous rodents have two breeding seasons a year, winter and spring, although those that hibernate (e.g., ground squirrels) only breed once a year in the spring (Hayssen 2008; Thorington and Ferrell 2006). Detailed studies of maternal caregiving in *Sciuridae* have been conducted in a number of species. Similar to other rodents that give birth to altricial young, American red squirrel mothers (*Tamiasciurus hudsonicus)* have a short gestation of 35 days but then care for their pups (average of 3 per litter) for a long ~70 days of lactation (McAdam et al. 2007). Because tree squirrel nests (called dreys) are obviously elevated, there are few details known about what mothers do while with the pups, although it has been said that tree squirrel mothers have an exceptionally strong retrieval drive (Muul 1970). Captive southern flying squirrels (*Glaucomys volans)* will indiscriminately retrieve both related and unrelated young until late lactation, after which they will only retrieve their own pups and attack others (Muul 1970). It is also known that the time that red squirrel mothers spend with their young unsurprisingly declines as postpartum time advances (Dantzer et al. 2011). Detailed studies of maternal caregiving in ground squirrels are equally undetailed, as their underground nests are also not amenable to viewing, although captive Richardson's ground squirrels will quickly retrieve pups placed outside the nest box and abruptly stop doing so at 20–25 days postpartum when the pups are more mobile (Michener 1971). Postpartum captive 13-lined ground squirrels will also quickly retrieve young pups that are experimentally displaced (Droge 1976).

Chipmunks may be unusual in that a number of species are known to have both arboreal and subterranean nests, and some have been observed to interact with their young at both sites (Broadbooks 1977). Siberian chipmunks (*Eutamias sibiricus lineatus)* have a 30-day pregnancy and give birth to litters of four to five pups (Kawamichi and Kawamichi 1993). These hibernating rodents do not use their tree nests for parturition or early interaction with pups but remain underground. Early postpartum dams spend almost all their time in the burrow with the pups, but this naturally declines, and mothers then visit their offspring only a few times a day for up to about 60 days. They remain with them all night until postpartum day 35 though (Kawamichi 1989; Kawamichi and Kawamichi 1984). Siberian chipmunk mothers also change their burrows one or more times during lactation, including sometimes leading the mobile young from a burrow to a tree nest (Kawamichi 1989). Similar to chipmunks, we know little about maternal caregiving in prairie dogs. They co-nurse with related females, and alloparenting is thought to be common (Hoogland 1995). In a single mother that was non-systematically observed with her litter in a laboratory setting, she was seen to use a solid nest chamber that she did not line with available cotton, and she most often nursed in a supine position, used her mouth to put displaced pups back in the nest box (although pups were almost always attached to teats), and licked the pups when returning to them (Pizzimenti and McClaneghan 1974). Although nest building was not seen in this captive animal, female prairie dogs in a natural setting bring potential nest material to the entrance of a burrow before the pups are born and through lactation (Rioja-Paradela et al. 2008).

Family *Gliridae* include 29 living species of dormice, divided in three subfamilies. Some species are highly territorial and live solitary lives outside of mating. Mating couples may remain together and co-parent in captivity though (Nowak 1999). Dormouse mothers give birth to 2–10 pups after a typical gestation of 21–30 days (Nowak 1999; Ruf et al. 2006). The fat dormouse (*Glis glis*) mostly nests alone with her pups, although some choose to communally nest with close relatives, with rates of communal nesting varying dramatically by year and particularly likely at times of high population density (Pilastro 1992; Marin and Pilastro 1994). In one study, survival of fat dormouse pups to weaning did not differ between solo and group nesting however, so group nesting may benefit fitness of the mother's smaller, younger sisters who delayed breeding more than it does for herself (Pilastro et al. 1996).

Hystricomorpha

The suborder *Hystricomorpha* contains 19 families and 291 species within the infraorders *Ctenodactylomorphi* and *Hystricognathi*. The families *Anomaluridae* (African scaly-tailed squirrels), *Pedetidae* (spring hares), and *Zenkerelidae* (Cameroon scaly-tail) originally included in this suborder were recently reclassified to form the fifth rodent suborder, called *Anomaluromorpha* and discussed separately below (Carleton and Musser 2005; Coster et al. 2015). Rodents of the *Hystricomorpha* suborder appear to be more diversified in their reproductive strategies and social structure than any other rodent group. They range from being entirely solitary (e.g., porcupines, tuco-tucos) to socially monogamous (Dassie rat, some gundies, Old World porcupines) to highly eusocial (e.g., naked mole rats and Damaraland mole rats). The reproductive success of many Hystricognaths is enhanced by a postpartum estrus that enables females to be almost continuously pregnant. The infraorder *Ctenodactylomorphi* includes two extant families within the superfamily *Ctenodactyloidea*: the small *Ctenodactylidae* family (Gundis) of gregarious rodents from northern Africa and the *Diatomyidae* family represented by a single living species, *Laonastes aenigmamus* (Laotian rock rat), native to Laos in Southeast Asia. Gundis live in colonies of up to 100+ individuals in all manner of rocky desert habitats. Gundis breeding patterns vary from facultative monogamy to uni-male polygyny to multi-male polygyny (Nutt 2007). Female gundis have long gestations and give birth to one to three highly precocial young that within an hour of their birth are roaming, sunbathing, and feeding on chewed leaves (George 1978). The Laotian rock rat (also known as Kha-nyou) inhabits rocky limestone areas dotted by small patches of forest in Laos and Vietnam. They use small rock crevices for their dens. The Laotian rock rat is an example of the "Lazarus effect" as it was rediscovered in 2005 after a long gap of having been considered extinct (Dawson et al. 2006; Jenkins et al. 2005). Unfortunately, very little is known about their reproductive biology except that similar to other *Hystricognathi*, Laonastes females give birth to a single precocial young (Jenkins et al. 2005).

The infraorder *Hystricognathi* (true hystricognaths) is a large group of rodents, with 18 families and more than 230 species found primarily in South America, although some exist in North America, Africa, and Asia. This group is divided into two parvorders, the *Phiomorpha* and the *Caviomorpha*, and includes some of the largest rodents including porcupines (~25 pounds/11 kilograms in most males) and capybaras (often >100 pounds/45 kilograms). It also includes some of the most unusual rodents, such as the eusocial naked mole rats and extremely precocial pacas, as well as some of the most familiar rodents that are commonly kept as pets, including guinea pigs and chinchillas.

The *Phiomorpha* parvorder families *Hystricidae* (Old World porcupines), *Bathyergidae* (African mole rats), *Heterocephalidae* (naked mole rats), *Petromuridae* (Dassie rats), and *Thryonomyidae* (cane rats) are all found in Africa and Asia. Old World porcupines are monogamous and usually pair for life (Farida et al. 2019; Sever and Mendelssohn 1988; Morris and Van Aarde 1985). They have one to four young once or twice a year after a pregnancy of ~90–112 days. The young are born quite developed, and the spines that are initially soft harden within a few hours of birth. Although porcupettes begin to eat solid food within 2 weeks, they are not fully weaned until 13–19 weeks after birth. They remain with the mother until they reach sexual maturity at around 9 months to 2.5 years of age, depending on species, and share the burrow system with their parents and siblings (prickle) from other litters until dispersal (Mori et al. 2014). Fathers also share parental duties, such as cub guarding, and help defend the family group from intruders (Sever and Mendelsson 1988, 1989; Mori et al. 2016). The *Phiomorpha* parvorder is also striking because it contains the only two known mammalian examples of eusociality. While most mole rat species are solitary, the Damaraland mole rat (*Fukomys damarensis*) and the naked mole rat (*Heterocephalus glaber)* are eusocial, displaying a cooperative breeding system in which most members have specialized tasks and forgo reproduction. Naked mole rats are the most social of the burrowing mole rats and form eusocial colonies of up to 300 individuals consisting of a single breeding female (the queen), one to three breeding males, and sexually immature adults of both sexes (subordinates) that provide alloparental care by retrieving, licking, and huddling with the pups (Jarvis 1981; Lacey and Sherman 1991). This investment in alloparental pup care is enhanced in subordinate adults after the queen has given birth, which is partly mediated through consumption of the queen's estrogen-rich feces (Watarai et al. 2018). Breeding female mole rats can have as many as 28 altricial pups per litter after a long 60- to 80-day pregnancy (Sherman et al. 1999). The family *Petromuridae* consists of only one species, Dassie rats, that are endemic to rocky outcrops in southwestern Africa. Dassie rats (also called rock rats, but different from Loatian rock rats) are colonial, live in small groups of several adults and young, monogamous, and biparental (paternal care includes allogrooming and vigilance against predators). Reproduction is seasonal, during the spring months, with litters of one to two highly precocial young. Multiple generations remain in the parental home range. The greater and lesser cane rats are the only two living species of the family *Thryonomyidae*. They are found throughout Africa in marshy areas and along river and lake banks. Depending on the season, greater cane rats are solitary or communal. Lesser cane rats live in small family groups. Females give birth to litters of two to four precocial young once or twice per year after ~152 days of pregnancy (Stier et al. 1991).

The *Caviomorpha* parvorder unites all New World hystricognaths, including the extinct *Heptaxodontidae* (giant hutias) and 12 extant families. New World porcupines (family *Erethizontidae*) are arboreal and nocturnal. They are generally solitary (Mertz 2003), and females carry their young for a gestation period ~ 195–210 days, depending on species, and give birth to one to three precocial young at a time. Similar to the Old World porcupettes mentioned above, New World porcupettes are born with their eyes open, tail strongly prehensile, and claws well developed. At birth, they have soft quills, which harden in a few days. Porcupettes will stay with their mother for about 6 months, which is beyond nutritional weaning.

The family *Cuniculidae* (pacas) contains three species of ground-dwelling rodents that are the second largest rodent occurring in South and Central America. Pacas are monogamous and mate for life. Typically, mothers give birth to one to two offspring, following a gestation of ~114–119 days. Pacas are extremely precocial at birth, being born with fur and open eyes, and are able to run, swim, and eat solid food within a day later. Weaning begins after 90 days, but the young start to follow their mothers early and can do so for up to a year. Paca mothers mind their highly mobile offspring carefully. The young are kept in nests inside holes too small for both predators and the mother to enter, and the mother uses a low rolling vocalization to invite the young out of the hole for daily interaction and nursing (Lima et al. 2018).

The family *Dasyproctidae* (agoutis and acouchis) contains 13 species. Agoutis range from southern Mexico to northern Argentina and Paraguay and make their dens in burrows in stream banks; between boulders, roots, and bushes; or in tree hollows. They are group living and diurnal and are monogamous for life. After a long 104–120 pregnancy, they give birth to litters of one to two precocial young in which they invest a great deal of time and parental care. Young are well coordinated and able to run within an hour of birth. Acouchis live in the Amazon basin and exist in two species: red acouchi (*Myoprocta acouchy*) and green acouchi (*Myoprocta pratti*). Both are probably nonmonogamous and uniparental. Their gestation lasts ~ 99 days, and one to three relatively precocial young are born per litter. Even so, the pups remain sheltered inside a burrow until they are several weeks old. The mother returns to the nest burrow to nurse the young and produces a unique "purring" vocalization to help maintain contact with the offspring (Kleiman 1972). Weaning is extended, lasting 2–3 months, and young also remain with the mother for several weeks or months postweaning to benefit from her protection (Kleiman 1972).

The family *Caviidae* consists of ~23 species in five genera of rodents native to South America, including the largest extant rodent – the capybara – which approaches up to 200 pounds/90 kilograms. The family includes two subfamilies: the *Caviinae*, guinea pigs, wild cavies, and the capybara, and the *Dolichotinae*, the Patagonian hares or maras. They are unusually diurnal and are herbivorous. Capybaras (*Hydrochoerus hydrochaeris*) are semiaquatic rodents that present a complex sociality (Burton and Burton 2002). They are polygynous and live in long-lasting groups of 10–30 (and up to 100) individuals, including one male, several females that cooperatively care for and nurse all offspring, and one or more subordinate males (Herrera et al. 2011; Macdonald et al. 2007). Capybara gestation is ~130–150 days and usually produces a litter of four extremely precocious young once or twice a year (Ojasti 1973). Parturition occurs on land, and the young are fully mobile at birth and can follow the mother and eat grass but will continue to nurse for up to 16 weeks. The males do not usually provide paternal care but will tolerate the presence of the young (Ojasti 1973).

Wild cavies (rock cavy and mountain cavies) give birth to one or two young per litter several times per year after a pregnancy of ~75 days. The young are able to eat solid food within days of birth but continue to nurse for up to 7 weeks. The milk produced for these young is very dense in energy. This may be due to the lack of water availability in their environment. Both male and female rock cavies care for their young (Tasse 1986). Wild guinea pigs may have up to four litters throughout the year. They have an extended pregnancy (62 days) and on average one to two precocious young per litter (Kraus et al. 2005). Guinea pig females dedicate anywhere from 25 to 35 days to each litter; however, pups can eat solid food as early as 3 days after birth (Eisenberg 1989; Kober et al. 2008). Males invest little in the survival of their offspring but protect females and pups for a short period after parturition (Eisenberg 1989; Asher et al. 2004).

Maras are the sole extant representatives of the subfamily *Dolichotinae*, common in the Patagonian steppes of Argentina, but also live in Paraguay and Bolivia (Campo et al. 2020). Maras are monogamous for life but rear their young in communal warrens. The burrow can be shared with as many as 15 different pairs and their offspring. Because males are fiercely protective of their mates, only one mara pair occupies the burrow at any given time. They use a crèche system, where one pair of adults keeps watch of all the young in the crèche. Their pregnancy lasts around 100 days, and they have from one to three precocious offspring each year that can start grazing within the first day of birth. Offspring are born out in the open but are quickly transferred into a communal burrow. Females take on almost all direct care of their young, while males serve as sentries, protecting the den from potential predators. Young maras will nurse for much longer than most other rodent species, ~75 days, before being weaned. The mother will visit the den for about an hour at a time because she must split time with other nursing mothers utilizing the same den (Macdonald et al. 2007).

The family *Dinomyidae* now contains only a single living species, the pacarana, native to South America. They are nocturnal and vegetarian, live in burrows, prone to climb trees, and typically found in family groups of four or five. Very little is known about their reproductive biology, except that pregnancy lasts up to ~283 days and that females often give birth to two precocial young (Macdonald 1984).

Chinchillidae (chinchillas and viscachas) consists of six species in three genera, found along the central and southern Andes and throughout most of Patagonia in southern South America. Members of this family often jump bipedally, but mostly, they move on all four limbs. All species are colonial, living in groups that range from a few individuals to hundreds, and they are mostly monogamous. Females give birth two or three times a year to one to four precocial kits after a gestation period of about 3–4 months. Newborn chinchillas are born fully furred with a complete set of teeth, and their eyes open within 24 h after birth. Lactation can last up to 8 weeks even though the kits are able to eat solids from birth. Both parents tend to their young, and groups of chinchilla parents may communally raise their young (Macdonald 1984; Wilson and Reeder 1993).

The family *Abrocomidae* has only nine known extant species. Chinchilla rats (or chinchillones) are found in the Andes of South America from southern Peru to northern Chile. Their common name reflects the resemblance to chinchillas although they have a body structure more like a short-tailed rat. They are rock specialists that inhabit rocky areas and thickets up to around 5000 m in elevation. Very little is known about their reproduction. The length of gestation ranges from 105 to 118 days with one to three precocial young per litter (Taraborelli et al. 2015). Kits are born fully furred with a complete set of teeth and able to eat solids from birth, and their eyes open within 24 h. They are herbivorous and nocturnal.

The family *Octodontidae* is restricted to southwestern South America and consists of fourteen species in seven genera (degus, rock rats, and viscacha rats). This family is diverse in terms of ecology and morphology, with desert (*Octomys* and *Tympanoctomys*), semi-fossorial, and fossorial lifestyles (*Octodon*, *Aconaemys*, and *Spalacopus*). The best-known species is the common degu, *Octodon degus*, endemic to central Chile. They are diurnal and highly social and live in communally dug burrows. Group sizes range from 2 to 12 members, and females exhibit communal care (Hayes et al. 2009). Common degus are seasonal breeders, with pups born following a 90-day gestation in early to mid-spring. Litters contain an average of six pups, but size can range from 1 or 2 to up to 12 young. Common degu pups are born relatively precocial, fully furred, and with eyes open at birth and are able to recognize their biological mother at an early age (Fuchs et al. 2010). Male common degus take part in protecting and raising their pups (Aspillaga-Cid et al. 2021).

The family *Ctenomyidae* (tuco-tucos) includes ~60 species belonging to the only living genus of the family, *Ctenomys*, endemic to central and southern South America (Bidau et al. 2015). This genus is the second most diverse of the order Rodentia, being exceeded only by *Rattus*, with 69 species (Burgin et al. 2018). Tuco-tucos are diurnal and fossorial, spending up to 90% of their lives underground. Tuco-tucos occupy a diverse array of habitats, ranging from arid, coastal sand dunes to wet montane meadows. Members of the genus exhibit differing levels of sociality (Tomasco et al. 2019). Most species are solitary (e.g., *C. personi, C. mendocinus*), but some are semi-social (*C. rionegrensis*) or social (*C. sociabilis*). Tuco-tucos have relatively long gestations (~95 days) and give birth to one to three young per litter (Rosi et al. 1996; Pereira 2006). Newborn tuco-tucos are not as precocious as other neonatal hystricognath rodents with respect to the developmental stage at birth; their bodies are poorly furred, their eyes are still closed, and they most likely cannot thermoregulate. Solitary tuco-tuco mothers display a nonselective set of maternal behaviors, similar to that observed in traditional rodent species (Camín 2010; Pereira 2006). In the social tuco-tucos, all the adult females in a group are kin and nest communally (Izquierdo and Lacey 2008; Lacey and Wieczorek 2004).

The family *Capromyidae* (hutias) inhabit the Caribbean Islands/West Indies. This family contains about 26 species in eight genera, but most species are rare or extinct. Some species are terrestrial, while others are partly arboreal. Nocturnal hutias show variations in their social organization, with some species being solitary and others living in social family groups with communal nesting. The gestation lasts ~125 days, often producing one to six very precocial young (Witmer and Lowney 2007).

Echimyidae (spiny rats, coypu/nutrias) is the most diverse family of the South America hystricognath rodents with 78 species in 20 genera. They are found in tropical regions of Central and South America. Some are fully arboreal, and others are terrestrial, fossorial, or semiaquatic. Most species are solitary, but the broad-headed spiny rat (*Clyomys laticeps*) and the coypu (also known as nutria, *Myocastor coypus)* are colonial. Spiny rats live either individually, in small groups, or in large colonies. Little is known about the breeding habits of many species. In general, spiny rats breed throughout the year, and females give birth to four to six litters of two to four pups a year. Gestation period varies between 60 and 98 days. Coypus are semiaquatic and use a polygynous mating system within an organized social structure, containing about ten individuals, including adult females and males and their young (Gosling and Baker 1987). They are nonseasonal breeders with a gestation period of 126–141 days and average litter size of four to five. Coypus pups are precocial and born fully furred and with open eyes; they can eat vegetation within hours of birth. Mothers care exclusively for the young, and newborn coypus nurse for about 8 weeks, after which they leave their mothers (Gosling and Baker 1987).

Castorimorpha

The suborder *Castorimorpha* contains three families that include *Castoridae* (beavers), *Geomyidae* (pocket gophers), and *Dipodomyinae* (kangaroo rats and pocket mice). *Castoridae* consists of two species of beavers – the Eurasian beaver (*Castor fiber*) and the North American beaver (*Castor canadensis*). Both are nocturnal, semiaquatic (living near freshwater systems), and herbivorous and live in multigenerational family groups that can include a breeding pair, the current year's offspring (kits), and older adult offspring from the year before (yearlings) (Busher 2008). Beavers are monogamous, pairing together for up to 4–9 years, typically only repairing after the death of one of the partners (Busher 2008). Although they have been reported as genetically monogamous, with no evidence of extra pair copulations (EPC) (Syrůčková et al. 2014) in either species, other studies have found low percentages of EPCs (5–7%) in Eurasian beavers (Nimje et al. 2019). Nevertheless, both beaver species display qualities indicative of social monogamy including territorial defense, slow maturation of young (~2 years), and biparental care split relatively equally between mothers and fathers (Busher 2008).

Family *Geomyidae* is a superfamily consisting of seven genera and 35 species of fossorial (burrowing) gophers that have external cheek pouches known as "pockets" used for storing and transporting food (Russell 2012). Gophers are generally solitary except for breeding when males and females can sometimes be found sharing burrows (Davis et al. 1938; Smolen et al. 1980; Wight 1930). However, females are the sole providers of parental care and display maternal behaviors similar to other rodents that give birth to numerous, altricial young. The majority of maternal care involves positioning the young on their backs in order to nurse them, grooming pups, or moving pups further into tunnel systems if under threat (Hickman 1975). The breeding patterns and reproduction outcomes in gophers are quite variable (Andersen 1978; Barrington 1942; Brown 1971; Connior 2011; Hickman 1975; Smolen et al. 1980; Wight 1930; Wood 1949) though they have best been studied in the *Geomys* and *Thomomys* genera. From what we know, most gophers are capable of breeding year-round, but with one or two breeding peaks during the year, while some species show stricter seasonal breeding patterns, depending on habitat location. Variations in breeding times across the year are thought to be influenced by abundance of food, rainfall, temperature, and body metabolism (Wood 1949). Gestation lasts 18–51 days depending on the species, with some gophers having two litters in rapid succession (e.g., *T. bulbivorus*, *T. quadratus*; Wight 1930), whereas other species (e.g., *G. pinetis*) never show successive pregnancies, suggesting no postpartum estrus (Brown 1971). Litters generally consist of one to five pups, with average litter sizes of two to three.

Family *Heteromyidae* comprises three subfamilies including *Dipodomyinae* (kangaroo rats and mice), *Heteromyinae* (spiny pocket mice), and *Perognathinae* (pocket mice). Together, they contain six genera of burrow dwelling heteromyids generally found in the desert regions of North America but occasionally found in forested areas. Desert heteromyids share many similar features with other nocturnal rodents but also show unique adaptations to their xeric habitats. For example, they can go for long periods without drinking water and extract their required water supply through metabolism of their primarily seed diet (Reichman 1983). Heteromyids are solitary creatures that aggressively defend their mound, which provides a burrow for shelter, food storage, and reproduction, with some species only having one mound per animal and others owning up to five to seven mounds (Reichman 1983). Aggression between males and females ceases during female estrus in the breeding season (during long days in spring/summer), and mating pairs will sometimes cohouse in the same burrow for one night during this time, returning to separate burrows the following night (Eisenberg and Isaac 1963; Ostwald et al. 1972; Reichman 1983). Animals generally only produce one to two litters a year and have small litter sizes (two to three pups) that wean after only three weeks (Kenagy and Bartholomew 1985; Reichman 1983).

Several species of *Dipodomys* (kangaroo rats) and *Perognathus* (pocket mice) have been successfully brought into captivity and their reproductive behavior, parturition, and maternal behaviors are described in extensive detail (e.g., Butterworth 1964; Eisenberg and Isaac 1963; Hayden et al. 1966; Kenagy and Barnes 1984; Kenagy and Bartholomew 1985). In *Dipodomys*, females may assist with the birthing process by pulling at the fetal membrane, consuming the placenta, cleaning pups, and chewing the umbilical cord, while other species are more passively involved and allow the movement of the pups to break the umbilical cord and will cover new pups with sand to dry them off (Butterworth 1964). Similar to the commonly described laboratory rodents, maternal care involves picking up and retrieving the pups, either further into the burrow or, if the female possesses multiple burrows, moving pups from one to the other if under threat. Mothers show heightened aggression toward intruders, including foot stomping, a defensive behavior used to deter snakes from their territory (Randall and Matocq 1997). At three weeks of age, the mother prohibits nursing and juveniles disperse from the nest. While the harsh habitats and considerable challenges with maintaining captive colonies of heteromyids can prove challenging for research, many questions relating to ecological adaptations and parental care – such as how do desert dwelling kangaroo rats and mice maintain lactation when the conservation of water is so limited – could provide fascinating insights into what is likely a unique physiological maternal condition.

Anomaluromorpha

The suborder *Anomaluromorpha* contains nine living species in four genera and three families: *Anomaluridae* (African scaly-tailed squirrels, flying mice), *Pedetidae* (springhares), and *Zenkerelidae* (Cameroon scaly-tails). Little is known about the family life of these African rodents. The *Anomaluridae* are found in the forests of central Africa, and most species in this family are arboreal and can glide from tree to tree. Depending on the species, they either live alone with their young or live in colonies, often making dens in hollow trees or in nests made of leaves. Given that gliding is presumably influenced by weight gain of pregnancy, these and other gliders have fewer offspring per litter than do non-gliders (Fokidis and Risch 2008). The solitary *Pedetidae* springhares are found in semiarid regions of southern Africa (Peinke and Brown 2005). Both species of springhares (*P. capensis* and *P. surdaster*) are nocturnal and live in elaborate underground burrow systems. Springhares were recently reported to be biofluorescent, representing the first documented biofluorescence of an Old World eutherian mammal (Olson et al. 2021). Lastly, little is known about the reproductive behavior of the *Zenkerelidae* Cameroon scaly-tails as there are only few reported observations of this species (Adejumo et al. 2020; Dinets 2017).

3 Neural Basis of Maternal Caregiving in Rodents

In all rodent species, maternal behavior is essential for the survival and well-being of the offspring. Young have different needs as they grow and develop, and mothers coordinate and adjust their social interactions to match those needs (Grota and Ader 1969; Pereira and Ferreira 2006; Pereira et al. 2008; Reisbick et al. 1975; Rosenblatt 1975). Most of our knowledge on the neurobiological basis of maternal behavior has been obtained from *Myomorpha* species, mainly rats and mice.

Maternal behavior is a complex behavior that engages a wide array of cognitive, motivational, and affective functions and thus is necessarily supported by a distributed network of cortical, striatal, and limbic brain structures (Kohl et al. 2018; Lonstein et al. 2015; Numan 2020; Pereira and Ferreira 2016). These include structures such as the orbitofrontal, prefrontal, and sensory cortices, sensory and affective components of the amygdala, the bed nucleus of the stria terminalis (BNST), the medial preoptic area (mPOA), hypothalamic and brainstem areas, the periaqueductal gray (PAG), the nucleus accumbens, and its dopaminergic input from the ventral tegmental area (VTA) (Fig. 4). Some of these structures are considered critical for the onset, maintenance, and/or regulation of key aspects of parenting (i.e., recognition, discrimination of and attraction toward young-relevant stimuli, effort-related functions), whereas others regulate attention, memory, empathy, decision-making, and other processes (e.g., alterations in food intake, stress reactivity, and aggression) that are recruited to support parenting (Lonstein et al. 2015; Numan 2020). The nucleus accumbens receives convergent inputs from all these cortical and subcortical structures (de Olmos and Heimer 1999) and is considered a corticolimbic-motor interface within this network (Mogenson et al. 1980). As a major modulatory component of this circuitry, accumbens dopamine is considered to play a critical role in modulating motivational aspects of parenting (Champagne et al. 2004; Grieb et al. 2020; Hansen et al. 1991, 1993; Numan 2020; Pereira and Morrell 2011). The major target of nucleus accumbens medium spiny neurons is the ventral pallidum (VP), which itself subsequently sends GABAergic axons to motor-related structures, including substantia nigra pars reticulata, ventral tegmental area, as well as several hypothalamic and brainstem nuclei, to modulate motivational

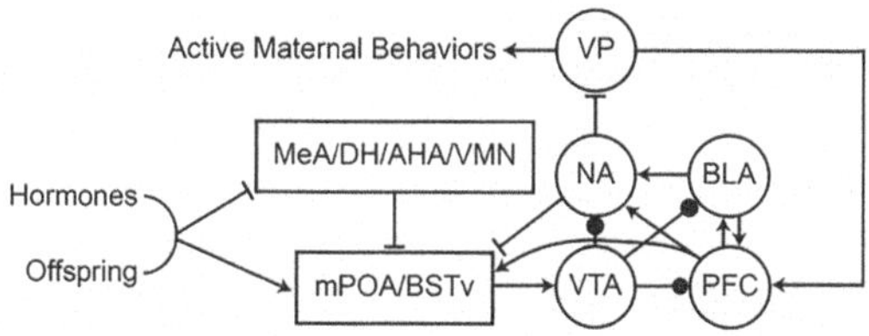

Fig. 4 Proposed neural model for the display of active maternal behaviors. The hormones of pregnancy and sensory cues of offspring suppress inhibitory input from the MeA/DH/AHA/VMN to the mPOA/BSTv while simultaneously stimulating mPOA/BSTv output to the VTA. This VTA activity elicits dopamine release in the NA, as well as the PFC and BLA. DA release in the NA inhibits VP output, thus promoting active components of maternal behavior including retrieval. The NA, VP, PFC, and BLA can modulate this pathway by their connections to the MPOA/vBST or NA. *AHA* anterior hypothalamic area, *BLA* basolateral amygdala, *BSTv* ventral bed nucleus of the stria terminalis, *DH* dorsal hypothalamus, *DA* dopamine, *MeA* medial amygdala, *mPOA* medial preoptic area, *NA* nucleus accumbens, *PFC* prefrontal cortex, *VP* ventral pallidum, *VTA* ventral tegmental area. Lines ending in arrows = excitatory input; lines ending in vertical bars = inhibitory input; lines ending in circles = DAergic signaling. (Modified with permission from Lonstein et al. 2015)

aspects of behavioral output (Groenewegen et al. 1993; Mogenson et al. 1980; Swerdlow and Koob 1987).

Neonates provide multisensory signals that are clearly communicative to mothers and facilitate approach and corresponding caregiving responses. Among the brain structures involved in parenting, the mPOA is considered a primary integrative locus, not only in the laboratory rodents that have been studied but also numerous other mammalian orders (Lonstein et al. 2015; Numan 2020; Pereira and Ferreira 2016). Functional inhibition of the mPOA results in robust deficits in caregiving behaviors in laboratory mice, rat, and hamster mothers (Lee et al. 2000; Miceli and Malsbury 1982; Numan et al. 1977; Pereira and Morrell 2009; Tsuneoka et al. 2013), as well as in the less-studied California mouse (*P. californicus*) (Lee and Brown 2002). The mPOA receives converging offspring-related information from multiple sensory modalities (Risold et al. 1997; Simerly and Swanson 1986). In rodents, olfaction is a crucial sensory modality for social communication and, in combination with auditory, visual, and somatosensory stimuli, facilitates maternal understanding of offspring's physiological and emotional needs that shape corresponding caregiving decisions (Elyada and Mizrahi 2015; Fraser and Shah 2014; Schiavo et al. 2020). Note that olfaction in some species, including laboratory rats, inhibits maternal caregiving, and the originally noxious odors emanating from the pups must undergo a change in valence (naturally occurring in response to pregnancy hormones) before mothers respond positively to neonates. Activity mapping by visualizing Fos expression and calcium imaging studies showed increased activity of mPOA neurons during maternal interaction with offspring in early postpartum laboratory mice, laboratory rats, and prairie voles (Fleming and Walsh 1994; Kalinichev et al. 2000; Katz et al. 1999; Kelly et al. 2017; Kuroda et al. 2007; Lonstein and De Vries 2001; McHenry et al. 2017; Tsuneoka et al. 2013). In addition, the mPOA is a primary neural site where pregnancy-related hormones, including gonadal and placental steroids, prolactin, and oxytocin, act to synchronize maternal responsiveness to offspring-related stimuli at parturition (Bakowska and Morrell 1997; Brown et al. 2017; Insel 1990; Mann and Bridges 2002; Numan et al. 1999; Pi and Grattan 1999; Scott et al. 2000). Furthermore, the mPOA has widespread connections throughout the brain, including projections to the VTA that allow for offspring-related information to regulate NA dopamine and modulate corresponding caregiving decisions (Fang et al. 2018; Kohl et al. 2018; McHenry et al. 2017; Numan and Stolzenberg 2009; Simerly and Swanson 1986; Swanson 1976).

Many of the mPOA projection neurons recruited during maternal interaction with offspring are GABAergic and target cortical, midbrain, and hypothalamic and brainstem structures to modulate cognitive, motivational, and affective aspects of parenting (Kohl et al. 2018; Lonstein et al. 2015; Numan 2020). Modern genetic tools have allowed understanding of the neurochemical modulation of mPOA output circuitry by identifying the neuronal phenotypes that are engaged during maternal interaction with offspring. For instance, mPOA to VTA neurons containing neurotensin and/or expressing estrogen receptor alpha have been shown to positively modulate social approach and pup retrieval behavior (Fang et al. 2018; McHenry et al. 2017). Other work shows that activating GABA cells in the mPOA inhibits anxiety and promotes positive responses to pups, while the opposite is true for activation of mPOA glutamate neurons (Zhang et al. 2021). This is consistent with studies showing that ablating a particular subpopulation of mPOA GABAergic neurons that also express the neuropeptide galanin produces severe deficits in maternal behavior, while activating them facilitates and/or increases maternal responsiveness (Kohl et al. 2018; Wu et al. 2014).

Many of these structures and connections undergo significant modulation and plasticity mediated by the interplay between maternal hormones and social experience while parenting develops and adjusts in new mothers. There is an increasing body of evidence showing that remarkable and long-lasting adaptations occur in the maternal brain, including adult neurogenesis, extracellular matrix plasticity (PNNs), dendritic/synaptic remodeling, and synaptic plasticity (Albin-Brooks et al. 2017; Barrière et al. 2020; Kinsley and Lambert 2008; Lau et al. 2020; Uriarte et al. 2020). The neuroendocrine events accompanying parturition promote modifications in the processing of behaviorally relevant cues within the maternal circuitry, ultimately initiating and coordinating a new mother's responsiveness to her newborn's sensory cues and signals that allow for rapid recognition, and facilitate early mother-young interactions (Fleming et al. 1993; Lonstein et al. 2015). Considerable research has highlighted the role of sensory neuroplasticity in the rodent auditory, olfactory, and somatosensory systems in the transition to an early development of motherhood (Liu et al. 2006; Marlin et al. 2015; Vinograd et al. 2017). For instance, in rats and mice, hormonal-mediated plasticity in olfactory systems mediates the critical changes in odor processing that alters the hedonic valence of offspring' scent, from aversive to highly attractive (Fleming et al. 1989; Kinsley and Bridges 1990; Vinograd et al. 2017). Thereafter, the interplay between hormonal effects and the sensory experience of interacting with the young, as they grow and develop, is instrumental to subsequently modulate the maternal circuitry to facilitate ongoing social interactions (Stern 1990, 1996; Stern and Johnson 1990). These alterations account for the remarkable behavioral plasticity of mothers.

4 Endocrinology of Maternal Caregiving in Rodents

The endocrine determinants of maternal caregiving in rodents are also, not surprisingly, best understood for laboratory animals such as rats, mice, and hamsters. This research demonstrates that the hormonal factors that establish maternal care at parturition are the same as those necessary for successful pregnancy and parturition. That is, the hormonal fluctuations across female reproduction that fulfill the peripheral requirements of gestating, giving birth, and lactating are co-opted to act on the brain to first strongly initiate caregiving at parturition and then continue to influence maternal behaviors through the preweaning period.

Some of the earliest work on this topic found that transferring blood plasma from recently parturient rats to nulliparous female rats could facilitate retrieval of pups in the latter (Terkel and Rosenblatt 1968, 1972). This suggested the presence of blood-borne substances, such as hormones, that were responsible for the onset of maternal behavior in naturally pregnant and parturient females. Plasma concentration profiles of the most relevant steroid (estrogens, progesterone), protein (prolactin), and peptide (oxytocin) hormones measured across pregnancy and parturition are now well characterized in rats and a number of other rodents; these profiles are quite similar among them, as well as similar to numerous non-rodent mammals studied. It is important to note the caveat that most laboratory studies of the steroid and peptide hormones involved in maternal caregiving involve primiparous female rodents that are not inseminated during the postpartum estrus. That is, the endocrine state most commonly studied for mothering in the laboratory is probably rare in feral rodents, which are more likely to be multiparous and gestating a new litter while caring for any current existing one. It is known that parity and simultaneous pregnancy and lactation affect the endocrine state and behavior of a number of rodents and other animals but are rarely studied (Uriarte et al. 2008, 2014; Pose et al. 2019; Del Ferkin 2006; Naguib et al. 2010; Nephew et al. 2010).

Steroid Hormones and Maternal Care: Estrogens and Progesterone

Circulating concentrations of *estrogens* are generally low for the first two weeks of the three-week pregnancy in laboratory rats after which estradiol levels rise considerably and peak soon after parturition, coincident with the postpartum estrus (Garland et al. 1987) (Fig. 5). These levels remain low until the middle of lactation

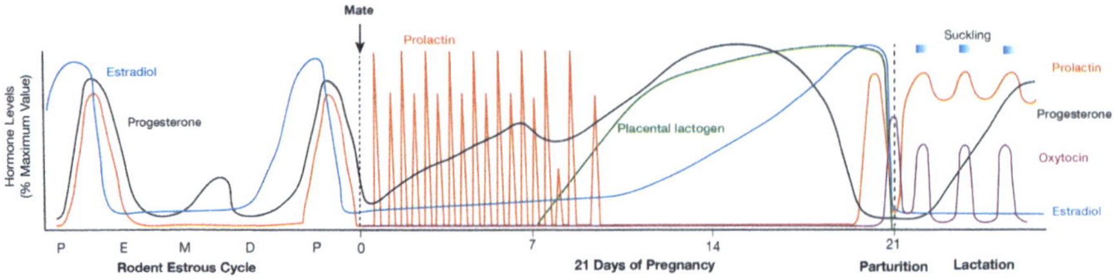

Fig. 5 Schematic representation of plasma estradiol, progesterone, and prolactin concentrations across pregnancy and peripartum period in laboratory rats. (Modified with permission from Smiley et al. 2019)

and then rise until weaning and resumption of cyclicity approaches (Taya and Greenwald 1982). Ovarian corpora lutea are the source of estradiol during pregnancy and lactation in laboratory rats (Niswender et al. 1994). The corpora lutea created at pregnancy is responsible for initial estradiol secretion although those created during the postpartum estrus take over during lactation (Taya and Greenwald 1982). During early pregnancy, the corpora lutea produce both the steroidogenic substrate for estrogen synthesis, and estradiol itself, but during the second half of pregnancy, the placenta becomes the source of androgens necessary for estradiol produced by the corpora lutea (Gibori and Sridaran 1981).

Maternal caregiving in laboratory rats begins during late pregnancy, a few days before parturition even begins (Mayer and Rosenblatt 1984), so it is thought that the rising *estradiol* at this time is essential for priming the brain for the rapid onset of maternal caregiving in naturally parturient laboratory rats. Note that this prepartum onset of caregiving is not universal even in laboratory rodent species, such as hamsters and prairie voles (Buntin et al. 1984; Hayes and DeVries 2007). Some expression of maternal care in laboratory rats can be achieved by as little as 10–13 days of natural pregnancy followed by termination (Rosenblatt and Siegel 1975), and similar effects are observed after treating ovariectomized females with exogenous estradiol for a similar duration of time (Bridges et al. 1974; Bridges 1984). In fact, even a single high dose of estradiol benzoate can reduce the latency for ovariectomized and hypophysectomized nulliparous rats to begin showing caregiving behaviors (Siegel and Rosenblatt 1975). Conversely, preventing the aromatization of androgens to estradiol prevents a rapid onset of maternal behaviors in hormone-treated, ovariectomized rats (Bridges and Russell 1981). However, the ability to induce maternal care in initially non-maternal nulliparous rats or mice through a few days of exposure to young pups (a process termed *maternal sensitization*; Rosenblatt 1967) is unaffected in a number of studies by the absence of estradiol or its receptor (Stolzenberg and Rissman 2011; Gallagher et al. 2019; Rosenblatt 1969) although other studies find that inhibiting estrogen synthesis or receptor signaling does impair it in female mice (Ogawa et al. 1998; Murakami 2016). Thus, while the onset of maternal caregiving can be quickly established both naturally and experimentally by exposure to estrogens, pup-directed caregiving can still sometimes be displayed (with a relatively long latency compared to estrogen-stimulated caregiving) in the absence of estrogen signaling.

Estrogens presumably influence many brain sites simultaneously to instill the full, robust expression of maternal caregiving at parturition, but the only individual brain site where modulating estradiol signaling can influence parenting in female rodents is the mPOA. When cannulae filled with crystalline estradiol benzoate (EB) were implanted for 2 days into the mPOA of late-pregnant rats that were hysterectomized and ovariectomized (to provide some natural hormone priming), maternal behaviors were shown almost immediately toward test pups; controls with cholesterol implanted into the mPOA took a median of 2 days to show these same behaviors (Numan et al. 1977; also see Felton et al. 1999). Similar studies later showed that estradiol implants into the mPOA could also reduce the latency to retrieve pups by ovariectomized nulliparous female rats (Fahrbach and Pfaff 1986). Conversely, implanting the selective estrogen receptor modulator tamoxifen into the mPOA of late-pregnant rats somewhat impairs the onset of maternal caregiving when the litters are delivered by Caesarian section (Ahdieh et al. 1987), which itself can hinder the initial onset of caregiving in some strains of laboratory rats (Stern 1985).

Estrogen receptors (ERs) in the brain exist in two isoforms – alpha and beta – that are distributed widely but differentially across the brain (Simerly et al. 1990; Mitra et al. 2003). The mPOA has very high expression of ERα, and it is much higher than ERβ (Shughrue et al. 1997), suggesting the former is more involved in estrogenic stimulation of maternal behaviors. The mPOA shows significant elevations in ER receptor binding, and specifically ERα mRNA and/or immunoreactivity, across various time points of pregnancy in laboratory rats (Giordano et al. 1990; Wagner and Morrell 1995, 1996) and mice (Koch 1989). Furthermore, female rats that display more maternal care have higher ERα expression in their mPOA compared to females showing less care (Champagne et al. 2003). ERβ also fluctuates across reproduction in laboratory rats, with the number of cells in the mPOA expressing this isoform lower in late-pregnant and lactating rats compared to proestrus virgins, although late-pregnant females have very high ERβ expression per cell (Greco et al. 2003). In support of a particular role for ERα in maternal care, short-hairpin RNA interference of ERα gene expression in the mPOA starting prepartum completely eliminated postpartum responsiveness to pups in mice (Ribiero et al. 2012) as did genetic deletion of ERα-expressing cells in the mPOA (Wei et al. 2018). On the other hand, optogenetically stimulating ERα-expressing cells in the mPOA that project to the VTA activated retrieval (Fang et al. 2018). Some interesting findings about the role of ERα in maternal caregiving is that even when maternal behavior is expressed during the postpartum maintenance phase of caregiving, when estrogens are thought to have very little role in pup-directed behaviors, ERα-expressing cells in the mPOA are still activated in parous rats interacting with pups (Lonstein et al. 2000), and optogenetically inhibiting these cells impaired retrieval in parous mice (Wei et al. 2018). It is unknown if it is the ERα expression on these mPOA cells, or some other signaling mechanisms in these cells, that matter for these effects, but it is intriguing to consider that perhaps ERα-expressing cells in the mPOA are being activated in a ligand-independent manner (i.e., in the absence of circulating estradiol) by pup cues to continue promoting motherhood through the postpartum period (see Maggi 2011).

Estrogens do not act alone to instill the rapid onset of caregiving in pregnant and parturient rats. Within days of insemination, plasma *progesterone* concentrations are rapidly rising and are exceedingly high through day 19 of pregnancy, after which they fall to very low levels at parturition (Garland et al. 1987; Grota and Eik-Nes 1967; Rosenblatt et al. 1994; Sanyal 1978) (Fig. 5). Progesterone alone has no effect on maternal responsiveness (Doerr et al. 1981), but when administered with estradiol, it has a biphasic effect on the onset of maternal caregiving in rats. More specifically, progesterone first synergizes with estradiol to rapidly promote motherhood but needs to be withdrawn for that influence to be revealed. Indeed, while estradiol administered alone can reduce the latency for ovariectomized females to begin retrieving, the dose and duration of estradiol can be reduced if progesterone is also provided (Bridges 1984), but if this progesterone is not withdrawn, the behavior does not emerge (Bridges and Russell 1981). In addition, if exogenous progesterone is injected after a Cesarean section, the onset of caregiving is prevented (Bridges and Feder 1978; Bridges et al. 1978; Numan 1978). These findings can be contrasted with a single study on Syrian hamsters in which exogenous progesterone at the end of pregnancy did not interfere with the onset of maternal caregiving (Siegel and Greenwald 1975). Interestingly, progesterone in rats not only can inhibit the onset of maternal behavior after it serves its initial facilitatory role, but it can continue to inhibit the behavior even after it is well established through lactation. There is a mostly understudied rise in progesterone beginning a few days after parturition in rats, and preventing this rise via postpartum ovariectomy has been seen to decrease pup licking early postpartum (de souza et al. 2010) but increase licking and nursing later postpartum (Grieb et al. 2017).

Few studies have examined changes in progestin receptor expression across female reproduction in the rat brain. In the mPOA, progestin receptor immunoreactivity is high in nulliparous females, falls soon after mating, rises again before and after parturition, and falls during early lactation (Numan et al. 1999; Grieb et al. 2017). Despite these results indicating that the sensitivity of the mPOA to progesterone fluctuates across reproduction, Numan (1978) long ago showed that implants of progesterone into the mPOA did not inhibit the onset of maternal behavior in Cesarean-sectioned and ovariectomized female rats that were also treated with exogenous estradiol. It was concluded that either the mPOA was not a site where progesterone inhibits maternal care or multiple sites need to be exposed to progesterone for it to have its negative effects.

Protein and Peptide Hormones and Maternal Care: Prolactin and Oxytocin

The protein hormone *prolactin* is probably best known to be required for milk secretion by the mammary glands but can also assist estrogen and progesterone's pro-maternal effects. Starting soon after insemination, prolactin is released by the anterior pituitary gland in twice-daily surges that maintain the first half of pregnancy, but by the middle of gestation, lactogens secreted by the placenta(e) rise and inhibit the pituitary gland to become the primary source of this hormone (Grattan 2015). Placental lactogens then begin falling a few days before parturition, resulting in a prepartum prolactin surge from the now disinhibited pituitary (Grattan 2015) (Fig. 5). After parturition, prolactin levels are high until postpartum day 10 and then decline (Taya and Sasamoto 1980; Mattheij et al. 1985). Suckling or non-suckling stimulation of the nipples maintains high postpartum concentrations of prolactin in mothers, and even maternally sensitized, nulliparous female rats may show elevated circulating prolactin when interacting with pups (Stern and Siegel 1978).

Administering exogenous prolactin does not alone stimulate maternal care in ovariectomized and hypophysectomized virgin rats (Baum 1978), and consistent with that finding, preventing the late-pregnancy surge in prolactin in mated females by ovariectomizing them on pregnancy day 17 (thereby preventing the immediate prepartum rise in ovarian estrogens from stimulating pituitary prolactin release) does not inhibit the later display of caregiving (Bridges et al. 1974). However, when given against a background of chronic exogenous estradiol and progesterone treatment, injecting prolactin or implanting pituitary grafts ectopically in the body cavity helps stimulate maternal behavior in ovariectomized and hypophysectomized virgin rats (Bridges et al. 1985; Bridges and Ronsheim 1990). Conversely, injecting an inhibitor of endogenous pituitary prolactin release (i.e., bromocriptine) disrupts the onset of responsiveness to pups in estradiol plus progesterone-treated virgin rats (Bridges and Ronsheim 1990).

Intracerebroventricular (ICV) infusion of prolactin will also stimulate the onset of maternal behavior in nulliparous virgin female rats treated with estradiol and progesterone and in pregnancy-terminated females (Bridges et al. 1990). Prolactin acts on the brain through two isoforms of the prolactin receptor – a long form and a short form. Expression of the long form, but not the short form, rises in the mPOA across pregnancy to peak at parturition and then falls through lactation (Bakowska and Morrell 1997; Mann and Bridges 2002; Pi and Grattan 1999). This pattern of prolactin receptor expression in rats is very similar to their circulating estradiol across reproduction, which is a strong upregulator of prolactin receptor expression (Pi et al. 2003). Eliminating prolactin receptors in the brain and body by null mutation of the prolactin gene impairs pup responsiveness in both nulliparous and primiparous mice (Lucas et al. 1998), and prolactin receptors in the mPOA is one of the brain sites where prolactin acts to stimulate maternal care. Infusing ovine prolactin or placental lactogens into the mPOA greatly shortens the latency of steroid-primed nulliparous rats to retrieve pups (Bridges et al. 1990, 1997; Bridges and Freemark 1995), and conditional knockout of mPOA prolactin receptors prevents the onset of maternal caregiving at parturition (Brown et al. 2017).

Oxytocin is a peptide hormone that has been well studied for a role in caregiving behaviors. Because there is no specific transporter mechanism that can carry oxytocin released by the posterior pituitary into the blood back into the brain (as there is for prolactin; Ermisch et al. 1985), the oxytocin affecting maternal caregiving and other behaviors is thought to arise from intracerebrally projecting oxytocinergic cells in the hypothalamus and elsewhere (Veenema and Neumann 2008). Consistent with the uterine muscles needing to refrain from contracting and prematurely expelling the fetuses, plasma oxytocin concentrations are low for the first 2 weeks of pregnancy in laboratory rats but quadruple by day 17. Oxytocin levels continue to rise as each pup is born and then fall. These low concentrations persist unless there is suckling by the young to elicit brief boluses of oxytocin release into the blood, resulting in milk ejection (Crowley 2011).

Perhaps surprising to some readers, the body of work on oxytocin and motherhood demonstrates that oxytocin can facilitate – but is not essential – for the onset or maintenance of caregiving behaviors in laboratory rats or mice (Yoshihara et al. 2017). On the positive side, this conclusion comes from research demonstrating that a single ICV oxytocin administration could induce maternal behavior in female rats but not if they were in the low-estradiol diestrus state of their cycle if they were gonadally intact and not if they were not given exogenous estradiol if they were ovariectomized (Pedersen and Prange 1979; Pedersen et al. 1982). It was also found that ICV injection of an oxytocin antiserum or an oxytocin antagonist delays the onset of maternal behaviors in pregnancy-terminated or nulliparous rats given estradiol (Fahrbach et al. 1984, 1985; Pedersen et al. 1985). Other studies, however, suggested that oxytocin's effects on the onset of caregiving were more complicated and perhaps depended on the animals' sensory capabilities or stress level before and during testing (Wamboldt and Insel 1987; Fahrbach et al. 1986; Rubin et al. 1983).

Oxytocin receptor expression and autoradiographic binding in many sites in the female brain are sensitive to the female's reproductive state and are particularly elevated in response to high concentrations of circulating estrogens (e.g., Bale et al. 1995). Studies are somewhat conflicting about where in the brain and when during reproduction OTR expression or binding peaks, but elevations during late pregnancy and/or the peripartum period have been reported in the ventromedial hypothalamus (VMH), BNST, lateral septum, olfactory bulbs, VTA, and mPOA (Bale et al. 1995; Meddle et al. 2007; Pedersen et al. 1994; Insel 1990). The mPOA and VTA are currently the only known targets for oxytocin's effects on maternal behavior. Medial preoptic area infusion of oxytocin facilitates maternal responding (Fahrbach et al. 1985), whereas mPOA infusions of an oxytocin antagonist during parturition greatly delay or completely eliminate retrieval of pups and delay the latency to crouch over pups in rats (Pedersen et al. 1994). Oxytocin receptor antagonism also prevents the experience-related enhancements in retrieval in nulliparous mice (Okabe et al. 2017). Notable impairment in the onset of maternal behavior is also found after parturitional OT receptor antagonism in the rat VTA, but the effects are not as pronounced as those found when the mPOA is targeted (Pedersen et al. 1994).

Despite all of this evidence suggesting the importance of oxytocin for maternal caregiving, more recent studies knocking out the genes for oxytocin or its receptor clearly demonstrate that the absence of oxytocin signaling results in surprisingly minor detriments to the onset of maternal caregiving, at least in laboratory mice (Nishimori et al. 2008; Macbeth et al. 2010; Rich et al. 2014; Pedersen et al. 2006). Oxytocin receptor antagonism or gene knockout also produces relatively minor deficits in pup licking and nursing when assessed during the postpartum maintenance phase of caregiving (Pedersen and Boccia 2003; Champagne et al. 2001). A recent review of this topic concluded that more attention to methodological details and avoiding hyperbolic reporting of experimental results would be important for defining oxytocin's role in parental care (Yoshihara et al. 2017).

Other rodent families have not been extensively studied for hormone-behavior relationships in maternal females. Given that, the endocrinology of pregnancy and the peripartum period in *Sciuridae* appears to be at least partly similar to laboratory-studied *Mymorpha* (and most mammals in general). As seen in studies of numerous species of squirrels (Miller et al. 2021; Tait et al. 1981; Holekamp et al. 1988; Nunes et al. 2000; Yadav and Haldar 2009) and numerous marmots (Keeley et al. 2012; Concannon et al. 1984; Exner et al. 2003), there is a tremendous rise in progesterone or its metabolites in blood or feces across pregnancy, followed by a precipitous decline starting soon before parturition. Mated female prairie dogs (*Cynomys ludovicianus*) also show this pattern of progesterone secretion, but it occurs even in females that aborted their litters, indicating that the placentae are not the hormonal source of progesterone in this and probably other species (Foreman and Garris 1984). Furthermore, a number of *Sciuridae* have a second rise in progesterone after parturition (Holekamp et al. 1988; Holekamp and Talamantes 1991; Exner et al. 2003), which also occurs during mid-lactation in laboratory rats (Hansen et al. 1983). However, studies of pregnancy concentrations of estrogens in *Sciuridae* show more diversity, with significantly higher total estrogens in pregnant *Funambulus pennantii* (Yadav and Haldar 2009) and *Spermophilus beecheyi* compared to non-mated females (but in the latter not as high as during mating; Holekamp and Talamantes 1991), similarly high estrogen levels in pre-mated and pregnant *S. beldingi* (Nunes et al. 2000), somewhat higher levels in pregnant than nonpregnant *Marmota vancouverensis* (Keeley et al. 2012), and similarly low levels found for females in any reproductive state in *Marmota marmota* (Exner et al. 2003). Relationships to behavior are scant although building and maintaining the nest is most consistently high in relation to the high progesterone (and somewhat less consistently with the high estradiol and testosterone) levels of midpregnancy in Belding's ground squirrels (Nunes et al. 2000), but in American red squirrels, fecal androgen metabolites are highest when postpartum maternal nest attendance is lowest (Dantzer et al. 2011). These studies suggest that androgens and hormones other than those discussed in detail above that are widely studied in laboratory-raised *Myomorpha* would be valuable to study in most rodent species.

In family *Castoridae,* studies on hormone plasma levels and transcript levels in endocrine organs (e.g., brain, adrenals, gonads) have focused on seasonal patterns and sex differences in beavers (Bogacka et al. 2017; Chojnowska et al. 2015, 2017; Czerwinska et al. 2015, 2017). To date, hormones have not been directly linked with parental behavior, but rather to breeding/pregnancy times, with females showing higher blood plasma levels of testosterone and progesterone during April (pregnancy times), relative to July (end of lactation period) and November (sexual quiescence), and higher follicle-stimulating hormone (FSH) levels in July and November as they gear toward mating (Chojnowska et al. 2015).

Unfortunately, very little is known about gestational and peripartum hormones in *hystricomorpha* species, but the scarce available information on agoutis, guinea pigs, chinchillas, naked mole rats, degus, and porcupines generally agrees with *Myomorpha* (Antonini et al. 1976; Ebensperger et al. 2010; Edwards et al. 2021; Guimarães et al. 2016; Mikkelsen et al. 2017; van Aarde and Potgieter 1986).

5 Patterns of Paternal Caregiving

While maternal care is obligatory in rodents, paternal care is the exception. Paternal care typically corresponds with monogamous mating systems and has evolved independently in response to conditions where the male's contribution is essential for offspring survival. The fascinating evolution of the emergence of paternal care in mammals, including rodents, has been discussed extensively elsewhere (e.g., Clutton-Brock 1991; Gromov 2020; Kölliker 2012; Royle et al. 2012). As with maternal care, the study of paternal care has relied on only a few rodent species from family *Myomorpha*, mainly prairie voles, mice, gerbils, and hamsters. This is due both to the limited number of rodent species which display paternal care in addition to the feasibility of breeding and studying these animals in a laboratory setting. In rodents, paternal behavior is generally similar to that of maternal behavior (with the exception of true nursing behavior and physiological ability to lactate) and may include pup-directed behaviors such as retrieving, licking, and huddling over pups or indirect parental behaviors such as nest building, defense of young, or providing provisions for the mother. Some males, such as *Phodopus campbelli* (referred to as dwarf or Djungarian hamsters), will even assist with the delivery of pups and consume amniotic fluid and the placenta (Jones and Wynne-Edwards 2000). These behavioral similarities to females beg the question as to whether paternal behaviors are controlled by the same neurobiological and hormonal mechanisms as maternal behaviors. Compared to maternal caregiving, the investigation of paternal caregiving is still in its infancy. While many similarities are indeed present, and will be discussed below, the mechanisms driving paternal care are usually more variable, species-specific, and highly dependent on social, experiential, and other external factors that interact with neuroendocrine mechanisms. As these topics have been thoroughly reviewed many times elsewhere (e.g., Bales and Saltzman 2016; Guoynes and Marler 2020; Horrell et al. 2018; Saltzman et al. 2017; Saltzman and Ziegler 2014), we will highlight only some of the major advances and concepts that have emerged from the literature on rodents thus far.

In the absence of pregnancy and lactation, males generally require a mating stimulus and/or cohousing with the pregnant female to promote the onset of paternal behavior. However, the requirements to become paternal vary by species. For example, exposure to soiled bedding from the pregnant female partner alone can stimulate paternal care in California mice (*Peromyscus californicus;* Gubernick 1990; Gubernick and Alberts 1989), whereas some other mouse species will not show paternal care unless mating with ejaculation occurs (Vom Saal 1985). In the absence of mating and cohabitation with a pregnant female, repeated exposure to pups over the course of several days can induce paternal behavior, as in females, through the phenomenon discussed above known as sensitization. However, for the purposes of this chapter, we will mainly focus on what is known about the paternal behavior of sires (i.e., mated males). Other proximate factors driving paternal care include developmental experience, dominance status, housing conditions, previous paternal experience, intrauterine position, age, and the testing procedure (e.g., Brown 1993; Ghiraldi and Svare 1986; McCarthy and vom Saal 1986; Vom Saal 1983). In some species, paternal care is not observed in the wild, but when males are raised in captivity, they will share the nest with the female and may provide some paternal behavior toward the young. In addition, it should be noted that some of these examples of paternal behavior may be an artifact of laboratory conditions and may not substantially affect pup outcomes (e.g., physical or behavioral development) or future reproductive success (Shilton and Brooks 1989).

6 Neural Basis of Paternal Caregiving

The research on maternal care has revealed that two main actions must occur: (1) the stimulation of circuits which promote pup-directed behaviors and (2) inhibition of circuits that are involved in young avoidance/aggression. In females, these actions are modulated by hormones that change over pregnancy, parturition, and lactation to modify these circuits. So far, the evidence shows that males also require activation of circuits that promote caregiving behaviors and inhibition of circuits which interfere with care (e.g., pup-directed aggression) and that these circuits are also sensitive to hormonal modulation. Compared to the extent of knowledge we currently have on the neuroendocrine regulation of maternal behavior, the circuits and mechanisms underlying paternal behavior are not nearly as well understood.

In females, olfactory cues from young are relayed to the medial amygdala (MeA) and then to the BNST and mPOA. In the presence of pregnancy-related hormones, this circuit promotes attraction, approach, and other maternal responses toward pups. As such, these areas have been the main target of investigation in paternal males. As with maternal behavior, the mPOA has been repeatedly shown to be activated following paternal interactions (as indicated by Fos immunoreactivity) during paternal interactions in many rodent species (de Jong et al. 2009, 2010; Kirkpatrick et al. 1994a; Lambert et al. 2013; Smiley et al. 2021; Tsuneoka et al. 2015). Some of the earliest studies on paternal care revealed that lesions to the mPOA caused major deficits to paternal behaviors (Lee and Brown 2002; Rosenblatt et al. 1996; Sturgis and Bridges 1997), much like they do to maternal behavior. The technological advancement in transgenic tools, particularly in laboratory mice, has allowed for more precise identification of the mPOA cell types and circuits involved in paternal care. For instance, studies have shown that optogenetic stimulation of particular subsets of neuropeptide-synthesizing or steroid hormones-sensitive mPOA neurons can modulate caregiving behaviors in paternal (as well as maternal) laboratory mice

(Wu et al. 2014; Wei et al. 2018). Inhibition or ablation of these neurons decreased parental behaviors in both sexes, whereas optogenetic activation of these neurons in virgin male mice (which are normally infanticidal) decreased pup-directed aggression and increased pup grooming (a paternal response). Through a series of elegant experiments, Kohl et al. (2018) showed that discrete populations of mPOA galanin neurons (most of which also contain the inhibitory neurotransmitter GABA) projected to different brain regions, including the MeA, PAG, VTA, and PVN. Each of these distinct projections was then shown to regulate a different aspect of paternal behavior including motor control of pup-directed behaviors, motivation to approach pups, and neuromodulation of other peptides during these interactions. Together, these studies showed how one population of neurons in the mPOA can coordinate multiple aspects of parental behavior simultaneously.

Immunohistochemical studies using Fos expression have identified other brain regions associated with paternal care including the caudal dorsal raphe nucleus, lateral habenula, lateral septum, paraventricular nucleus of the thalamus, nucleus reuniens of the thalamus, and amygdala (de Jong et al. 2009, 2010; Kirkpatrick et al. 1994a; Smiley et al. 2021). In California mice, lesions of the basolateral nucleus of the amygdala result in similar paternal deficits as mPOA lesions in first-time fathers, including longer latencies to retrieve pups and less pup grooming and time spent in contact with pups (Lee and Brown 2007). In virgin male prairie voles (*Microtus ochrogaster*), which normally show alloparental care, lesions to corticomedial, but not basolateral amygdala, reduce the amount of time males spent in contact with a foster pup (Kirkpatrick et al. 1994c). In sum, so far, it appears that fathers require similar neural circuitry as mothers to promote paternal care, but the research is currently limited to the mPOA and some subdivisions of the amygdala.

With the exception of a few species, such as prairie voles which show alloparental care as virgins, many virgin male rodents are infanticidal and will attack and kill unrelated pups outside of the time they may have sired offspring. This behavior is hypothesized to be an adaptive reproductive strategy in which males can reinstate estrous cycling in lactating females, which allows them to mate and sire offspring of his own with her. Virgin females of some rodent species are typically not infanticidal (but see Shilton and Brooks 1989, for example, where both male and female collared lemmings display infanticide); therefore, the requirement for the suppression of pup-attacking behavior in order to show parental care is often unique to males. In male laboratory mice, mating with ejaculation has been shown to be the necessary stimulus that suppresses infanticidal behavior (Vom Saal 1985). Bilateral excitotoxic lesions in the rhomboid nucleus of the BNST (BNSTrh) resulted in a delayed latency to show infanticidal behavior in virgin male mice (Tsuneoka et al. 2015). However, it did not eliminate infanticide, suggesting that other brain regions are involved. Perplexingly, studies using Fos expression have shown that the MeA and mPOA are activated following both paternal and infanticidal interactions with pups in virgin male mice (Chen et al. 2019; Tachikawa et al. 2013). To resolve how the MeA neurons could be involved in two opposing behaviors – paternal care and infanticide – Chen et al. (2019)

used fiber photometry to record intracellular calcium activity in GABA-expressing neurons in the MeA of freely behaving male mice interacting with pups. MeA-GABA neurons were active during pup grooming in both virgin males and females which showed spontaneous parental behaviors but showed an approximately fourfold increased activity when males were attacking pups. Optogenetic stimulation of GABAergic neurons in the MeA of virgin males at higher laser intensities induced pup-attacking behavior, while optogenetic stimulation at lower frequencies induced pup-grooming behavior, indicating that the MeA can influence both infanticidal and paternal responses in an activity-dependent manner.

Like maternal care, both paternal and infanticidal behavior are highly regulated in both positive and negative ways by olfactory input from pups. Genetic ablation of the vomeronasal organ (VNO), which specially detects social odors, eliminates infanticidal behavior in virgin male laboratory mice. Instead, VNO-deficient virgin male mice will show the full display of paternal behavior toward foster pups (Tachikawa et al. 2013). In male rats, which normally do not show paternal behavior but can be sensitized into acting paternal through repeated pup exposure, olfactory bulbectomy reduced the number of days to show pup retrieval (Fleischer et al. 1981). However, in virgin male prairie voles, which are not normally aggressive toward pups, bilateral olfactory bulbectomy increased pup-directed attacking behavior in roughly half of the males tested (Kirkpatrick et al. 1994b) and increased attacking behavior in virgin female mice, which are also normally alloparental toward pups (Neckers et al. 1975). Therefore, in species that show pup-directed aggression, olfactory inputs from pups need to be suppressed, whereas olfactory inputs appear to facilitate parental responses in species which already show these alloparental behaviors as virgins.

In sum, while many brain regions that are involved in maternal care have been implicated in paternal care, only the mPOA, some subdivisions of the amygdala, and olfactory regions have been extensively studied and have a causal role in promoting paternal care. Although the circuits that control infanticide are not well understood, they appear to be distinct from those promoting paternal behaviors and may require separate regulatory mechanisms to suppress attack behaviors. What is clear is that both of these processes must take place simultaneously in order to become paternal and that these changes are likely triggered through mating and/or other stimuli associated with the pregnant female partner.

7 Endocrinology of Paternal Caregiving

Analogous to females, mated males also undergo some systematic changes in circulating hormones that are stimulated by copulation, cohabitation with pregnant females, and interactions with young. These hormones can then act on the neural circuitry underlying paternal behavior to shift behaviors away from aggression and toward the onset of caregiving.

Steroid Hormones and Paternal Care

Early studies of testosterone in male mammals showed that testosterone levels drop when males become fathers, leading to the conclusion that high levels of testosterone were incompatible with paternal behaviors. However, the role of testosterone-regulating rodent paternal behavior is not as straightforward as this dogma suggests. In several rodent species, including *Phodopus campbelli* (Reburn and Wynne-Edwards 1999), Mongolian gerbils (*Meriones unguiculatus*; Juana et al. 2010), and volcano mice (*Neotomodon alstoni*; Luis et al. 2012), males show increases in testosteroneover the female's gestation period, which remain unchanged during the pup-rearing period. Gonadectomy prior to puberty + testosterone replacement in adulthood facilitates paternal behavior in C57/B6 mice (Kercmar et al. 2014) and volcano mice (Luis et al. 2012), but whether these actions are mediated through the aromatization to estradiol has not been tested. In adult virgin male gerbils, which are normally aggressive toward pups, castration + testosterone, + estradiol, or dihydrotestosterone (DHT; a non-aromatizable androgen) caused males to show paternal responses toward pups, with testosterone and estradiol having a greater effect than DHT (Martínez et al. 2015). However, when adult male gerbils were castrated, cohoused with pregnant females, and then tested during the postpartum period, testosterone reduced paternal behavior (Clark and Galef 1999).

In California mice, testosterone has been shown to facilitate paternal behavior through the conversion to estradiol. Castration in adult males reduces paternal behavior with replacement with either testosterone or estradiol restoring paternal behavior (Trainor and Marler 2001, 2002). This was further confirmed by using aromatase inhibitors + testosterone or + DHT to show that, indeed, the conversion to estradiol is required for paternal care and, specifically, via local production of estradiol in the mPOA (Trainor et al. 2003). Estrogen signaling through the ERβ receptor also mediates hippocampal neurogenesis, which is important for reducing anxiety in father mice during the mid-pup-rearing period when paternal care starts shifting from passive (e.g., huddling) to active (e.g., retrieving pups that leave the nest) (Glasper et al. 2016; Hyer et al. 2016, 2017). However, a role for estrogen signaling in paternal behaviors has received mixed support in the hamster and vole models. Although male *P. campbelli* have high baseline levels of circulating estradiol throughout their mate's pregnancy and lactation period, and at levels similar to breeding females, castration in adult males had no effect on paternal behavior (Hume and Wynne-Edwards 2005), with no evidence to support local brain aromatization to estrogen or an upregulation of ERα receptors in key brain regions for parental care (Hume and Wynne-Edwards 2006; Timonin et al. 2008). In hamsters, aromatase inhibitors given during adolescence (postnatal day (pnd) 18) reduced subsequent paternal behavior in adults (Timonin and Wynne-Edwards 2008). However, postnatal exposure to ERα agonists (pnd 8–14) resulted in increased pup-directed aggression (Perry et al. 2015). Finally, mixed findings for a role of gonadal hormones have been produced in prairie voles with one study reporting castration decreased paternal behavior (Wang and De Vries 1993) and another reporting no effects (Lonstein and De Vries 1999). However, neonatal castration significantly reduced the percentage of males who showed alloparental care in adulthood (Lonstein and De Vries 2000).

Progesterone has not been as well studied as testosterone and E2. In hamsters, progesterone rises at the end of their mate's pregnancy (Schum and Wynne-Edwards 2005) and declines in California mouse fathers (Trainor et al. 2003). Progesterone receptors are downregulated in father California mice (Perea-Rodriguez et al. 2015), and importantly, progesterone-receptor knockout mice do not exhibit infanticidal behavior as virgins (Schneider et al. 2009). Treatment with progesterone increased pup-attacking behavior in mice, but not other aggressive behaviors, while blocking progesterone receptors enhanced paternal responses (Schneider et al. 2003).

In sum, it is difficult to generalize any certain patterns of steroid hormone effects on paternal care as they appear to be species- and brain site-specific although developmental (i.e., organizational) effects of estrogen may be important for paternal behavior expression in adulthood. Drawing from the limited data from male mice, progesterone interferes with paternal behavior, and progesterone activity must be suppressed in order for males to show caregiving.

Peptide and Protein Hormones and Paternal Care

As discussed above, oxytocin generally has some positive effects on maternal caregiving. Oxytocin neuron activation, serum oxytocin levels, and oxytocin receptor expression are similarly positively related to paternal care in a number of rodent species (e.g., Kenkel et al. 2012, 2014; Li et al. 2015; Perea-Rodriguez et al. 2015). Given these positive associations, it is surprising that very few studies have manipulated oxytocin during rodent paternal care. Recently, it was shown that an oxytocin receptor antagonist infused directly into the mPOA of father Mandarin voles (*Lasiopodomys mandarinus*, another monogamous vole species) reduced paternal responses and increased the latency to initiate paternal behavior toward pups (Yuan et al. 2017). Chemogenetic stimulation of oxytocin-expressing neurons in the PVN projecting to either the VTA or the NA stimulated paternal behavior in Mandarin voles, particularly licking and grooming responses, while inhibiting these pathways reduced these behaviors (He et al. 2021). In laboratory mice, CD38, a membrane glycoprotein which facilitates oxytocin secretion from neurons in the hypothalamus, has been shown to be an important factor for paternal care. CD38 knockout mice do not retrieve pups but will do so if injected with OT. Furthermore, introducing CD38 into the NA using a lentiviral vector recovers pup retrieval behavior in most males, with full recovery if males are also injected with oxytocin (Akther et al. 2013).

The role of another closely related peptide, arginine-vasopressin (AVP), has been primarily investigated in male voles. In meadow voles (*Microtus pennsylvanicus*), which are nonmonogamous but can show facultative paternal care under winter conditions, central infusions of AVP into the lateral ventricle reduced pup-directed aggression and increased paternal responses in previously non-paternal male virgins (Parker and Lee 2001). In prairie voles, sires have a reduced density of AVP fibers in the lateral septum (LS) and lateral habenula, compared to virgins or males cohoused in pregnant females (Bamshad et al. 1993, 1994), potentially reflecting increased synthesis and release during parenting. Indeed, central infusions of AVP in the lateral septum promoted, while AVP antagonists inhibited paternal behavior in prairie voles (Wang et al. 1994a). However, castration that eliminates AVP immunoreactivity in this area either has detrimental effects (Wang and De Vries 1993) or no effects on alloparental behaviors in virgin males (Lonstein 2002), suggesting that AVP action in the lateral septum alone is probably insufficient for paternal responses. AVP-containing neurons in the PVN are also activated when males are exposed to pups (Kenkel et al. 2012), with fathers having increased AVP mRNA levels compared to virgins (Wang et al. 2000). However, in California mice, no such differences in AVP mRNA levels exist (de Jong et al. 2013). In fact, California mice fathers show reduced levels of AVP V1a receptor mRNA in the BNST, compared to virgins (Perea-Rodriguez et al. 2015). In virgin male prairie voles, both OT and AVP receptor antagonists are required to reduce alloparental responses (Bales et al. 2004).

Another important hormone for paternal care is prolactin. In several rodent species, including Mongolian gerbils (Brown et al. 1995), California mice (Gubernick and Nelson 1989), and dwarf hamsters (*P. Campbelli*) (Reburn and Wynne-Edwards 1999), circulating prolactin levels are higher in fathers compared to virgins. In some species, this increase in prolactin may be in response to pregnant female stimuli as male gerbils and California mice that cohabited with a female during gestation had higher circulating prolactin levels compared to males cohoused with pregnant females (Brown 1993; Gubernick and Nelson 1989). However, pharmacologically suppressing circulating prolactin had no effect on paternal behavior in *P. Campbelli* fathers (Brooks et al. 2005). In laboratory mice, on the other hand, a clear role for prolactin has been established. A recent study showed that the differences in parental responses to pups in male rats (which are not normally parental) and mice (which show mating-induced paternal care) can be attributed to differences in the dynamics of prolactin secretion (Stagkourakis et al. 2020). Prolactin is under inhibitory control of dopamine neurons in the arcuate nucleus (i.e., TIDA neurons). In rats, these neurons oscillate at slow rates to sustain dopamine release, leading to lower circulating levels of prolactin. In contrast, TIDA neurons in mice oscillate at faster rates, leading to dopamine depletion and, hence, higher prolactin levels (relative to rats); this, in turn, facilitates paternal care via acting on prolactin receptors in the mPOA (Stagkourakis et al. 2020). Optogenetically stimulating TIDA neurons in rats to create faster mouselike oscillations increases prolactin secretion and can induce paternal responses in males. From these results, it is likely that most males have a circuitry that can be activated to show paternal care, but that males require the appropriate hormonal stimulation to induce the onset of this behavior.

In sum, the peptide hormones oxytocin and AVP, and the protein hormone prolactin, are generally positively related to paternal behavior. However, the casual evidence establishing a role for these peptides has been limited to only a handful of studies. Both oxytocin and prolactin have been shown to exert its effects on paternal care through receptors in the mPOA, whereas AVP has some effects in the LS. Although it is clear that hormones are important for paternal care, hormonal changes also occur in species-specific ways, with different species requiring different hormonal stimulation to show these behaviors. In many species, but not all, these hormonal changes are facilitated by mating and cohabitation with pregnant females.

8 Conclusions

Rodents make up roughly a quarter of all mammalian species and have undoubtedly provided us with some of the most detailed descriptions of parental care and information about the physiological and neurobiological underpinnings of this incredibly important reproductive behavior. Most work on the neuroendocrinology of parental care has focused on a limited subset of species – laboratory rats, mice, voles, hamsters, and gerbils – all within the suborder *Myomorpha*. This work has greatly influenced our understanding of the parental brain and has proven to be one of the most translatable models for understanding human parenting. We hope that the discussion above is convincing enough that this research is actually quite limited and is not completely representative of all rodents. In fact, many other species of rodents within the other four rarely-studied suborders (*Sciuromorpha, Hystricomorpha, Castorimorpha, and Anomaluromorpha*) show interesting unique parental behaviors that have evolved from their particular environments and other constraints.

There are substantial opportunities for comparative analyses among closely related species that display different parental strategies within *Myomorpha* (and likely other rodent families) that allows us to discover which features of parental care are common and which are unique adaptations. This has been demonstrated well through studies comparing the monogamous, biparental prairie vole with the promiscuous, uniparental meadow vole (Bamshad et al. 1993; Fowler et al. 2005; McGuire et al. 1992; McGuire and Novak 1984; Oliveras and Novak 1986; Wang et al. 1994b), as well comparing the biparental *Phodopus campbelli* with the uniparental *Phodopus sungorus* (Ma et al. 2005; Reburn and Wynne-Edwards 1999; Schum and Wynne-Edwards 2005). Some species, such as *Peromyscus polionotus* and *Peromyscus maniculatus*, which display opposite mating/parental strategies, can interbreed and/or tolerate cross fostering (Bendesky et al. 2017), which are powerful tools to understand the genetic contributions to monogamy and biparental care, as well as the influence of different parental types on offspring phenotypes.

With regard to the endocrine and neural basis of motherhood, it is clear that the hormones that prepare the body for pregnancy and lactation also act on the brain circuitry that promotes maternal care behaviors. However, the hormonal correlates of maternal behavior have been studied in a handful of laboratory rodents that all give birth to altricial young and thus show quite similar patterns of caregiving behavior. Most of these animals also have a postpartum estrus that involves its own unique endocrine state in the peripartum period, which is not universal across rodents. Most females are also studied after their first parturition, which is certainly not representative of many free-living rodents even of these well-studied species that are simultaneously pregnant while caring for young.

Although paternal care is displayed in only a small number of rodent species, paternal care can nonetheless be equally important for offspring survival as maternal care in natural settings. Paternal care is far less studied compared to maternal care but appears to require similar neural circuits as maternal care, and these are also sensitive to hormonal modulation. While males undergo systemic changes in hormone secretion patterns between mating and weaning of offspring, these changes tend to be more variable and are more species-specific, relative to females. Males are also especially susceptible to external social and environmental factors, which greatly influences where they show paternal care. It is also important to study paternal care in wild populations as rodent paternal behavior is sometimes an artifact of laboratory/captive conditions.

In sum, for the field studying parental caregiving to gain a fuller understanding of how these behaviors evolved and how generalizable the mechanisms underlying these behaviors are, there needs to be a wider range of rodents studied across the different rodent families. We recommend that researchers consider alternative models of rodent parental care, especially when developing new research programs investigating the hormonal or neural basis of these behaviors. Only by studying both similarities and differences between species can we generalize concepts among species, which will lead to great predictive models and translational outcomes for understanding parenting in other mammals, including humans.

References

Adejumo DO, Adeyanju TA, Adeyanju TE. Range extension and first confirmed record of the Flightless Anomalure Zenkerella insignis (Matschie, 1898) (Mammalia: Rodentia: Anomaluridae) in Nigeria. J Threat Taxa. 2020;12(13):16900–3.

Ahdieh HB, Mayer AD, Rosenblatt JS. Effects of brain antiestrogen implants on maternal behavior and on postpartum estrus in pregnant rats. Neuroendocrinology. 1987;46(6):522–31.

Akther S, Korshnova N, Zhong J, Liang M, Cherepanov SM, Lopatina O, Komleva YK, Salmina AB, Nishimura T, Fakhrul AA, Hirai H, Kato I, Yamamoto Y, Takasawa S, Okamoto H, Higashida H. CD38 in the nucleus accumbens and oxytocin are related to paternal behavior in mice. Mol Brain. 2013;6:41.

Albin-Brooks C, Nealer C, Sabihi S, Haim A, Leuner B. The influence of offspring, parity, and oxytocin on cognitive flexibility during the postpartum period. Horm Behav. 2017;89:130–6.

Andersen DC. Observations on reproduction, growth, and behavior of the northern pocket gopher (*Thomomys talpoides*). J Mammal. 1978;59:418–22.

Antonini R, Turner TT, Pauerstein CJ. The hormonal control of the guinea pig corpus luteum during early pregnancy. Comp Study Fertil Steril. 1976;27(11):1322–5.

Asher M, Spinelli de Oliveira E, Sachser N. Social system and spatial organization of wild guinea pigs (Cavia aperea) in a natural population. J Mammal. 2004;85(4):788–96.

Aspillaga-Cid A, Vera DC, Ebensperger LA, Correa LA. Parental care in male degus (Octodon degus) is flexible and contingent upon female care. Physiol Behav. 2021;238:113487.

Bakowska JC, Morrell JI. Atlas of the neurons that express mRNA for the long form of the prolactin receptor in the forebrain of the female rat. J Comp Neurol. 1997;386(2):161–77.

Bale TL, Pedersen CA, Dorsa DM. CNS oxytocin receptor mRNA expression and regulation by gonadal steroids. Adv Exp Med Biol. 1995;395:269–80.

Bales KL, Saltzman W. Fathering in rodents: neurobiological substrates and consequences for offspring. Horm Behav. 2016;77:249–59.

Bales KL, Kim AJ, Lewis-Reese AD, Sue Carter C. Both oxytocin and vasopressin may influence alloparental behavior in male prairie voles. Horm Behav. 2004;45:354–61.

Bamshad M, Novak MA, De Vries GJ. Sex and species differences in the vasopressin innervation of sexually naive and parental prairie voles, Microtus ochrogaster and meadow voles, Microtus pennsylvanicus. J Neuroendocrinol. 1993;5:247–55.

Bamshad M, Novak MA, de Vries GJ. Cohabitation alters vasopressin innervation and paternal behavior in prairie voles (Microtus ochrogaster). Physiol Behav. 1994;56:751–8.

Barrière DA, Ella A, Szeremeta F, Adriaensen H, Même W, Chaillou E, Migaud M, Même S, Lévy F, Keller M. Brain orchestration of pregnancy and maternal behavior in mice. BioRxiv. 2020; https://doi.org/10.1101/2020.05.23.112045.

Barrington BA. Description of Birth and Young of the Pocket Gopher, Geomys Floridanus. J Mammal. 1942;23:428–30.

Baum MJ. Failure of pituitary transplants to facilitate the onset of maternal behavior in ovariectomized virgin rats. Physiol Behav. 1978;20(1):87–9.

Bendesky A, Kwon Y-M, Lassance J-M, Lewarch CL, Yao S, Peterson BK, He MX, Dulac C, Hoekstra HE. The genetic basis of parental care evolution in monogamous mice. Nature. 2017;544:434–9.

Benedetto L, Rivas M, Pereira M, Ferreira A, Torterolo P. A descriptive analysis of sleep and wakefulness states during maternal behaviors in postpartum rats. Arch Ital Biol. 2017;155(3):99–109.

Bidau CJ, Patton JL, Pardiñas UFJ, D'Elía G. Mammals of South America. JL Patton UFJ Pardiñas G. Delía Rodents. The University of Chicago Press London and Chicago, 2015;818–876.

Bogacka I, Paukszto Ł, Jastrzębski JP, Czerwińska J, Chojnowska K, Kamińska B, Kurzyńska A, Smolińska N, Giżejewski Z, Kamiński T. Seasonal differences in the testicular transcriptome profile of free-living European beavers (*Castor fiber L.*) determined by the RNA-Seq method. PLoS ONE. 2017;12:e0180323.

Bosch OJ. Maternal aggression in rodents: brain oxytocin and vasopressin mediate pup defence. Philos Trans R Soc B Biol Sci. 2013;368(1631):20130085.

Breed WG, Leigh CM. Reproductive biology of an old endemic murid rodent of Australia, the Spinifex hopping mouse, Notomys alexis: adaptations for life in the arid zone. Integr Zool. 2011;6:321–33.

Brewster J, Leon M. Relocation of the site of mother–young contact: maternal transport behavior in Norway rats. J Comp Physiol Psychol. 1980;94(1):69.

Bridges RS. A quantitative analysis of the roles of dosage, sequence, and duration of estradiol and progesterone exposure in the regulation of maternal behavior in the rat. Endocrinology. 1984;114(3):930–40.

Bridges RS, Feder HH. Inhibitory effects of various progestins and deoxycorticosterone on the rapid onset of maternal behavior induced by ovariectomy-hysterectomy during late pregnancy in rats. Horm Behav. 1978;10(1):30–9.

Bridges RS, Freemark MS. Human placental lactogen infusions into the medial preoptic area stimulate maternal behavior in steroid-primed, nulliparous female rats. Horm Behav. 1995;29(2):216–26.

Bridges RS, Ronsheim PM. Prolactin (PRL) regulation of maternal behavior in rats: bromocriptine treatment delays and PRL promotes the rapid onset of behavior. Endocrinology. 1990;126(2):837–48.

Bridges RS, Russell DW. Steroidal interactions in the regulation of maternal behaviour in virgin female rats: effects of testosterone, dihydrotestosterone, oestradiol, progesterone and the aromatase inhibitor, 1, 4, 6-Androstatriene-3, 17-Dione. J Endocrinol. 1981;90(1):31–40.

Bridges RS, Goldman BD, Bryant LP. Serum prolactin concentrations and the initiation of maternal behavior in the rat. Horm Behav. 1974;5(3):219–26.
Bridges RS, Rosenblatt JS, Feder HH. Serum progesterone concentrations and maternal behavior in rats after pregnancy termination: behavioral stimulation after progesterone withdrawal and inhibition by progesterone maintenance. Endocrinology. 1978;102(1):258–67.
Bridges RS, DiBiase R, Loundes DD, Doherty PC. Prolactin stimulation of maternal behavior in female rats. Science. 1985;227(4688):782–4.
Bridges RS, Numan M, Ronsheim PM, Mann PE, Lupini CE. Central prolactin infusions stimulate maternal behavior in steroid-treated, nulliparous female rats. Proc Natl Acad Sci. 1990;87(20):8003–7.
Bridges RS, Robertson MC, Shiu RP, Sturgis JD, Henriquez BM, Mann PE. Central lactogenic regulation of maternal behavior in rats: steroid dependence, hormone specificity, and behavioral potencies of rat prolactin and rat placental lactogen I. Endocrinology. 1997;138(2):756–63.
Broadbooks HE. Tree nesting and maternal behavior of chipmunks. Southwest Nat. 1977;22:154–5.
Brooks PL, Vella ET, Wynne-Edwards KE. Dopamine agonist treatment before and after the birth reduces prolactin concentration but does not impair paternal responsiveness in Djungarian hamsters, *Phodopus campbelli*. Horm Behav. 2005;47(3):358–66.
Brown LN. Breeding Biology of the Pocket Gopher (*Geomys pinetis*) in Southern Florida. Am Midl Nat. 1971;85:45–53.
Brown RE. Hormonal and experiential factors influencing parental behaviour in male rodents: an integrative approach. Behav Process. 1993;30:1–27.
Brown RE, Murdoch T, Murphy PR, Moger WH. Hormonal responses of male gerbils to stimuli from their mate and pups. Horm Behav. 1995;29:474–91.
Brown RS, Aoki M, Ladyman SR, Phillipps HR, Wyatt A, Boehm U, Grattan DR. Prolactin action in the medial preoptic area is necessary for postpartum maternal nursing behavior. Proc Natl Acad Sci. 2017;114(40):10779–84.
Buntin JD, Jaffe S, Lisk RD. Changes in responsiveness to newborn pups in pregnant, nulliparous golden hamsters. Physiol Behav. 1984;32(3):437–9.
Burgin CJ, Colella JP, Kahn PL, Upham NS. How many species of mammals are there?. Journal of Mammalogy, 2018;99(1):1–14.
Burton M, Burton R. International wildlife encyclopedia, vol. 1. 3rd ed. Tarrytown: Marshall Cavendish; 2002.
Busher P. Chapter 24. Social organization and monogamy in the beaver. In: Rodent societies. Chicago: University of Chicago Press; 2008.
Butterworth BB. Parental relations and the behavior of juvenile kangaroo rats (Dipodomys) in captivity. Southwest Nat. 1964;8:213–20.
Camín S. Gestation, maternal behaviour, growth and development in the subterranean caviomorph rodent Ctenomys mendocinus (Rodentia, Hystricognathi, Ctenomyidae) Animal Biology. 2010;60(1):79–95.
Camp CL. Excavations of burrows of the rodent aplodontia: with observations on the habits of the animal, vol. 17. Berkeley: University of California Press; 1918.
Campo DH, Caraballo DA, Cassini GH, Lucero SO, Teta P. Integrative taxonomy of extant maras supports the recognition of the genera Pediolagus and Dolichotis within the Dolichotinae (Rodentia, Caviidae). J Mammal. 2020;101(3):817–34.
Carleton MD, Musser GG. Order Rodentia. In: Wilson DE, Reeder DM, editors. Mammal species of the world: a taxonomic and geographic reference. Baltimore: Johns Hopkins University Press; 2005. p. 745–52.
Champagne F, Diorio J, Sharma S, Meaney MJ. Naturally occurring variations in maternal behavior in the rat are associated with differences in estrogen-inducible central oxytocin receptors. Proc Natl Acad Sci. 2001;98(22):12736–41.
Champagne FA, Weaver IC, Diorio J, Sharma S, Meaney MJ. Natural variations in maternal care are associated with estrogen receptor α expression and estrogen sensitivity in the medial preoptic area. Endocrinology. 2003;144(11):4720–4.

Champagne F, Chretien P, Stevenson CW, Zhang TY, Gratton A, Meaney MJ. Variations in nucleus accumbens dopamine associated with individual differences in maternal behavior in the rat. J Neurosci. 2004;24(17):4113–23.

Champagne FA, Curley JP, Keverne EB, Bateson PP. Natural variations in postpartum maternal care in inbred and outbred mice. Physiol Behav. 2007;91(2–3):325–34.

Chen PB, Hu RK, Wu YE, Pan L, Huang S, Micevych PE, Hong W. Sexually dimorphic control of parenting behavior by the medial amygdala. Cell. 2019;176:1206–1221.e18.

Chojnowska K, Czerwinska J, Kaminski T, Kaminska B, Panasiewicz G, Kurzynska A, Bogacka I. Sex- and seasonally related changes in plasma gonadotropins and sex steroids concentration in the European beaver (Castor fiber). Eur J Wildl Res. 2015;61:807–11.

Chojnowska K, Czerwinska J, Kaminski T, Kaminska B, Kurzynska A, Bogacka I. Leptin plasma concentrations, leptin gene expression, and protein localization in the hypothalamic-pituitary-gonadal and hypothalamic-pituitary-adrenal axes of the European beaver (*Castor fiber*). Theriogenology. 2017;87:266–75.

Clark MM, Galef BG. A testosterone-mediated trade-off between parental and sexual effort in male mongolian gerbils (Meriones unguiculatus). J Comp Psychol Wash DC. 1999;1983(113):388–95.

Clutton-Brock TH. The evolution of parental care. Princeton: Princeton University Press; 1991.

Concannon P, Baldwin B, Tennant B. Serum progesterone profiles and corpora lutea of pregnant, postpartum, barren and isolated females in a laboratory colony of woodchucks (*Marmota monax*). Biol Reprod. 1984;30(4):945–51.

Connior MB. Geomys bursarius (Rodentia: Geomyidae). Mamm Species. 2011;43:104–17.

Coster, P.M.C., Beard, K.C., Salem, M.J., Chaimanee, Y., Jaeger, JJ., 2015. New fossils from the Paleogene of central Libya illuminate the evolutionary history of endemic African anomaluroid rodents. Front Earth Sci https://doi.org/10.3389/feart.2015.00056.

Cramblet HM, Ridenhour RL. Parturition in *Aplodontia*. J Mammal. 1956;37(1):87–90.

Crowley WR. Neuroendocrine regulation of lactation and milk production. Compr Physiol. 2011;5(1):255–91.

Czerwińska J, Chojnowska K, Kamiński T, Bogacka I, Panasiewicz G, Smolińska N, Kamińska B. Plasma-glucocorticoids and ACTH levels during different periods of activity in the European Beaver (*Castor fiber L.*). Folia Biol (Praha). 2015;63:229–34.

Czerwińska J, Chojnowska K, Kaminski T, Bogacka I, Smolinska N, Kaminska B. Prepro-orexin and orexin expression in the hypothalamic–pituitary–adrenal and hypothalamic–pituitary–gonadal axes of free-living Eurasian beavers (*Castor fiber L.*) depends on season. J Mammal. 2017;98:895–905.

Dantzer B, McAdam AG, Palme R, Humphries MM, Boutin S, Boonstra R. Maternal androgens and behaviour in free-ranging North American red squirrels. Anim Behav. 2011;81(2):469–79.

Davis WB, Ramsey RR, Arendale JM Jr. Distribution of pocket gophers (*Geomys breviceps*) in relation to soils. J Mammal. 1938;19:412–8.

Dawson MR, Marivaux L, Li C, Beard KC, Metais G. Laonastes and the "Lazarus Effect" in recent mammals. Science. 2006;311(5766):1456–8.

Day CS, Galef BG. Pup cannibalism: one aspect of maternal behavior in golden hamsters. J Comp Physiol Psychol. 1977;91(5):1179.

de Jong TR, Chauke M, Harris BN, Saltzman W. From here to paternity: neural correlates of the onset of paternal behavior in California mice (*Peromyscus californicus*). Horm Behav. 2009;56:220–31.

de Jong TR, Measor KR, Chauke M, Harris BN, Saltzman W. Brief pup exposure induces Fos expression in the lateral habenula and serotonergic caudal dorsal raphe nucleus of paternally experienced male California mice (*Peromyscus californicus*). Neuroscience. 2010;169:1094–104.

de Jong TR, Harris BN, Perea-Rodriguez JP, Saltzman W. Physiological and neuroendocrine responses to chronic variable stress in male California mice (*Peromyscus californicus*): influence of social environment and paternal state. Psychoneuroendocrinology. 2013;38:2023–33.

de Olmos JS, Heimer L. The concepts of the ventral striatopallidal system and extended amugdada. Ann N Y Acad Sci. 1999;877:1–32.

de Sousa FL, Lazzari V, de Azevedo MS, de Almeida S, Sanvitto GL, Lucion AB, Giovenardi M. Progesterone and maternal aggressive behavior in rats. Behav Brain Res. 2010;212(1):84–9.

Dewsbury DA, Hodges AW. Copulatory Behavior and Related Phenomena in Spiny Mice (*Acomys cahirinus*) and Hopping Mice (*Notomys alexis*). J Mammal. 1987;68:49–57.

Dinets V. First observations on the behavior of the flightless anomalure (Zenkerella insignis). Zoology. 2017;123:121–3.

Doerr HK, Siegel HI, Rosenblatt JS. Effects of progesterone withdrawal and estrogen on maternal behavior in nulliparous rats. Behav Neural Biol. 1981;32(1):35–44.

Droge MH. Maternal retrieval in the lactation thirteen-lined ground squirrel, Spermophilus tridecemlineatus. Doctoral dissertation, Texas Tech University; 1976.

Ebensperger LA, Ramírez-Otarola N, León C, Ortiz ME, Croxatto HB. Early fitness consequences and hormonal correlates of parental behaviour in the social rodent, Octodon degus. Physiol Behav. 2010;101(4):509–17.

Edwards PD, Arguelles DA, Mastromonaco GF, Holmes MM. Queen pregnancy increases group estradiol levels in cooperatively breeding naked mole-rats. Integr Comp Biol. 2021;61:1841–51.

Eisenberg J. Mammals of the neotropics. Chicago: The University of Chicago Press; 1989.

Eisennberg JF, Isaac DE. The reproduction of heteromyid rodents in captivity. J Mammal. 1963;44:61–7.

Elyada YM, Mizrahi A. Becoming a mother – circuit plasticity underlying maternal behavior. Curr Opin Neurobiol. 2015;35:49–56.

Ermisch A, Barth T, Rühle HJ, Skopkova J, Hrbas P, Landgraf R. On the blood-brain barrier to peptides: accumulation of labelled vasopressin, DesGlyNH2-vasopressin and oxytocin by brain regions. Endocrinol Exp. 1985;19(1):29–37.

Exner C, Wehrend A, Hospes R, Einspanier A, Hoffmann B, Heldmaier G. Hormonal and behavioural changes during the mating season and pregnancy in Alpine marmots (*Marmota marmota*). Reproduction. 2003;126(6):775–82.

Fahrbach SE, Pfaff DW. Effect of preoptic region implants of dilute estradiol on the maternal behavior of ovariectomized, nulliparous rats. Horm Behav. 1986;20(3):354–63.

Fahrbach SE, Morrell JI, Pfaff DW. Oxytocin induction of short-latency maternal behavior in nulliparous, estrogen-primed female rats. Horm Behav. 1984;18(3):267–86.

Fahrbach SE, Morrell JI, Pfaff DW. Possible role for endogenous oxytocin in estrogen-facilitated maternal behavior in rats. Neuroendocrinology. 1985;40(6):526–32.

Fahrbach SE, Morrell JI, Pfaff DW. Effect of varying the duration of pre-test cage habituation on oxytocin induction of short-latency maternal behavior. Physiol Behav. 1986;37(1):135–9.

Fang YY, Yamaguchi T, Song SC, Tritsch NX, Lin D. A hypothalamic midbrain pathway essential for driving maternal behaviors. Neuron. 2018;98(1):192–207.

Farida WR, Sari AP, Inayah N, Nugroho HA. Observations of behavioral development on common Porcupines (Hystrix brachyura) undergoing domestication. IOP Conf Ser Earth Environ Sci. 2019;308:012076.

Feldhamer GA, Drickamer LC, Vessey SH, Merritt JF. Population growth and regulation. In: Mammalogy: adaptation, diversity, and ecology, vol. 1999. Boston: WCB/McGraw-Hill; 1999. p. 406–16.

Feldhamer GA, Rochelle JA, Rushton CD. Mountain beaver (Aplodontia rufa). In: Wild mammals of North America: biology, management, and conservation. Baltimore: Johns Hopkins; 2003. p. 179–87.

Felton TM, Linton LN, Rosenblatt JS, Morrell JI. Estrogen implants in the lateral habenular nucleus do not stimulate the onset of maternal behavior in female rats. Horm Behav. 1999;35(1):71–80.

Ferkin MH. Similarities between female meadow voles mating during post-partum oestrus and raising two concurrent litters and females raising only one litter. Reprod Fertil Dev. 2006;18(7):751–6.

Fleischer S, Kordower JH, Kaplan B, Dicker R, Smerling R, Ilgner J. Olfactory bulbectomy and gender differences in maternal behaviors of rats. Physiol Behav. 1981;26:957–9.

Fleming AS, Walsh C. Neuropsychology of maternal behavior in the rat: c-fos expression during mother-litter interactions. Psychoneuroendocrinology. 1994;19(5–7):429–43.

Fleming AS, Cheung U, Myhal N, Kessler Z. Effects of maternal hormones on "timidity" and attraction to pup-related odors in female rats. Physiol Behav. 1989;46(3):449–53.

Fleming AS, Corter C, Franks P, Surbey M, Schneider B, Steiner M. Postpartum factors related to mother's attraction to newborn infant odors. Dev Psychobiol. 1993;26(2):115–32.

Fokidis HB, Risch TS. The burden of motherhood: gliding locomotion in mammals influences maternal reproductive investment. J Mammal. 2008;89(3):617–25.

Foreman D, Garris D. Plasma progesterone levels and corpus luteum morphology in the female prairie dog (*Cynomys ludovicianus*). Gen Comp Endocrinol. 1984;55(2):315–22.

Fowler CD, Johnson F, Wang Z. Estrogen regulation of cell proliferation and distribution of estrogen receptor-alpha in the brains of adult female prairie and meadow voles. J Comp Neurol. 2005;489:166–79.

Fraser EJ, Shah NM. Complex chemosensory control of female reproductive behaviors. PLoS One. 2014;9:e90368.

Friedman MI, Bruno JP, Alberts JR. Physiological and behavioral consequences in rats of water recycling during lactation. J Comp Physiol Psychol. 1981;95(1):26.

Fuchs T, Iacobucci P, MacKinnon KM, Panksepp J. Infant-mother recognition in a social rodent (octodon degus). J Comp Psychol. 2010;124(2):166–75.

Gallagher JM, Nephew BC, Poirier G, King JA, Bridges RS. Estrogen receptor-alpha knockouts and maternal memory in nulliparous rats. Horm Behav. 2019;110:40–5.

Garland HO, Atherton JC, Baylis C, Morgan MRA, Milne CM. Hormone profiles for progesterone, oestradiol, prolactin, plasma renin activity, aldosterone and corticosterone during pregnancy and pseudopregnancy in two strains of rat: correlation with renal studies. J Endocrinol. 1987;113(3):435–44.

George W. Reproduction in female gundis (Rodentia: Ctenodactylidae). J Zool. 1978;185:57–71.

Ghiraldi LL, Svare B. Postpubertal isolation decreases infanticide and elevates parental care in C57BL/6J male mice. Physiol Behav. 1986;36:59–62.

Gibori G, Sridaran R. Sites of androgen and estradiol production in the second half of pregnancy in the rat. Biol Reprod. 1981;24(2):249–56.

Giordano AL, Ahdieh HB, Mayer AD, Siegel HI, Rosenblatt JS. Cytosol and nuclear estrogen receptor binding in the preoptic area and hypothalamus of female rats during pregnancy and ovariectomized, nulliparous rats after steroid priming: correlation with maternal behavior. Horm Behav. 1990;24(2):232–55.

Glasper ER, Hyer MM, Katakam J, Harper R, Ameri C, Wolz T. Fatherhood contributes to increased hippocampal spine density and anxiety regulation in California mice. Brain Behav. 2016;6.

González-Mariscal G, Melo AI. Neuroscience in the 21st century. New York: Springer Science and media; 2013.

Gosling LM, Baker SJ. Planning and monitoring an attempt to eradicate coypus from Britain. Symp Zool Soc Lond. 1987;58:99–113.

Grattan DR. 60 years of neuroendocrinology: the hypothalamo-prolactin axis. J Endocrinol. 2015;226(2):T101–22.

Gréco B, Lubbers LS, Blaustein JD. Estrogen receptor β messenger ribonucleic acid expression in the forebrain of proestrous, pregnant, and lactating female rats. Endocrinology. 2003;144(5):1869–75.

Grieb ZA, Tierney SM, Lonstein JS. Postpartum inhibition of ovarian steroid action increases aspects of maternal caregiving and reduces medial preoptic area progesterone receptor expression in female rats. Horm Behav. 2017;96:31–41.

Grieb ZA, Vitale EM, Morrell JI, Lonstein JS, Pereira M. Decreased mesolimbic dopaminergic signaling underlies the waning of maternal caregiving across the postpartum period in rats. Psychopharmacology. 2020;237(4):1107–19.

Groenewegen HJ, Berendse HW, Haber SN. Organization of the output of the ventral striatopallidal system in the rat: ventral pallidal efferents. Neuroscience. 1993;57(1):113–42.

Gromov V. Interactions in family pairs and parental care in captive steppe lemming (*Lagurus lagurus*). Contemp Probl Ecol. 2010;3:133–9.

Gromov VS. Paternal care in rodents: Ultimate causation and proximate mechanisms. Russ J Theriol. 2020;19:1–20.

Grota LJ, Eik-Nes KB. Plasma progesterone concentrations during pregnancy and lactation in the rat. J Reprod Fertil. 1967;13(1):83–91.

Grota LJ, Ader R. Continuous recording of maternal behaviour in *Rattus norvegicus*. Anim Behav. 1969;17(4):722–9.

Gubernick DJ. A maternal chemosignal maintains paternal behaviour in the biparental california mouse, Peromyscus californicus. Anim Behav. 1990;39:936–42.

Gubernick DJ, Alberts JR. Postpartum maintenance of paternal behaviour in the biparental California mouse, Peromyscus californicus. Anim Behav. 1989;37:656–64.

Gubernick DJ, Nelson RJ. Prolactin and paternal behavior in the biparental California mouse, *Peromyscus californicus*. Horm Behav. 1989;23:203–10.

Guimarães DA, Ohashi OM, Singh M, Vale W. Profile of plasmatic progesterone on pregnancy, and the postpartum estrus of Dasyprocta prymnolopha (Rodentia: Dasyproctidae). Rev Biol Trop. 2016;64(4):1519–26.

Guoynes C, Marler C. Paternal behavior from a neuroendocrine perspective. Neurosci: Oxf. Res. Encycl; 2020.

Hansen S, Sodersten P, Eneroth P. Mechanisms regulating hormone release and the duration of dioestrus in the lactating rat. J Endocrinol. 1983;99(2):173–80.

Hansen S, Harthon C, Wallin E, Lofberg L, Svensson K. The effects of 6-OHDA-induced dopamine depletions in the ventral or dorsal striatum on maternal and sexual behavior in the female rat. Pharmacol Biochem Behav. 1991;39:71–7.

Hansen S, Bergvall AH, Nyiredi S. Interaction with pups enhances dopamine release in the ventral striatum of maternal rats: a microdialysis study. Pharmacol Biochem Behav. 1993;45(3):673–6.

Hasler J, Banks E. Reproductive performance and growth in captive collared lemmings (*Dicrostonyx groenlandicus*). Can J Zool. 1975;53:777–87.

Hayden P, Gambino JJ, Lindberg RG. Laboratory breeding of the little pocket mouse, *Perognathus longimembris*. J Mammal. 1966;47:412–23.

Hayes UL, De Vries GJ. Role of pregnancy and parturition in induction of maternal behavior in prairie voles (*Microtus ochrogaster*). Horm Behav. 2007;51(2):265–72.

Hayes LD, Chesh AS, Castro RA, Tolhuysen LO, Burger JR, Bhattacharjee J, Ebensperger LA. Fitness consequences of group living in the degu Octodon degus, a plural breeder rodent with communal care. Anim Behav. 2009;78:131–9.

Hayssen V. Reproductive effort in squirrels: ecological, phylogenetic, allometric, and latitudinal patterns. J Mammal. 2008;89(3):582–606.

He Z, Zhang L, Hou W, Zhang X, Young LJ, Li L, Liu L, Ma H, Xun Y, Lv Z, Li Y, Jia R, Li J, Tai F. Paraventricular nucleus oxytocin subsystems promote active paternal behaviors in mandarin voles. J Neurosci. 2021;41:6699–713.

Hennessy MB, Jenkins R. A descriptive analysis of nursing behavior in the guinea pig (Cavia porcellus). J Comp Psychol. 1994;108(1):23.

Herrera EA, Salas V, Congdon ER, Corriale MJ, Tang-Martinez Z. Capybara social structure and dispersal patterns: variations on a theme. J Mammal. 2011;92(1):12–20.

Hickman GC. The maternal behavior of a Mexican pocket gopher (*Pappogeomys castanops*). Southwest Nat. 1975;20:142–4.

Holekamp KE, Talamantes F. Seasonal variation in circulating testosterone and oestrogens of wild-caught California ground squirrels (*Spermophilus beecheyi*). Reproduction. 1991;93(2):415–25.

Holekamp KE, Nunes S, Talamantes F. Patterns of progesterone secretion in free-living California ground squirrels (*Spermophilus beecheyi*). Biol Reprod. 1988;39(5):1051–9.

Hoogland JL. The black-tailed prairie dog: social life of a burrowing mammal. Chicago: University of Chicago Press; 1995.

Horrell ND, Hickmott PW, Saltzman W. Neural regulation of paternal behavior in mammals: sensory, neuroendocrine, and experiential influences on the paternal brain. SpringerLink. 2018:1–50.

Hume JM, Wynne-Edwards KE. Castration reduces male testosterone, estradiol, and territorial aggression, but not paternal behavior in biparental dwarf hamsters (*Phodopus campbelli*). Horm Behav. 2005;48:303–10.

Hume JM, Wynne-Edwards KE. Paternal responsiveness in biparental dwarf hamsters (*Phodopus campbelli*) does not require estradiol. Horm Behav. 2006;49:538–44.

Hyer MM, Hunter TJ, Katakam J, Wolz T, Glasper ER. Neurogenesis and anxiety-like behavior in male California mice during the mate's postpartum period. Eur J Neurosci. 2016;43:703–9.

Hyer MM, Khantsis S, Venezia AC, Madison FN, Hallgarth L, Adekola E, Glasper ER. Estrogen-dependent modifications to hippocampal plasticity in paternal California mice (*Peromyscus californicus*). Horm Behav. 2017;96:147–55.

Insel TR. Regional changes in brain oxytocin receptors post-partum: time-course and relationship to maternal behaviour. J Neuroendocrinol. 1990;2(4):539–45.

Izquierdo G, Lacey EA. Effects of group size on nest attendance in the communally breeding colonial tuco-tuco (Ctenomys sociabilis). Mamm Biol. 2008;73:438–43.

Jarvis JU. Eusociality in a mammal: cooperative breeding in naked mole-rat colonies. Science. 1981;212(4494):571–3.

Jenkins PD, Kilpatrick CW, Robinson MF, Timmins RJ. Morphological and molecular investigations of a new family, genus and species of rodent (Mammalia: Rodentia: Hystricognatha) from Lao PDR. Systematics and Biodiversity. 2 (4): 419–454. Corrigendum: Volume 2(4): 419–454. Syst Biodivers. 2005 3(3):343.

Jones JS, Wynne-Edwards KE. Paternal hamsters mechanically assist the delivery, consume amniotic fluid and placenta, remove fetal membranes, and provide parental care during the birth process. Horm Behav. 2000;37:116–25.

Juana L, Bárbara V-G, Martín M-T, Agustín C, Guillermo R-B, Guadalupe O. Neither testosterone levels nor aggression decrease when the male Mongolian gerbil (*Meriones unguiculatus*) displays paternal behavior. Horm Behav. 2010;57:271–5.

Kalinichev M, Rosenblatt JS, Nakabeppu Y, Morrell JI. Induction of c-fos-like and fosB-like immunoreactivity reveals forebrain neuronal populations involved differentially in pup-mediated maternal behavior in juvenile and adult rats. J Comp Neurol. 2000;416(1):45–78.

Katz LF, Ball GF, Nelson RJ. Elevated Fos-like immunoreactivity in the brains of postpartum female prairie voles, *Microtus ochrogaster*. Cell Tissue Res. 1999;298(3):425–35.

Kawamichi M, Kawamichi T. Maternal care and independence of young Siberian chipmunks. Honyurui Kagaku [Mammal Sci]. 1984;48:3–17.

Kawamichi M. Nest structure dynamics and seasonal use of nests by Siberian chipmunks (Eutamias sibiricus). J Mammalogy 1989;71:44–57.

Kawamichi T, Kawamichi M. Gestation period and litter size of Siberian chipmunk Eutamias sibiricus lineatus in Hokkaido, northern Japan. J Mammal Soc Jpn. 1993;18(2):105–9.

Keeley T, Goodrowe KL, Graham L, Howell C, MacDonald SE. The reproductive endocrinology and behavior of Vancouver Island marmot (*Marmota vancouverensis*). Zoo Biol. 2012;31(3):275–90.

Kelly AM, Hiura LC, Saunders AG, Ophir AG. Oxytocin neurons exhibit extensive functional plasticity due to offspring age in mothers and fathers. Integr Comp Biol. 2017;57(3):603–18.

Kenagy GJ, Barnes BM. Environmental and endogenous control of reproductive function in the Great Basin pocket mouse *Perognathus parvus*. Biol Reprod. 1984;31:637–45.

Kenagy GJ, Bartholomew GA. Seasonal reproductive patterns in five coexisting California desert rodent species. Ecol Monogr. 1985;55:372–96. https://doi.org/10.2307/2937128.

Kenkel WM, Paredes J, Yee JR, Pournajafi-Nazarloo H, Bales KL, Carter CS. Neuroendocrine and behavioural responses to exposure to an infant in male prairie voles. J Neuroendocrinol. 2012;24:874–86.
Kenkel WM, Suboc G, Sue Carter C. Autonomic, behavioral and neuroendocrine correlates of paternal behavior in male prairie voles. Physiol Behav. 2014;128:252–9. https://doi.org/10.1016/j.physbeh.2014.02.006.
Kercmar J, Snoj T, Tobet SA, Majdic G. Gonadectomy prior to puberty decreases normal parental behavior in adult mice. Horm Behav. 2014;66:667–73. https://doi.org/10.1016/j.yhbeh.2014.09.007.
Kinsley CH, Bridges RS. Morphine treatment and reproductive condition alter olfactory preferences for pup and adult male odors in female rats. Dev Psychobiol. 1990;23(4):331–47.
Kinsley GH, Lambert KG. Reproduction-induced neuroplasticity: natural behavioural and neuronal alterations associated with the production and care of offspring. J Neuroendocrinol. 2008;20(4):515–25.
Kirkpatrick B, Kim JW, Insel TR. Limbic system fos expression associated with paternal behavior. Brain Res. 1994a;658:112–8. https://doi.org/10.1016/S0006-8993(09)90016-6.
Kirkpatrick B, Williams JR, Slotnick BM, Carter CS. Olfactory bulbectomy decreases social behavior in male prairie voles (*M. ochrogaster*). Physiol Behav. 1994b;55:885–9. https://doi.org/10.1016/0031-9384(94)90075-2.
Kirkpatrick B, Carter CS, Newman SW, Insel TR. Axon-sparing lesions of the medial nucleus of the amygdala decrease affiliative behaviors in the prairie vole (*Microtus ochrogaster*): behavioral and anatomical specificity. Behav Neurosci. 1994c;108:501–13.
Kleiman DG. Maternal behaviour of the green acouchi (Myoprocta pratti Pocock), a South American caviomorph rodent. Behaviour. 1972;43(1/4):48–84.
Kober M, Trillmich F, Naguib M. Vocal mother-offspring communication in guinea pigs: females adjust maternal responsiveness to litter size. Front Zool. 2008;5:13.
Koch M. Immunocytochemical localization and quantitation of estrogen-binding cells in the male and female (virgin, pregnant, lactating) mouse brain. Brain Res. 1989;489(1):101–12.
Kohl J, Babayan BM, Rubinstein ND, Autry AE, Marin-Rodriguez B, Kapoor V, Miyamishi K, Zweifel LS, Luo L, Uchida N, Dulac C. Functional circuit architecture underlying parental behaviour. Nature. 2018;556:326–31.
Kölliker M. The evolution of parental care. New York: Oxford University Press; 2012.
Kraus C, Trillmich F, Kunkele J. Reproduction and growth in a precocial small mammal, Cavia magna. J Mammal. 2005;86(4):763–72.
Kuroda KO, Meaney MJ, Uetani N, Fortin Y, Ponton A, Kato T. ERK-FosB signaling in dorsal MPOA neurons plays a major role in the initiation of parental behavior in mice. Mol Cell Neurosci. 2007;36(2):121–31.
Lacey EA, Sherman PW. Social organization of naked mole-rat colonies: evidence for a division of labor. In: Sherman PW, Jarvis JUM, Alexander RD, editors. The biology of the naked mole-rat. Princeton: Princeton University Press; 1991. p. 275–337.
Lacey EA, Wieczorek JR. Kinship in colonial tuco-tucos: evidence from group composition and population structure. Behav Ecol. 2004;15:988–96.
Lambert KG, Franssen CL, Hampton JE, Rzucidlo AM, Hyer MM, True M, Kaufman C, Bardi M. Modeling paternal attentiveness: distressed pups evoke differential neurobiological and behavioral responses in paternal and nonpaternal mice. Neuroscience. 2013;234:1–12.
Lau BYB, Krishnan K, Huang ZJ, Shea SD. Maternal experience-dependent cortical plasticity in mice is circuit- and stimulus-specific and requires MECP2. J Neurosci. 2020;40:1964–19.
Laurie EMO, Goodrich ES. The reproduction of the house-mouse (*Mus musculus*) living in different environments. Proc R Soc Lond Ser B Biol Sci. 1946;133:248–81.
Lee AW, Brown RE. Medial preoptic lesions disrupt parental behavior in both male and female California mice (*Peromyscus californicus*). Behav Neurosci. 2002;116:968–75.

Lee AW, Brown RE. Comparison of medial preoptic, amygdala, and nucleus accumbens lesions on parental behavior in California mice (*Peromyscus californicus*). Physiol Behav. 2007;92:617–28. https://doi.org/10.1016/j.physbeh.2007.05.008.

Lee A, Clancy S, Fleming AS. Mother rats bar-press for pups: effects of lesions of the MPOA and limbic sites on maternal behavior and operant responding for pup reinforcement. Behav Brain Res. 2000;108:215–31.

Li Y, Lian Z, Wang B, Tai F, Wu R, Hao P, Qiao X. Natural variation in paternal behavior is associated with central estrogen receptor alpha and oxytocin levels. J Comp Physiol A Neuroethol Sens Neural Behav Physiol. 2015;201:285–93.

Lima SGC, Sousa-Lima RS, Tokumaru RS, Nogueira-Filho SLG, Nogueira SSC. Vocal complexity and sociality in spotted paca (Cuniculus paca). PLoS ONE. 2018;13(1):e0190961.

Liu RC, Linden JF, Schreiner CE. Improved cortical entrainment to infant communication calls in mothers compared with virgin mice. Eur J Neurosci. 2006;23:3087–97.

Lonstein JS. Effects of dopamine receptor antagonism with haloperidol on nurturing behavior in the biparental prairie vole. Pharmacol Biochem Behav. 2002;74:11–9.

Lonstein JS, De Vries GJ. Comparison of the parental behavior of pair-bonded female and male prairie voles (*Microtus ochrogaster*). Physiol Behav. 1999;66:33–40.

Lonstein JS, De Vries GJ. Influence of gonadal hormones on the development of parental behavior in adult virgin prairie voles (*Microtus ochrogaster*). Behav Brain Res. 2000;114:79–87. https://doi.org/10.1016/S0166-4328(00)00192-3.

Lonstein JS, De Vries GJ. Social influences on parental and nonparental responses toward pups in virgin female prairie voles (Microtus ochrogaster). J Comp Psychol. 2001;115(1):53–61.

Lonstein JS, Gammie SC. Sensory, hormonal, and neural control of maternal aggression in laboratory rodents. Neurosci Biobehav Rev. 2002;26(8):869–88.

Lonstein JS, Gréco B, De Vries GJ, Stern JM, Blaustein JD. Maternal behavior stimulates c-fos activity within estrogen receptor alpha-containing neurons in lactating rats. Neuroendocrinology. 2000;72(2):91–101.

Lonstein JS, Pereira M, Morrell JI, Marler CA. Parenting behavior. In: Knobil and Neill's physiology of reproduction. New York: Academic; 2015. p. 2371–437.

Lovejoy BP, Black HC. Growth and weight of the mountain beaver, *Aplodontia rufa pacifica*. J Mammal. 1974;55(2):364–9.

Lucas BK, Ormandy CJ, Binart N, Bridges RS, Kelly PA. Null mutation of the prolactin receptor gene produces a defect in maternal behavior. Endocrinology. 1998;139(10):4102–7.

Luis J, Cadena C, Zedillo B, Reyes J, Ramos G, Martínez M. Testosterone replacement induced paternal behaviour in the Mexican volcano mouse Neotomodon alstoni (*Rodentia Muridae*). Ethol Ecol Evol. 2012;24:275–83.

Ma E, Lau J, Grattan DR, Lovejoy DA, Wynne-Edwards KE. Male and female prolactin receptor mRNA expression in the brain of a biparental and a uniparental hamster, phodopus, before and after the birth of a litter. J Neuroendocrinol. 2005;17:81–90.

Macbeth AH, Stepp JE, Lee HJ, Young WS III, Caldwell HK. Normal maternal behavior, but increased pup mortality, in conditional oxytocin receptor knockout females. Behav Neurosci. 2010;124(5):677.

Macdonald D. The encyclopedia of mammals. New York: Facts on File Publications; 1984.

Macdonald DW, Herrera EA, Taber AB, Moreira JR. Social organization and resource use in capybaras and maras. Chapter 33. In: Wolff JO, Sherman PW, editors. Rodent societies: an ecological and evolutionary perspective. Chicago: University of Chicago Press; 2007. p. 393–402.

Maggi A. Liganded and unliganded activation of estrogen receptor and hormone replacement therapies. Biochim Biophys Acta (BBA)-Mol Basis Dis. 2011;1812(8):1054–60.

Mann PE, Bridges RS. Prolactin receptor gene expression in the forebrain of pregnant and lactating rats. Mol Brain Res. 2002;105(1–2):136–45.

Marin G, Pilastro A. Communally breeding dormice, Glis glis, are close kin. Anim Behav. 1994;47(6):1485–7.

Marlin BJ, Mitre M, D'amour JA, Chao MV, Froemke RC. Oxytocin enables maternal behaviour by balancing cortical inhibition. Nature. 2015;520(7548):499–504.

Martin P. Movements and activities of the mountain beaver (*Aplodontia rufa*). J Mammal. 1971;52(4):717–23.

Martínez A, Ramos G, Martínez-Torres M, Nicolás L, Carmona A, Cárdenas M, Luis J. Paternal behavior in the Mongolian gerbil (*Meriones unguiculatus*): estrogenic and androgenic regulation. Horm Behav. 2015;71:91–5.

Mattheij JA, Swarts HJ, van Mourik S. Plasma prolactin in the rat during suckling without prior separation from pups. Eur J Endocrinol. 1985;108(4):468–74.

Mayer AD, Rosenblatt JS. Prepartum changes in maternal responsiveness and nest defense in *Rattus norvegicus*. J Comp Psychol. 1984;98(2):177.

McAdam AG, Boutin S, Sykes AK, Humphries MM. Life histories of female red squirrels and their contributions to population growth and lifetime fitness. Ecoscience. 2007;14(3):362–9.

McCarthy MM, vom Saal FS. Infanticide by virgin CF-1 and wild male house mice (*Mus musculus*): effects of age, prolonged isolation, and testing procedure. Dev Psychobiol. 1986;19:279–90.

McFadyen-Ketchum SA, Porter RH. Transmission of food preferences in spiny mice (Acomys cahirinus) via nose-mouth interaction between mothers and weanlings.Behavioral Ecology and Sociobiology, 1989;24(1):59–62.

McGuire B, Novak M. A comparison of maternal behaviour in the meadow vole (*Microtus pennsylvanicus*), prairie vole (*M. ochrogaster*) and pine vole (*M. pinetorum*). Anim Behav. 1984;32:1132–41.

McGuire B, Russell KD, Mahoney T, Novak M. The effects of mate removal on pregnancy success in prairie voles (*Microtus ochrogaster*) and meadow voles (*Microtus pennsylvanicus*). Biol Reprod. 1992;47:37–42.

McGuire B, Vermeylen F, Bemis WE. Suckling behaviour in three species of voles. Behaviour. 2011;148:551–73.

McHenry JA, Otis JM, Rossi MA, Robinson JE, Kosyk O, Miller NW, McElligott ZA, Budygin EA, Rubinow DR, Stuber GD. Hormonal gain control of a medial preoptic area social reward circuit. Nat Neurosci. 2017;20(3):449–58.

Meddle SL, Bishop VR, Gkoumassi E, van Leeuwen FW, Douglas AJ. Dynamic changes in oxytocin receptor expression and activation at parturition in the rat brain. Endocrinology. 2007;148(10):5095–104.

Mertz L. New world porcupines. In: Grzimek B, editor. Thomson Gale's Grzimek's animal life encyclopedia, vol. 13. New York/Cincinnati/Chicago: Van Nostrand Reinhold Co; 2003.

Miceli MO, Malsbury CW. Sagittal knife cuts in the near and far lateral preoptic area-hypothalamus disrupt maternal behaviour in female hamsters. Physiol Behav. 1982;28(5):856–67.

Michener GR. Maternal behaviour in Richardson's ground squirrel, *Spermophilus richardsonii richardsonii*: retrieval of young by lactating females. Anim Behav. 1971;19(4):653–6.

Mikkelsen E, Lauridsen H, Nielsen PM, Qi H, Nørlinger T, Andersen MD, Uldbjerg N, Laustsen C, Sandager P, Pedersen M. The chinchilla as a novel animal model of pregnancy. R Soc Open Sci. 2017;4(4):161098.

Miller A, Jentz E, Duncan C. Progestogen metabolites for use in pregnancy monitoring of 13-lined ground squirrels (*Ictidomys tridecemlineatus*). Reprod Fertil. 2021;2(2):81–8.

Mitra SW, Hoskin E, Yudkovitz J, Pear L, Wilkinson HA, Hayashi S, Pfaff DW, Ogawa S, Rohrer SP, Schaeffer JM, McEwen BS. Immunolocalization of estrogen receptor β in the mouse brain: comparison with estrogen receptor α. Endocrinology. 2003;144(5):2055–67.

Mogenson GJ, Jones DL, Yim CY. From motivation to action: functional interface between the limbic system and the motor system. Prog Neurobiol. 1980;14:69–97.

Mori E, Lovari S, Sforzi A, Romeo G, Pisani C, Massolo A, Fattorini L. Patterns of spatial overlap in a monogamous large rodent, the crested porcupine. Behav Process. 2014;107:112–8.

Mori E, Menchetti M, Lucherini M, Sforzi A, Lovari S. Timing of reproduction and paternal cares in the crested porcupine. Mamm Biol. 2016;81:345–9.

Morris DJ, Van Aarde RJ. Sexual behavior of the female porcupine Hystrix africaeaustralis. Horm Behav. 1985;19(4):400–12.
Murakami G. Distinct effects of estrogen on mouse maternal behavior: the contribution of estrogen synthesis in the brain. PLoS One. 2016;11(3):e0150728.
Muul I. Intra-and inter-familial behaviour of *Glaucomys volans* (Rodentia) following parturition. Anim Behav. 1970;18:20–5.
Naguib M, Kober M, Trillmich F. Mother is not like mother: concurrent pregnancy reduces lactating guinea pigs' responsiveness to pup calls. Behav Process. 2010;83(1):79–81.
Naughton D. The natural history of Canadian mammals. Toronto: University of Toronto Press; 2012.
Neckers LM, Zarrow MX, Myers MM, Denenberg VH. Influence of olfactory bulbectomy and the serotonergic system upon intermale aggression and maternal behavior in the mouse. Pharmacol Biochem Behav. 1975;3:545–50. https://doi.org/10.1016/0091-3057(75)90170-7.
Nephew BC, Byrnes EM, Bridges RS. Vasopressin mediates enhanced offspring protection in multiparous rats. Neuropharmacology. 2010;58(1):102–6.
Nimje PS, Tinnesand HV, Buesching C, Sæbø M, Senn H, Zedrosser A, Rosell F. Almost faithful: SNP markers reveal low levels of extra-pair paternity in the Eurasian beavers. PeerJ. 2019;7:e27866v1.
Nishimori K, Takayanagi Y, Yoshida M, Kasahara Y, Young LJ, Kawamata M. New aspects of oxytocin receptor function revealed by knockout mice: sociosexual behaviour and control of energy balance. Prog Brain Res. 2008;170:79–90.
Niswender GD, Juengel JL, McGuire WJ, Belfiore CJ, Wiltbank MC. Luteal function: the estrous cycle and early pregnancy. Biol Reprod. 1994;50(2):239–47.
Nolte DL, Farley JP, Campbell DL, Epple GM, Mason JR. Potential repellents to prevent mountain beaver damage. Crop Prot. 1993;12(8):624–6.
Nowak RM. Walker's Mammals of the World. 6th ed. Baltimore: Johns Hopkins University Press; 1999.
Novak RM. 5th edn. The Johns Hopkins University Press; Baltimore, MD: 1991. Walker's mammals of the world.
Numan M. Progesterone inhibition of maternal behavior in the rat. Horm Behav. 1978;11(2):209–31.
Numan M. The parental brain: mechanisms, development, and evolution. New York: Oxford University Press; 2020.
Numan M, Stolzenberg DS. Medial preoptic area interactions with dopamine neural systems in the control of the onset and maintenance of maternal behavior in rats. Front Neuroendocrinol. 2009;30(1):46–64.
Numan M, Rosenblatt JS, Komisaruk BR. Medial preoptic area and onset of maternal behavior in the rat. J Comp Physiol Psychol. 1977;91(1):146.
Numan M, Roach JK, del Cerro MCR, Guillamón A, Segovia S, Sheehan TP, Numan MJ. Expression of intracellular progesterone receptors in rat brain during different reproductive states, and involvement in maternal behavior. Brain Res. 1999;830(2):358–71.
Nunes S, Muecke EM, Ross HE, Bartholomew PA, Holekamp KE. Food availability affects behavior but not circulating gonadal hormones in maternal Belding's ground squirrels. Physiol Behav. 2000;71(5):447–55.
Nutt KJ. Genetic reconstruction of breeding patterns in gundis (Rodentia: Ctenodactylidae). Behav Ecol Sociobiol. 2007;61(11):1651–63.
Ogawa S, Eng V, Taylor J, Lubahn DB, Korach KS, Pfaff DW. Roles of estrogen receptor-α gene expression in reproduction-related behaviors in female mice. Endocrinology. 1998;139(12):5070–81.
Ojasti J. Estudio biologico del chigüire o capibara. Caracas: Ed. Fondo Nacional de Investigaciones Agropecuarias; 1973.
Okabe S, Tsuneoka Y, Takahashi A, Ooyama R, Watarai A, Maeda S, Honda Y, Nagasawa M, Mogi K, Nishimori K, Kuroda M. Pup exposure facilitates retrieving behavior via the oxytocin neural system in female mice. Psychoneuroendocrinology. 2017;79:20–30.

Oliveras D, Novak M. A comparison of paternal behaviour in the meadow vole *Microtus pennsylvanicus*, the pine vole *M. pinetorum* and the prairie vole *M. ochrogaster*. Anim Behav. 1986;34:519–26.

Olson ER, Carlson MR, Ramanujam VMS, Sears L, Anthony SE, Anich PS, Ramon L, Hulstrand A, Jurewicz M, Gunnelson AS, Kohler AM, Martin JG. Vivid biofluorescence discovered in the nocturnal Springhare (Pedetidae). Sci Rep. 2021;11:4125.

Ostwald R, Wilken K, Simons J, Highstone H, Cimino S, Shimondle S. Influence of photoperiod and partial contact on estrus in the desert pocket mouse, *Perognathus penicillatus*. Biol Reprod. 1972;7:1–8.

Paco A. Repetitive DNA sequences in rodentia genomes. Its involvement in chromosome architecture reshuffling and in genome functionality. Doctoral thesis, Universidade de Trás-os-Montes e Alto Douro, Portugal; 2014.

Parker KJ, Lee TM. Central vasopressin administration regulates the onset of facultative paternal behavior in *Microtus pennsylvanicus* (meadow voles). Horm Behav. 2001;39:285–94. https://doi.org/10.1006/hbeh.2001.1655.

Pedersen CA, Boccia ML. Oxytocin antagonism alters rat dams' oral grooming and upright posturing over pups. Physiol Behav. 2003;80(2–3):233–41.

Pedersen CA, Prange AJ. Induction of maternal behavior in virgin rats after intracerebroventricular administration of oxytocin. Proc Natl Acad Sci. 1979;76(12):6661–5.

Pedersen CA, Ascher JA, Monroe YL, Prange AJ. Oxytocin induces maternal behavior in virgin female rats. Science. 1982;216(4546):648–50.

Pedersen CA, Caldwell JD, Johnson MF, Fort SA, Prange AJ Jr. Oxytocin antiserum delays onset of ovarian steroid-induced maternal behavior. Neuropeptides. 1985;6(2):175–82.

Pedersen CA, Caldwell JD, Walker C, Ayers G, Mason GA. Oxytocin activates the postpartum onset of rat maternal behavior in the ventral tegmental and medial preoptic areas. Behav Neurosci. 1994;108(6):1163.

Pedersen CA, Vadlamudi SV, Boccia ML, Amico JA. Maternal behavior deficits in nulliparous oxytocin knockout mice. Genes Brain Behav. 2006;5(3):274–81.

Peinke DM, Brown CR. Burrow utilization by springhares (*Pedetes capensis*) in the Eastern Cape. S Afr Afr Zool. 2005;40(1):37–44.

Perea-Rodriguez JP, Takahashi EY, Amador TM, Hao RC, Saltzman W, Trainor BC. Effects of reproductive experience on central expression of progesterone, oestrogen α, oxytocin and vasopressin receptor mRNA in male California mice (*Peromyscus californicus*). J Neuroendocrinol. 2015;27:245–52.

Pereira M. Nonselective maternal bonding but pup recognition in the subterranean rodent Ctenomys pearsoni. J Comp Psychol. 2006;120(4):411–5.

Pereira M, Ferreira A. Demanding pups improve maternal behavioral impairments in sensitized and haloperidol-treated lactating female rats. Behav Brain Res. 2006;175(1):139–48.

Pereira M, Ferreira A. Neuroanatomical and neurochemical basis of parenting: dynamic coordination of motivational, affective and cognitive processes. Horm Behav. 2016;77:72–85.

Pereira M, Morrell JI. The changing role of the medial preoptic area in the regulation of maternal behavior across the postpartum period: facilitation followed by inhibition. Behav Brain Res. 2009;205(1):238–48.

Pereira M, Morrell JI. Functional mapping of the neural circuitry of rat maternal motivation: effects of site-specific transient neural inactivation. J Neuroendocrinol. 2011;23(11):1020–35.

Pereira M, Seip KM, Morrell JI. Maternal motivation and its neural substrate across the postpartum period. In: Bridges RS, editor. The neurobiology of the parental mind. San Diego: Elsevier Academic Press; 2008. p. 39–59.

Perry AN, Sue Carter C, Cushing BS. Effects of postnatal estrogen manipulations on juvenile alloparental behavior. Horm Behav. 2015;75:11–7. https://doi.org/10.1016/j.yhbeh.2015.07.019.

Pfeiffer EW. The reproductive cycle of the female mountain beaver. J Mammal. 1958;39(2):223–35.

Phifer-Rixey M, Nachman MW. Insights into mammalian biology from the wild house mouse *Mus musculus*. elife. 2015;4:e05959.

Pi XJ, Grattan DR. Increased prolactin receptor immunoreactivity in the hypothalamus of lactating rats. J Neuroendocrinol. 1999;11(9):693–705.

Pi X, Zhang B, Li J, Voogt JL. Promoter usage and estrogen regulation of prolactin receptor gene in the brain of the female rat. Neuroendocrinology. 2003;77(3):187–97.

Pilastro A. Communal nesting between breeding females in a free-living population of fat dormouse (*Glis glis L.*). Ital J Zool. 1992;59(1):63–8.

Pilastro A, Missiaglia E, Marin G. Age-related reproductive success in solitarily and communally nesting female dormice (Glis glis). J Zool. 1996;239(3):601–8.

Pizzimenti JJ, McClenaghan LR Jr. Reproduction, growth and development, and behavior in the Mexican prairie dog, *Cynomys mexicanus* (Merriam). Am Midl Nat. 1974;92:130–45.

Pose S, Zuluaga MJ, Ferreño M, Agrati D, Bedó G, Uriarte N. Raising overlapping litters: differential activation of rat maternal neural circuitry after interacting with newborn or juvenile pups. J Neuroendocrinol. 2019;31(9):e12701.

Potapov MA, Zadubrovskaya IV, Zadubrovskii PA, Potapova OF, Evsikov VI. Mating systems in the steppe lemming (*Lagurus lagurus*) and narrow-skulled vole (*Microtus gregalis*) from the Northern Kulunda Steppe. Russ J Ecol. 2012;43:40–4.

Randall JA, Matocq MD. Why do kangaroo rats (*Dipodomys spectabilis*) footdrum at snakes? Behav Ecol. 1997;8:404–13. https://doi.org/10.1093/beheco/8.4.404.

Reburn CJ, Wynne-Edwards KE. Hormonal changes in males of a naturally biparental and a uniparental mammal. Horm Behav. 1999;35:163–76. https://doi.org/10.1006/hbeh.1998.1509.

Reichman OJ. Behavior of desert heteromyids. Gt Basin Nat Mem. 1983:77–90.

Reisbick S, Rosenblatt JS, Mayer AD. Decline of maternal behavior in the virgin and lactating rat. J Comp Physiol Psychol. 1975;89(7):722–32.

Ribeiro AC, Musatov S, Shteyler A, Simanduyev S, Arrieta-Cruz I, Ogawa S, Pfaff DW. siRNA silencing of estrogen receptor-α expression specifically in medial preoptic area neurons abolishes maternal care in female mice. Proc Natl Acad Sci. 2012;109(40):16324–9.

Rich ME, deCárdenas EJ, Lee HJ, Caldwell HK. Impairments in the initiation of maternal behavior in oxytocin receptor knockout mice. PLoS One. 2014;9(6):e98839.

Rioja-Paradela TM, Scott-Morales LM, Cotera-Correa M, Estrada-Castillón E. Reproduction and behavior of the Mexican prairie dog (*Cynomys mexicanus*). Southwest Nat. 2008;53(4):520–3.

Risold PY, Thompson RH, Swanson LW. The structural organization of connections between hypothalamus and cerebral cortex. Brain Res Brain Res Rev. 1997;19;24(2-3):197–254.

Rosenblatt JS. Nonhormonal basis of maternal behavior in the rat. Science. 1967;156(3781):1512–3.

Rosenblatt JS. The development of maternal responsiveness in the rat. Am J Orthopsychiatry. 1969;39(1):36–56.

Rosenblatt JS. Prepartum and postpartum regulation of maternal behaviour in the rat. CIBA Found Symp. 1975;33:17–37.

Rosenblatt JS. Psychobiology of maternal behavior: Contributions to the clinical understanding of maternal behavior among humans. Acta Paediatrica. 1994;83:3–8.

Rosenblatt JS, Siegel HI. Hysterectomy-induced maternal behavior pregnancy in the rat. J Comp Physiol Psychol. 1975;89(7):685.

Rosenblatt JS, Hazelwood S, Poole J. Maternal behavior in male rats: effects of medial preoptic area lesions and presence of maternal aggression. Horm Behav. 1996;30:201–15. https://doi.org/10.1006/hbeh.1996.0025.

Rosi MI, Puig S, Videla F, Cona MI, Roig VG. Ciclo reproductivo y estructura etaria de Ctenomys mendocinus (Rodentia, Ctenomyidae) del Piedemonte de Mendoza, Argentina. Ecologfa Austral 1996;6:87–93.

Royle, NJ, Smiseth, PT, Kölliker, M. The Evolution of Parental Care. OUP Oxford; 2012.

Rubin BS, Menniti FS, Bridges RS. Intracerebroventricular administration of oxytocin and maternal behavior in rats after prolonged and acute steroid pretreatment. Horm Behav. 1983;17(1):45–53.

Ruf T, Fietz J, Schlund W, Bieber C. High survival in poor years: life history tactics adapted to mast seeding in the edible dormouse. Ecology. 2006;87(2):372–81.

Russell RJ. Evolution and classification of the pocket gophers of the subfamily *Geomyinae*. 2012.
Saltzman W, Ziegler TE. Functional significance of hormonal changes in mammalian fathers. J Neuroendocrinol. 2014;26:685–96. https://doi.org/10.1111/jne.12176.
Saltzman W, Harris BN, De Jong TR, Perea-Rodriguez JP, Horrell ND, Zhao M, Andrew JR. Paternal care in biparental rodents: intra- and inter-individual variation. Integr Comp Biol. 2017;57 https://doi.org/10.1093/icb/icx047.
Sanyal MK. Secretion of progesterone during gestation in the rat. J Endocrinol. 1978;79(2):179–90.
Sarli J, Lutermann H, Alagaili AN, Mohammed OB, Bennett NC. Seasonal reproduction in the Arabian spiny mouse, *Acomys dimidiatus* (Rodentia: Muridae) from Saudi Arabia: the role of rainfall and temperature. J Arid Environ. 2016;124:352–9. https://doi.org/10.1016/j.jaridenv.2015.09.008.
Schiavo JK, Valtcheva S, Bair-Marshall CJ, Song SC, Martin KA, Froemke RC. Innate and plastic mechanisms for maternal behaviour in auditory cortex. Nature. 2020;587:426–31.
Schiml PA, Hennessy MB. Light-dark variation and changes across the lactational period in the behaviors of undisturbed mother and infant guinea pigs (Cavia porcellus). J Comp Psychol. 1990;104(3):283.
Schneider JE, Wade GN. Effects of maternal diet, body weight and body composition on infanticide in Syrian hamsters. Physiol Behav. 1989;46(5):815–21.
Schneider JS, Burgess C, Horton TH, Levine JE. Effects of progesterone on male-mediated infant-directed aggression. Behav Brain Res. 2009;199:340–4. https://doi.org/10.1016/j.bbr.2008.12.019.
Schum JE, Wynne-Edwards KE. Estradiol and progesterone in paternal and non-paternal hamsters (*Phodopus*) becoming fathers: conflict with hypothesized roles. Horm Behav. 2005;47:410–8. https://doi.org/10.1016/j.yhbeh.2004.11.009.
Schneider JS, Stone MK, Wynne-Edwards KE, Horton TH, Lydon J, O'Malley B, Levine JE. Progesterone receptors mediate male aggression toward infants. Proceedings of the National Academy of Sciences of the United States of America, 2003;100(5):2951–2956.
Scott CJ, Tilbrook AJ, Simmons DM, Rawson JA, Chu S, Fuller PJ, Ing NH, Clarke IJ. The distribution of cells containing estrogen receptor-alpha (ERalpha) and ERbeta messenger ribonucleic acid in the preoptic area and hypothalamus of the sheep: comparison of males and females. Endocrinology. 2000;141(8):2951–62.
Sever Z, Mendelssohn H. Copulation as a possible mechanism to maintain monogamy in porcupines, Hystrix indica. Anim Behav. 1988;36:1541–2.
Sever Z, Mendelssohn H. Paternal behavior in porcupines. Isr J Zool. 1989;36:172–3.
Sherman PW, Braude S, Jarvis JUM. Litter sizes and mammary numbers of naked mole-rats: breaking the one-half rule. J Mammal. 1999;80(3):720–33.
Shilton CM, Brooks RJ. Paternal care in captive collared lemmings (*Dicrostonyx richardsoni)* and its effect on development of the offspring. Can J Zool. 1989;67:2740–5. https://doi.org/10.1139/z89-388.
Shughrue PJ, Lane MV, Merchenthaler I. Comparative distribution of estrogen receptor-α and-β mRNA in the rat central nervous system. J Comp Neurol. 1997;388(4):507–25.
Siegel HI, Greenwald GS. Prepartum onset of maternal behavior in hamsters and the effects of estrogen and progesterone. Horm Behav. 1975;6(3):237–45.
Siegel HI, Rosenblatt JS. Estrogen-induced maternal behavior in hysterectomized-ovariectomized virgin rats. Physiol Behav. 1975;14(4):465–71.
Simerly RB, Swanson LW. The organization of neural inputs to the medial preoptic nucleus of the rat. J Comp Neurol. 1986;246:312–42.
Simerly RB, Swanson LW, Chang C, Muramatsu M. Distribution of androgen and estrogen receptor mRNA-containing cells in the rat brain: an in situ hybridization study. J Comp Neurol. 1990;294(1):76–95.
Smiley, KO, Brown, RSE, and Grattan, DR. Prolactin action in the brain is required for paternal care in the mouse. BioRxiv 2021. https://doi.org/10.1101/2021.12.30.474525.

Smiley, KO, Ladyman, SR, Gustafson, P, Grattan, DR, Brown, RSE. Neuroendocrinology and Adaptive Physiology of Maternal Care. In: Neuroendocrine Regulation of Behavior. Springer International Publishing, 2019;161–210.

Smolen MJ, Genoways HH, Baker RJ. Demographic and Reproductive Parameters of the Yellow-Cheeked Pocket Gopher (*Pappogeomys castanops*). J Mammal. 1980;61:224–36. https://doi.org/10.2307/1380043.

Stagkourakis S, Smiley KO, Williams P, Kakadellis S, Ziegler K, Bakker J, Brown RSE, Harkany T, Grattan DR, Broberger C. A neuro-hormonal circuit for paternal behavior controlled by a hypothalamic network oscillation. Cell. 2020;182:960–75.e15.

Stern JM. Parturition influences initial pup preferences at later onset of maternal behavior in primiparous rats. Physiol Behav. 1985;35(1):25–31.

Stern JM. Multisensory regulation of maternal behavior and masculine sexual behavior: a revised view. Neurosci Biobehav Rev. 1990;14(2):183–200.

Stern JM. Somatosensation and maternal care in Norway rats. In: Advances in the study of behavior, vol. 25. London: Academic; 1996. p. 243–94.

Stern JM, Johnson SK. Ventral somatosensory determinants of nursing behavior in Norway rats. I. Effects of variations in the quality and quantity of pup stimuli. Physiol Behav. 1990;47(5):993–1011.

Stern JM, Siegel HI. Prolactin release in lactating, primiparous and multiparous thelectomized and maternal virgin rats exposed to pup stimuli. Biol Reprod. 1978;19(1):177–82.

Stier CH, Mensah GA, Gall CF. Elevage d'aulacodes (Thryonomys swinderianus) pour la production de viande. World Anim Rev. 1991;69:44–9.

Stolzenberg DS, Rissman EF. Oestrogen-independent, experience-induced maternal behaviour in female mice. J Neuroendocrinol. 2011;23(4):345–54.

Sturgis JD, Bridges RS. N-Methyl-DL-aspartic acid lesions of the medial preoptic area disrupt ongoing parental behavior in male rats. Physiol Behav. 1997;62:305–10. https://doi.org/10.1016/S0031-9384(97)88985-8.

Swanson LW. An autoradiographic study of the efferent connections of the preoptic region in the rat. J Comp Neurol 1976;167(2):227–256.

Swerdlow NR, Koob GF. Lesions of the dorsomedial nucleus of the thalamus, medial prefrontal cortex and pedunculopontine nucleus: effects on locomotor activity mediated by nucleus accumbens-ventral pallidal circuitry. Brain Res. 1987;412(2):233–43.

Syrůčková A, Saveljev EP, Frosch C, Durka W, Savelyev AA, Munclinger P. Genetic relationships within colonies suggest genetic monogamy in the Eurasian beaver (*Castor fiber*). Mammal Res. 2014;60:139–47.

Tachikawa KS, Yoshihara Y, Kuroda KO. Behavioral transition from attack to parenting in male mice: a crucial role of the vomeronasal system. J Neurosci. 2013;33:5120–6.

Tait AJ, Pope GS, Johnson, E. Progesterone concentrations in peripheral plasma of non-pregnant and pregnant grey squirrels (*Sciurus carolinensis*). J Endocrinol. 1981;89(1):107–16.

Taraborelli PA, Sassi PL, Dacar MA, Moreno P, Ojeda RA. Abrocoma schistacea (Rodentia: Abrocomidae). Mamm Species. 2015;47(921):45–50.

Tasse J. Maternal and paternal care in the Rock Cavy, Kerodon rupestris, a South American hystricomorph rodent. Zoo Biol. 1986;5:27–43.

Taya K, Greenwald GS. Peripheral blood and ovarian levels of sex steroids in the lactating rat. Endocrinol Jpn. 1982;29(4):453–9.

Taya K, Sasamoto S. Initiation of follicular maturation and ovulation after removal of the litter from the lactating rat. J Endocrinol. 1980;87(3):393–400.

Terkel J, Rosenblatt JS. Maternal behavior induced by maternal blood plasma injected into virgin rats. J Comp Physiol Psychol. 1968;65(3p1):479.

Terkel J, Rosenblatt JS. Humoral factors underlying maternal behavior at parturition: cross transfusion between freely moving rats. J Comp Physiol Psychol. 1972;80(3):365.

Thorington RW Jr, Ferrell KE. Squirrels: the animal answer guide. Baltimore: JHU Press; 2006.

Timonin ME, Wynne-Edwards KE. Aromatase inhibition during adolescence reduces adult sexual and paternal behavior in the biparental dwarf hamster *Phodopus campbelli*. Horm Behav. 2008;54:748–57.

Timonin ME, Cushing BS, Wynne-Edwards KE. In three brain regions central to maternal behaviour, neither male nor female *Phodopus* dwarf hamsters show changes in oestrogen receptor alpha distribution with mating or parenthood. J Neuroendocrinol. 2008;20:1301–9.

Tomasco IH, Sánchez L, Lessa EP, Lacey EA. Genetic analyses suggest burrow sharing by río negro tuco-tucos (Ctenomys rionegrensis). Mastozool Neotrop. 2019;26(2):430–9.

Trainor BC, Marler CA. Testosterone, paternal behavior, and aggression in the monogamous California Mouse (*Peromyscus californicus*). Horm Behav. 2001;40:32–42.

Trainor BC, Marler CA. Testosterone promotes paternal behaviour in a monogamous mammal via conversion to oestrogen. Proc R Soc Lond B Biol Sci. 2002;269:823–9.

Trainor BC, Bird IM, Alday NA, Schlinger BA, Marler CA. Variation in aromatase activity in the medial preoptic area and plasma progesterone is associated with the onset of paternal behavior. Neuroendocrinology. 2003;78:36–44.

Tsuneoka Y, Maruyama T, Yoshida S, Nishimori K, Kato T, Numan M, Kuroda KO. Functional, anatomical, and neurochemical differentiation of medial preoptic area subregions in relation to maternal behavior in the mouse. J Comp Neurol. 2013;521(7):1633–63.

Tsuneoka Y, Tokita K, Yoshihara C, Amano T, Esposito G, Huang AJ, Yu LM, Odaka Y, Shinozuka K, McHugh TJ, Kuroda KO. Distinct preoptic-BST nuclei dissociate paternal and infanticidal behavior in mice. EMBO J. 2015;34:2652–70.

Uriarte N, Ferreira A, Rosa XF, Sebben V, Lucion AB. Overlapping litters in rats: effects on maternal behavior and offspring emotionality. Physiol Behav. 2008;93(4–5):1061–70.

Uriarte N, Fernández MV, Agrati D, Zuluaga MJ, Ferreño M, Ferreira A. Maternal and affective behaviors of lactating rats reared in overlapping litters. J Physiol Paris. 2014;108(2-3):221–230.

Uriarte N, Ferreño M, Méndez D, Nogueira J. Reorganization of perineuronal nets in the medial Preoptic Area during the reproductive cycle in female rats. Sci Rep. 2020;10:Article number: 5479.

van Aarde RJ, Potgieter HC. Circulating progesterone, progesterone-binding proteins and oestradiol-17 beta concentrations in the pregnant Cape porcupine, Hystrix africaeaustralis. J Reprod Fertil. 1986;76(2):561–7.

Veenema AH, Neumann ID. Central vasopressin and oxytocin release: regulation of complex social behaviours. Prog Brain Res. 2008;170:261–76.

Vinograd A, Fuchs-Shlomai Y, Stern M, Mukherjee D, Gao Y, Citri A, Davison I, Mizrahi A. Functional plasticity of odor representations during motherhood. Cell Rep. 2017;21(2):351–65.

Voloschin LM, Tramezzani JH. Milk ejection reflex linked to slow wave sleep in nursing rats. Endocrinology. 1979;105(5):1202–7.

Vom Saal FS. Variation in infanticide and parental behavior in male mice due to prior intrauterine proximity to female fetuses: elimination by prenatal stress. Physiol Behav. 1983;30:675–81. https://doi.org/10.1016/0031-9384(83)90162-2.

Vom Saal FS. Time-contingent change in infanticide and parental behavior induced by ejaculation in male mice. Physiol Behav. 1985;34:7–15. https://doi.org/10.1016/0031-9384(85)90069-1.

Wagner CK, Morrell JI. In situ analysis of estrogen receptor mRNA expression in the brain of female rats during pregnancy. Mol Brain Res. 1995;33(1):127–35.

Wagner CK, Morrell JI. Levels of estrogen receptor immunoreactivity are altered in behaviorally-relevant brain regions in female rats during pregnancy. Mol Brain Res. 1996;42(2):328–36.

Wamboldt MZ, Insel TR. The ability of oxytocin to induce short latency maternal behavior is dependent on peripheral anosmia. Behav Neurosci. 1987;101(3):439.

Wang Z, De Vries GJ. Testosterone effects on paternal behavior and vasopressin immunoreactive projections in prairie voles (*Microtus ochrogaster*). Brain Res. 1993;631:156–60.

Wang Z, Ferris CF, De Vries GJ. Role of septal vasopressin innervation in paternal behavior in prairie voles (*Microtus ochrogaster*). Proc Natl Acad Sci U S A. 1994a;91:400–4.

Wang Z, Smith W, Major DE, De Vries GJ. Sex and species differences in the effects of cohabitation on vasopressin messenger RNA expression in the bed nucleus of the stria terminalis in prairie voles (*Microtus ochrogaster*) and meadow voles (Microtus pennsylvanicus). Brain Res. 1994b;650:212–8. https://doi.org/10.1016/0006-8993(94)91784-1.

Wang ZX, Liu Y, Young LJ, Insel TR. Hypothalamic vasopressin gene expression increases in both males and females postpartum in a biparental rodent. J Neuroendocrinol. 2000;12:111–20.

Watarai A, Arai N, Miyawaki S, Okano H, Miura K, Mogi K, Kikusui T. Responses to pup vocalizations in subordinate naked mole-rats are induced by estradiol ingested through coprophagy of queen's feces. PNAS. 2018;115(37):9264–9.

Wei YC, Wang SR, Jiao ZL, Zhang W, Lin JK, Li XY, Li SS, Zhang X, Xu XH. Medial preoptic area in mice is capable of mediating sexually dimorphic behaviors regardless of gender. Nat Commun. 2018;9(1):1–15.

Wight HM. Breeding habits and economic relations of the Dalles Pocket Gopher. J Mammal. 1930;11:40–8.

Wilson DE, Reeder DM. Mammal species of the world: a taxonomic and geographic reference. 2nd ed. Washington, DC/London: Smithsonian Institution Press; 1993.

Witmer GW, Lowney M. Population biology and monitoring of the Cuban hutia at Guantanamo Bay, Cuba. Mammalia. 2007;71:115–21.

Wöhr M, Oddi D, D'Amato FR. Effect of altricial pup ultrasonic vocalization on maternal behavior. In: Handbook of behavioral neuroscience, vol. 19. Amsterdam: Elsevier; 2010. p. 159–66.

Wood JE. Reproductive Pattern of the Pocket Gopher (*Geomys breviceps brazensis*). J Mammal. 1949;30:36–44. https://doi.org/10.2307/1375195.

Wu Z, Autry AE, Bergan JF, Watabe-Uchida M, Dulac CG. Galanin neurons in the medial preoptic area govern parental behaviour. Nature. 2014;509:325–30.

Yadav R, Haldar C. Variation in gonadal steroid and melatonin during gestation of a diurnal seasonal breeder Indian palm squirrel Funambulus pennanti. J Endocrinol Reprod. 2009;13(1):1–6.

Yoshihara C, Numan M, Kuroda KO. Oxytocin and parental behaviors. In: Behavioral pharmacology of neuropeptides: oxytocin. Cham: Springer; 2017. p. 119–53.

Yuan W, Yang X-P, Yu P, Jia R, Tai F-D, Wei B, Liu X, Ma L-G. Different behavioral, neural and neuropeptide responses of fathers to their own and to alien pups in mandarin voles. J Comp Physiol A. 2017:1–13. https://doi.org/10.1007/s00359-017-1229-x.

Zhang GW, Shen L, Tao C, Jung AH, Peng B, Li Z, Zhang LI, Tao HW. Medial preoptic area antagonistically mediates stress-induced anxiety and parental behavior. Nat Neurosci. 2021;24(4):516–28.

Parental Behavior in Carnivores

Robyn Hudson, Péter Szenczi, and Oxána Bánszegi

Abstract The mammalian order *Carnivora* is generally defined as species that feed exclusively or to some degree by eating other animals. The *Carnivora* comprise around 280 species, divided into 16 families, 13 of which are terrestrial and 3 aquatic. Carnivores are spread across the entire planet, including the two polar regions and on land and sea. Consistent with such diverse ecologies, there is no typical pattern of parental care distinguishing carnivores from other mammals. Using examples from different taxonomic *families*, our aim is to illustrate the diversity of parental care in *Carnivora*. Major topics include parental care before and after birth of the young, paternal, and alloparental care and the process of weaning. Given the position of many carnivores at the apex of food chains, a greater understanding of their patterns of parental care as a vital part of reproductive biology is essential to conservation programs.

Keywords Maternal care · Paternal care · Helpers · Prenatal preparation · Parturition · Postnatal care · Weaning · Independence · Diversity

1 Introduction

The mammalian order *Carnivora* according to recent molecular phylogenetic studies is a monophyletic taxonomic group (Eizirik et al. 2010; Hassanin et al. 2021) made up of species generally characterized by a high proportion of vertebrates in their diet. The *Carnivora* comprise around 270–290 species (depending on sources), a number comparable to that of primates (256 species) but considerably less than

R. Hudson (✉) · O. Bánszegi (✉)
Instituto de Investigaciones Biomédicas, Universidad Nacional Autónoma de México, Mexico City, Mexico
e-mail: rhudson@biomedicas.unam.mx

P. Szenczi
CONACyT – Instituto Nacional de Psiquiatría Ramón de la Fuente Muñiz, Mexico City, Mexico

G. González-Mariscal (ed.), *Patterns of Parental Behavior*, Advances in Neurobiology 27, https://doi.org/10.1007/978-3-030-97762-7_2

the most numerous mammalian orders, *Chiroptera* (bats), with an estimated 977 species and *Rodentia* (rodents) with an estimated 2000 species. Nevertheless, the range in body mass of carnivores, from the 25 g least weasel *Mustela nivalis*[1] to a 4000–6000 kg male southern elephant seal *Mirounga leonina*, exceeds that of all other living mammalian orders (Mittermeier and Wilson 2009). *Carnivora* are commonly divided into 16 *families*, 13 of which are terrestrial and 3 aquatic (Table 1).

Members of this order typically have strong jaws and dentition characterized by large, daggerlike canine teeth adapted for catching and holding prey and the remaining teeth shaped for cutting, tearing at, and masticating meat. Other adaptations to the demands of a carnivore's hunting lifestyle typically include speed and agility, muscular strength, highly developed sensory acuity, and strongly convoluted brains indicative of the importance of learning and other cognitive abilities. Together, these characteristics are thought to account for the considerable charismatic appeal of many carnivores. Nevertheless, only the *Felidae*, *Phocidae,* and *Otariidae* (Table 1) are obligate carnivores, depending exclusively on a diet of fresh animal protein for their survival. All other carnivores are also scavengers or foragers to a greater or lesser degree, which include in their diet carrion, insects, and other invertebrates, fungi, and plant material such as berries, fruits, leaves, roots, and nuts. Exceptions among carnivores are the giant and the red panda *Ailuropoda melanoleuca* and *Ailurus fulgens*, respectively, which can be considered obligate herbivores feeding almost exclusively on bamboo, and the mainly frugivorous kinkajou *Potos flavus* and African palm civet *Nandinia binotata*. The bat-eared fox *Otocyon megalotis,* white-tailed mongoose *Ichneumia albicauda*, and aardwolf *Proteles cristata*, for example, are primarily insectivorous.

Hunting and/or scavenging is time-consuming. It can be dangerous, and success is often uncertain, obligating mothers, which in many carnivore species are the sole caretakers of the young, to leave them alone for extended periods. On the other hand, meat is calorie-rich, is easy to digest (as reflected by the relatively short and simple digestive tracts of carnivores compared to herbivores, for example), and can be brought to the mother and/or to the young. Some canid species transfer solid food to the mother and young by regurgitation, a canid innovation that allows mates or other caretakers to feed the mother at the den, whose ability to hunt may be compromised in late pregnancy and early lactation, and to introduce the young to solid food around the time of weaning (Macdonald and Sillero-Zubiri 2004).

A notable feature of carnivores is their worldwide distribution and the varied habitats they occupy. Carnivores can be found across the whole planet, including the two polar regions, aquatic environments, and the full range of terrestrial habitats: coastal areas, bushland, swampland, desert, forest, jungle, inland savannah and steppe, high-altitude mountains, urban environments, and at all latitudes. A further notable feature is their diversity of social and mating systems, ranging from species leading primarily solitary lives, such as most *Felidae* and *Viverridae*, to those forming pairs, family groups, or packs such as many *Canidae* or *Hyaenidae*, or even large complex societies such as members of the *Herpestidae* and *Mustelidae.*

[1] Latin names of species are given only at first mention, and they are listed in Table 1 according to the taxonomic *families* to which they are presently considered to belong.

Table 1 Carnivore *families* with some general characteristics and a list of species mentioned in the text. Group living propensity of the species refers mostly to their parental behaviors

Family	Extant species	Size range	Litter size	Species in the text	Latin name	Group living propensity
Mustelidae	57	20 g/45 kg	1–18	American mink	*Neogale vison*	Solitary
Weasels, otters, and badgers				Asian small-clawed otter	*Aonyx cinerea*	Group living
				Domestic ferret	*Mustela putorius furo*	
				European badger	*Meles meles*	Facultative groups
				Fisher	*Pekania pennanti*	Solitary
				Giant river otter	*Pteronura brasiliensis*	Group living
				Least weasel	*Mustela nivalis*	Solitary
				River otter	*Lontra canadensis*	Facultative groups
				Sea otter	*Enhydra lutris*	Facultative groups
				Tayra	*Eira barbara*	Solitary
				Wolverine	*Gulo gulo*	Solitary
Felidae	37	1–300 kg	1–10	Domestic cat	*Felis silvestris catus*	
Cats				African lion	*Panthera leo*	Group living
				Black-footed cat	*Felis nigripes*	Solitary
				Bobcat	*Lynx rufus*	Solitary
				Caracal	*Caracal caracal*	Solitary
				Cheetah	*Acinonyx jubatus*	Facultative groups
				Cougar	*Puma concolor*	Solitary
				Iberian lynx	*Lynx pardinus*	Solitary
				Leopard	*Panthera pardus*	Solitary
				Margay	*Leopardus wiedii*	Solitary
				Ocelot	*Leopardus pardalis*	Solitary
				Pallas's cat	*Otocolobus manul*	Solitary
				Tiger	*Panthera tigris*	Solitary

(continued)

Table 1 (continued)

Family	Extant species	Size range	Litter size	Species in the text	Latin name	Group living propensity
Canidae	35	1–50 kg	1–16	African wild dog	*Lycaon pictus*	Group living
Dogs				Arctic fox	*Vulpes lagopus*	Facultative groups
				Bat-eared fox	*Otocyon megalotis*	Group living
				Black-backed jackal	*Canis mesomelas*	Group living
				Domestic dog	*Canis lupus familiaris*	
				Maned wolf	*Chrysocyon brachyurus*	Pair living
				Raccoon dog	*Nyctereutes procyonoides*	Facultative groups
				Red fox	*Vulpes vulpes*	Pair living
				Gray wolf	*Canis lupus*	Group living
Viverridae	34	1–14 kg	1–6	Binturong	*Arctictis binturong*	Solitary
Civets, genets, and olyans						
Herpestidae	34	200 g/4.5 kg	1–7	Banded mongoose	*Mungos mungo*	Group living
Mongooses				Dwarf mongoose	*Helogale parvula*	Group living
				Meerkat	*Suricata suricatta*	Group living
				White-tailed mongoose	*Ichneumia albicauda*	Solitary
Mephitidae	12	200 g/4.5 kg	1–10	Eastern spotted skunk	*Spilogale putorius*	Solitary
Skunks and stink badgers						
Procyonidae	12	1–10 kg	1–7	Kinkajou	*Potos flavus*	Facultative groups
Raccoons						
Eupleridae	8	500 g/10 kg	1–6	Malagasy civet	*Fossa fossana*	Pair living
Malagasy mongooses and civets				Malagasy narrow-striped mongoose	*Mungotictis decemlineata*	Group living

(continued)

Table 1 (continued)

Family	Extant species	Size range	Litter size	Species in the text	Latin name	Group living propensity
Ursidae	8	25–700 kg	1–5	American black bear	*Ursus americanus*	Solitary
Bears				Brown bear	*Ursus arctos*	Solitary
				Giant panda	*Ailuropoda melanoleuca*	Solitary
				Polar bear	*Ursus maritimus*	Solitary
				Sloth bear	*Melursus ursinus*	Solitary
				Spectacled bear	*Tremarctos ornatus*	Solitary
				Sun bear	*Helarctos malayanus*	Solitary
Hyaenidae	4	8–70 kg	1–5	Aardwolf	*Proteles cristata*	Pair living
Hyenas				Brown hyena	*Hyena brunnea*	Group living
				Spotted hyena	*Crocuta crocuta*	Group living
Prionodontidae	2	400 g/1 kg	2–3			
Linsangs						
Nandiniidae	1	1–3 kg	1–4	African palm civet	*Nandinia binotata*	Solitary
African palm civet						
Ailuridae	1	3–6 kg	1–3	Red panda	*Ailurus fulgens*	Solitary
Red panda						
Phocidae	18	30–3000 kg	1	Southern elephant seal	*Mirounga leonina*	Solitary
Earless seals; true seals				Baikal seal	*Pusa sibirica*	Solitary
				Harbor seal	*Phoca vitulina*	Solitary
				Hooded seal	*Cystophora cristata*	Solitary
				Ringed seal	*Pusa hispida*	Solitary
				Weddell seal	*Leptonychotes weddellii*	Solitary
Otariidae	15	25–1000 kg	1	Antarctic fur seal	*Arctocephalus gazella*	Solitary
Eared seals, fur seals, and sea lions				Galápagos fur seal	*Arctocephalus galapagoensis*	Solitary
				New Zealand sea lion	*Phocarctos hookeri*	Solitary

(continued)

Table 1 (continued)

Family	Extant species	Size range	Litter size	Species in the text	Latin name	Group living propensity
				South American fur seal	*Arctocephalus australis*	Solitary
Odobenidae	1	400–1700 kg	1	Walrus	*Odobenus rosmarus*	Group living
Walrus						

Diversity in mating systems ranges from the polygynous southern elephant seal that shows strong harem-based female defense to the female-dominated polyandrous mating system of the spotted hyena *Crocuta crocuta* to the varying degrees of monogamy in the *Canidae*. These varied geographical and social ecologies have resulted in a great diversity in patterns of parental care in which even members of the same taxonomic family and species of similar size can differ markedly (Bekoff et al. 1984). Nevertheless, some generalities can be noted. Apart from seals and walrus, carnivore young can be considered altricial although with the exception of the American mink *Neogale vison* and the giant panda, most have some fur at birth, perhaps because most carnivores do not build nests although most give birth within the shelter of dens or burrows and because the young of some species are left alone for extended periods when mothers or other caretakers must leave to hunt or forage.

Following convention, we define parental care as any behavior that directly contributes to the well-being and survival of the young. Thus, for the purpose of this chapter, we have excluded territorial behavior and mate guarding, typically by males, which, while possibly helping ensure mothers a resource-rich environment and limiting infanticide by roaming males, might principally serve other purposes such as males' access to mating opportunities.

As we argue in this chapter, knowledge of parental care in carnivores is rather limited. It is based on detailed knowledge of only a handful of species, and this is often gained under the artificial conditions of laboratories, zoos or animal parks, and farm or household conditions. This is understandable given the considerable difficulty in observing most carnivores, particularly in nature. The young are typically born into dens or burrows, which are often difficult to access and often defended by a dangerous, well-armed mother and sometimes together with the father and other members of the social group. Moreover, many carnivores are primarily nocturnal or crepuscular, increasing the difficulty of observing parent-young interactions. Despite such difficulties, it is our aim to illustrate the diversity of carnivore parental behavior by using some of the best-documented examples from different taxonomic *families* and to show how limited our knowledge of this fundamental aspect of carnivore behavioral biology still is.

For this purpose, we have arranged the chapter in three main parts corresponding to the three main phases of carnivore parental care: prepartum preparation for arrival of the young, parturition and behaviors directly associated with this, and postpartum care, including weaning of the young and subsequent care to their complete independence. We will not discuss physiological mechanisms underlying parental behavior as so little is known about these in carnivores. Available information mainly relates to the endocrine regulation of mating and fertility in a few domestic or semi-domestic and laboratory species (dog, cat, ferret, mink). Also, we will not present in detail and only partly discuss accompanying differences in social systems that may be strongly influenced by ecological factors.

According to present knowledge, in the great majority of carnivores, parental care is provided by the mother alone. But before discussing the three main phases of parental care, we briefly consider the role of fathers (and in section "Alloparental care" of other members of the social group) in helping raise the young.

Paternal Care

Male parental care is unusual in mammals, occurring in only 5–10% of species (Woodroffe and Vincent 1994), and in the majority of carnivore species, females care for the young alone. Male care of offspring would be expected only when the benefits of helping the female outweigh the costs (Clutton-Brock 1991; Gross 2005). Hence, even though fathers may increase the survival or quality of their offspring by helping care for them or their mother, this usually entails a trade-off between time invested in care of the family and time lost in obtaining additional mating opportunities.

Among carnivores, male parental care is most common in the *Canidae* where it is reportedly present to some degree in all species (Kleiman and Malcolm 1981). Forms of paternal care are usually classified as direct or indirect. Direct care refers to interactions between males and their offspring, which can be reasonably understood to increase offspring fitness. Common forms of direct care are grooming and/or cleaning the young, carrying or retrieving them, providing warmth by huddling with them, providing food, defense against predators, and playing or other forms of socializing. Indirect care involves those behaviors that could benefit offspring even if there is no direct interaction between the father and young. Examples include den construction, giving alarm calls, and providing food for the mother. Arguably, the most social canid is the African wild dog *Lycaon pictus* in which females produce up to 16 pups per litter, thus requiring support from the father and even other pack members (Malcolm and Marten 1982). Fathers of feral domestic dogs *Canis lupus familiaris* reportedly guard the nest site and regurgitate food for the pups (Pal 2005). Gray wolf fathers *Canis lupus* also reportedly participate in guarding the den (Ruprecht et al. 2012) and feed the nursing mother, and when the cubs leave the den around weaning, the breeding male as well as other adults regurgitate food for them (Packard 2010). Even the maned wolf *Chrysocyon brachyurus*, thought to be solitary, has been observed to spend time sleeping together with its mate, and after birth of the cubs, the male stays in the vicinity, reducing his home range and activity levels, suggesting that he may contribute to parental care.

Male investment in some form has also been reported, at least occasionally, in three of the four species of *Hyaenidae* although it is apparently not found in the spotted hyena (Richardson 1987; Mills 1990; East et al. 2003). So-called raccoon dogs *Nyctereutes procyonoides* show long-term pair bonding, and males participate in rearing the young. In fact, they reportedly spend more time alone with the pups than females, guarding the litter at the den or in its close vicinity while females forage to satisfy their increased energy needs due to the high cost of lactation (Kauhala et al. 1998). Direct male care also occurs in the tropical otters: the giant river otter *Pteronura brasiliensis* and the Asian small-clawed otter *Aonyx cinerea* (Schmelz et al. 2017) and in families of the banded mongoose *Mungos mungo* (Rood 1974). Although male care in the form of playing with young and allowing them to take food items has been recorded in a number of felids in captivity (Kleiman and Malcolm 1981), it is difficult to know if this is an artifact of confined conditions as there are presently no reports of such activity in the wild. Direct paternal care has not been reported in any of the aquatic carnivores, and in fact, in species with large sexual dimorphism in body mass such as the southern elephant seal, males may severely injure or crush young pups to death – presumably sometimes even their own offspring – during battles with competing males for access to females for mating.

For any one topic, the examples given below represent only a fraction of the diversity of parental care in carnivores. Investigating in more detail any one aspect will surely reveal a richness and diversity beyond the bounds of this chapter.

2 Prepartum Behavior: Preparation for Arrival of the Young

For many mammals, parental care begins before birth of the young. Since altricial offspring require a particularly secure environment for early rearing (Case 1978), in many carnivores one of the most important prepartum behaviors is seeking out or actively constructing a nursery burrow or den. Such structures are vital to protect the young from harsh weather, and from aerial and land predators, which may include infanticidal males or other conspecifics (Ruggiero et al. 1998; Ross et al. 2010; Libal et al. 2011; Jackson et al. 2014; White et al. 2015). Den sites are critical resources that influence the survival of the young and ultimately the population dynamics of several species. The use of subterranean natal dens that can be readily defended is characteristic of almost all species in several families of carnivores (*Canidae, Hyaenidae, Mephitidae*) and prevalent in others (*Herpestidae, Mustelidae, Ursidae*) (Noonan et al. 2015). Many carnivores that use dens or burrows do not actively excavate them but occupy already existing structures such as hollow or fallen trees, rock crevices, or other naturally formed cavities or occupy burrows and tunnels constructed by other species. In contrast, mongooses, otters, hyenas, badgers, wolverines, and several canid species such as foxes, dingoes, coyotes, or wolves dig their own natal dens, a behavior not seen in any felids.

Only few species of carnivores have been reported to build nests of plant or other material within such nursery dens or cavities. Examples include the European badger *Meles meles* (Roper 1992), the American mink (Malmkvist and Palme 2008), and the red panda (Roka et al. 2015).

Properties of dens contributing to protection, including avoiding human disturbance, are reportedly more important for breeding females than habitat features such as prey density or structure of vegetation as reported for the Iberian lynx (Fernández and Palomares 2000), African wild dog (Jackson et al. 2014), gray wolf (Sazatornil et al. 2016), spotted hyena (Périquet et al. 2016), and the fisher *Pekania pennanti* (Matthews et al. 2019). Such shelters can also be important in helping to maintain a stable and adequate thermal environment for the altricial young (Reichman and Smith 1990). Dens of several species of the family *Felidae* such as Pallas's cat *Otocolobus manul*, Iberian lynx *Lynx pardinus*, and ocelot *Leopardus pardalis* contribute importantly to thermoregulation, as well as providing a refuge from other carnivores, including predatory conspecifics, and from humans (Fernández and Palomares 2000; Laack et al. 2005; Ross et al. 2010). Thermoregulatory factors in den selection were also found to be important in wolverines *Gulo gulo* (Magoun and Copeland 1998), and the Arctic fox *Vulpes lagopus* constructs and uses dens with southward-facing entrances to improve microclimate conditions (Smits et al. 1988). For most bear species, which have very altricial young, dens also provide a safe and sheltered environment for giving birth, nursing, and early cub growth in some species during winter hibernation (Oli et al. 1997; Seryodkin et al. 2003; Manchi and Swenson 2005; Zhang et al. 2007; Derocher et al. 2011; Faure et al. 2020).

Pinnipeds depend on and seek out solid substrates to give birth and, with the partial exception on the walrus *Odobenus rosmarus* (see section *Nursing*), for nursing the young. They utilize a wide variety of habitats, including pack ice, fast ice, and land. Phocid seal species give birth mostly on ice, whereas all 15 species of otariids give birth on land. The only surviving member of the *Odobenidae,* the walrus, is also an ice-breeding species (Bowen 1991). In several pinniped species, these sites are part of long-established breeding grounds with high individual philopatry, on beaches, rocky shores, or ice sheets; for example, the Galápagos fur seal *Arctocephalus galapagoensis*, southern elephant seal (Hindell and Little 1988), Antarctic fur seal *Arctocephalus gazella* (Hoffman and Forcada 2012), Weddell seal *Leptonychotes weddellii* (Cameron et al. 2007), and New Zealand sea lion *Phocarctos hookeri* (Chilvers and Wilkinson 2008). Because young seals are born without a substantial layer of subcutaneous fat and accumulate most of their lipid reserves after birth (Donohue et al. 2000), thermoregulatory factors are important in pupping site selection. However, the majority of seal pups are born into an environment that gives little protection against weather or predators, and predation is a significant source of pup mortality including in several species of ice-breeding pinnipeds (Bowen 1991). In this regard, the ringed seal *Pusa hispida* and the Baikal seal *Pusa sibirica* are exceptions as females of these species construct dens from snow under ice ridges or over breathing holes (Smith and Stirling 1975; Miyazaki 2009). Such dens have a main chamber and smaller tunnels off the central cavity and provide thermal protection and concealment from predators such as Arctic foxes and polar bears *Ursus maritimus*.

In addition, pregnant females may adjust their hunting range as parturition approaches. This is particularly the case for solitary carnivores such as pinnipeds, mustelids, and most felids where the mother raises the young alone. This change might be due to anatomical constraints of the pregnant mother, change in prey type or a focus on nest defense. Even after the young start to accompany their mother on hunting or foraging trips, their still limited motor abilities and endurance may also influence her behavior. For example, female North American cougars *Puma concolor* reduce their home range around and following parturition (Seidensticker et al. 1973; Maehr et al. 1989), as do leopards *Panthera pardus* (Seidensticker 1976), tigers *Panthera tigris* (Sunquist 1981), bobcats *Lynx rufus* (Nielsen and Woolf 2001), and the mustelid tayras *Eira barbara* (Presley 2000). Harbor seals *Phoca vitulina* restrict their foraging range during lactation (Thompson et al. 1994) while several large-bodied phocid seals fast throughout lactation completely (Schulz and Bowen 2005).

Thus, from the above, we may conclude that carnivore species show a wide range of prepartum behaviors relevant to the successful raising of their young. And furthermore, that the availability of suitable breeding sites such as dens, burrows, or ice flows and associated productive hunting grounds are vital for the conservation of many carnivore species (Squires et al. 2008).

3 Parturition

This is a critical phase in the reproductive cycle of all mammals in which females give birth to live young in various stages of maturation, and as noted above, including for carnivores, all the young of which are altricial or semi-altricial. Parturition involves varying degrees of physical, physiological, and external hazard, both for mothers and their young. While giving birth, mothers and young, particularly in solitary species, may be vulnerable to attack by predators as at this time, mothers are largely unable to defend themselves or their offspring.

Carnivore mothers typically give birth alone, and even in monogamous pairs, the female often does not allow the male to enter the breeding den (Naaktgeboren 1968). All pinniped species (Boness and Bowen 1996), the sea otter *Enhydra lutris* (Estes 1980), the feline margay *Leopardus wiedii* (Moreira 2001), and some members of the *Eupleridae* such as the Malagasy narrow-striped mongoose *Mungotictis decemlineata* and the Malagasy civet *Fossa fossana* give birth to only one offspring (Nowak 2005; Schneider and Kappeler 2016). In other *families,* the litter size can differ markedly, for example, in *Ursidae* between one and two young in the sun bear *Helarctos malayanus*, sloth bear *Melursus ursinus*, spectacled bear *Tremarctos ornatus*, and giant panda, to as many as five in the American black bear *Ursus americanus* and brown bear *Ursus arctos* (Garshelis 2004). In *families* such as the *Mustelidae*, *Felidae, Viverridae,* and *Hyaenidae,* between two and six young is most common, but up to 14 has been reported in the least weasel (Sundell 2003) and up to 18 in the domestic ferret *Mustela putorius furo* (Lindeberg 2008). In some *Canidae* such as the domestic dog, the Arctic fox, and the African wild dog, there may be up to 16 pups (McNutt and Silk 2008; Table 1).

Duration of birth including the expulsion of each young and the interval between each expulsion can vary considerably. Large individual differences have been observed lasting from 8 to 225 min for mothers giving birth to singletons, as in the South American fur seal *Arctocephalus australis* (Franco-Trecu et al. 2016), and even in polytocous species, the rhythm can be very different, for example, in foxes 30–120 min between young (Naaktgeboren 1968), gray wolf 9–90 min, domestic dog 6–212 min (Klarenbeek et al. 2007), and the tayra 17–30 min (Poglayen-Neuwall 1974). In the domestic cat *Felis silvestris catus,* the rhythm can also vary but is usually approximately 20 min (Hudson et al. 2009). Mainly in cats, but sometimes also in dogs, delivery between young may be interrupted for as much as 24–36 h (Lopate 2012, own observations).

With the arrival of each offspring, mothers usually bite through the umbilical cord, eat the placenta, and vigorously lick the young and surrounding area clean of birth fluids (Naaktgeboren 1968; Poglayen-Neuwall 1974). When delivery is complete, mothers typically lie on their side or back, exposing their nipples to the offspring although in the case of the domestic cat, early-born kittens may already have attached to nipples and started suckling while parturition is still in progress (Hudson et al. 2009). In general, the young, stimulated by the mother's vigorous licking, attach to a nipple within minutes and start to ingest colostrum and milk (Naaktgeboren 1968).

Orientation to the mammary region and attachment to nipples may be aided by emission of chemical cues from the mother's ventrum, a so-called nipple-search pheromone (cf. European rabbit *Oryctolagus cuniculus*: Hudson et al. 1990; domestic cat: Raihani et al. 2009), the emission of which appears to be associated with the female's reproductive cycle and under hormonal control (domestic cat: Raihani et al. 2009). In some litter-bearing species, the young rapidly develop a nipple order, with each offspring using only one or two particular nipples. This seems to be generally the case in felids (Pfeifer 1980; Hudson et al. 2009) although apparently not necessarily in canids (Hudson et al. 2016). Establishment of an order in nipple use has also been reported in black bears (Rogers et al. 2020) and the binturong *Arctictis binturong* (Schoknecht 1984).

Presumably to protect the young, immediately after parturition, mothers are reportedly more aggressive, especially around the den site. This has been seen in dogs (Pal et al. 1998) and hyenas (Kruuk 1973) but is possibly more general.

4 Postpartum Care and Transition to Independence

Nursing

In mammals, maternal care entails a large energetic cost. Lactation in particular can nearly triple a mother's caloric requirements (Gittleman and Thompson 1988; Oftedal and Gittleman 1989) while exposing her to greater risk of debilitation, injury, or even death, thereby reducing her fitness and future fecundity (König et al.

1988; Clutton-Brock et al. 1989; Koivula et al. 2003). Given the low-fat reserves in some taxa such as felids and some mustelids (Mustonen and Nieminen 2012), mothers of some species must soon leave their young to hunt, scavenge, or forage. The opposite is typical for bears and aquatic species, which often have extensive fat reserves. Bears remain in dens with their newborn cubs for weeks to months without eating or drinking, something some of them do, for example, polar bears, since they give birth during hibernation (Garshelis 2004). Some seal species draw on their fat reserves to remain with their pups for extensive periods of up to 50 days until being compelled to return to sea to hunt (Boness and Bowen 1996).

Among pinnipeds, this fasting strategy mostly occurs in the *Phocidae*, with a nursing-foraging cycle more typical for the *Otariidae*. Mothers of the latter family usually accumulate only a small amount of subcutaneous fat before they arrive to their breeding sites; thus, they fast for only 5–11 days after giving birth and then alternate foraging trips to sea with visits to land to nurse their pups, sometimes leaving them for 2–13 days. An extreme case, however, is presented by the hooded seal *Cystophora cristata*, which gives birth on an ice flow, an unstable environment, and nurses her pup almost continuously for 4 days, during which time the pup gains around 7 kg per 24 h and doubles its birth weight, after which the mother leaves it permanently and returns to sea (Bowen 1991). The walrus is the only carnivore with an aquatic nursing strategy. The young remain with their mother wherever she goes and are nursed at sea as well as on land (Bowen 1991).

Apart from providing milk, mothers of many species such as the domestic cat regularly lick the anogenital area of newborn young to stimulate urination and defecation, usually ingesting the excreta, thereby contributing to nest hygiene (see section *Den hygiene and safety*; Turner and Bateson 2014, own observations). Because of the bactericidal effect of saliva, periparturient licking by females of their mammary and anogenital areas is particularly adaptive since these are the body areas of the mother that could be contaminated by fecal-borne bacteria and which the newborns' mouths come into close contact during birth and suckling (Hart and Powell 1990). Newborn mammals, which are born with a sterile gut, do not have the intestinal bacterial flora that protect against opportunistic pathogens (Greene 1984).

Den Hygiene and Safety

Although dens and nests provide major benefits by protecting offspring from predators and harsh weather, they can also have costs in terms of hygiene due to a buildup of ectoparasites and by attracting predators, including infanticidal conspecifics, due to the accumulation of odors, worn trails, or repeated visits by caretakers. One strategy some species are thought to use to counteract such dangers is to move the nursery site from time to time although this might also be done to accommodate the changing needs of rapidly growing young. Thus, female ocelots are reported to use two to four den sites for each litter and to move kittens one to five times between them (Laack et al. 2005). Domestic cats and bobcats also frequently move their litters, especially if the mother is disturbed by unfamiliar males or humans (Feldman 1993, Nielsen and Woolf 2001, Turner and Bateson 2014, own observations). Giant panda mothers also regularly change dens although the reason is unknown as this is the only species of *Ursidae* that has been observed to do so (Garshelis 2004). Frequent den changes have also been reported in spotted and brown hyenas *Hyena brunnea*. The most likely reasons are thought to be human disturbance and a buildup of fleas at the den (Mills 1990). Limiting parasite infestations may also help explain moving the young in other carnivores although information on this is presently lacking.

Transport of the young is facilitated by the particular posture the offspring of some species adopt, notably felids such as the domestic cat and canids, which when carried in the mother's mouth by the nape of the neck reflexively curl up in a fetal-like posture, remaining motionless and completely silent (Turner and Bateson 2014; own observations).

Mother-Young Recognition

Since maternal care comes at a large energetic cost, evolutionary theory predicts that to maximize their fitness, mothers should preferentially care for their own young (Hamilton 1964) and thus should be able to distinguish them from alien offspring. If a mother indiscriminately cares for both her own and unrelated offspring, this could increase even further her energetic requirements and be detrimental to the development of her own young (Fleming and Rauscher 1978; Horrell and Bennett 1981; König et al. 1988; Mappes et al. 1995; Andersen et al. 2011).

This can be true in the opposite direction also. In many mammals, it is important for the early survival of the young that they quickly learn to recognize their own mother and to distinguish her from other conspecifics. The attempt to suckle from an alien mother, for example, may result not only in rejection but also in serious injury or even death (Le Boeuf et al. 1972; Wolski et al. 1980; Trillmich 1981; Harcourt 1992). Even for the young of solitary species that remain hidden in nests or dens, it can be vital that they remain quiet at the approach of predators or potentially infanticidal conspecifics and only respond positively to the approach of their mother (Sieber 1986, cf. Vaňková et al. 1997, Torriani et al. 2006 in ungulates).

Olfactory and vocal cues have been found to play a particularly important role in mother-offspring recognition. This is clearly the case and has been best studied in herd and colony-living species such as various pinnipeds (review in Insley et al. 2003; see also Charrier et al. 2010, Pitcher et al. 2010, Trimble and Insley 2010, Sauvé et al. 2015) where for mothers, identifying their own young among the throng of the colony is a daily task. Playback experiments in spotted hyenas have confirmed mutual vocal recognition between cubs and mothers/caretakers (Holekamp et al. 1999), and Hepper (1994) found mutual olfactory recognition in the domestic dog between mothers and young, which may last as long as 2 years after permanent separation.

However, less information is available on other taxonomic groups. This may be because the need for individual recognition between mothers and their offspring is less obvious for solitary than for social species. An exception is the domestic cat in which due to mothers readily permitting the handling and experimental manipulation of their newborn young by familiar caretakers, mutual olfactory recognition between mothers and young has also been found and that the young retain a memory of their mother's scent for more than a year after permanent separation (Bánszegi et al. 2017b; Jacinto et al. 2019; Szenczi et al. 2022). Mothers may also emit specific vocalizations to greet or call their young to follow and which the young rapidly learn to distinguish from similar calls from other mothers (Szenczi et al. 2016).

Alloparental Care

Care of the young may not only be by the mother or the breeding pair. Other, even unrelated members of a social group may also contribute to raising the young by helping guard, groom, carry, play with, nurse, or otherwise feed the offspring of others. The extent of cooperative care of young varies widely among carnivore species, ranging from joint breeding site defense to nursing and provisioning unrelated offspring (Clutton-Brock 2016). Both juveniles and adults may participate in alloparental behavior, while allolactators are frequently females who have lost their own young (Kleiman and Malcolm 1981; MacLeod et al. 2013).

Social carnivores in particular frequently show alloparental care, often accompanied by some degree of reproductive suppression in helpers (Montgomery et al. 2018). These include members of the canid, felid, herpestid, hyaenid, mustelid, and procyonid *families* (Rood 1978; Lukas and Clutton-Brock 2012; Federico et al. 2020). Among communal breeders, such as the African lion *Panthera leo*, the spotted and the brown hyenas, and the banded mongoose, most females breed during each reproductive cycle and participate in some degree of alloparental care although temporarily nonbreeding females and males may also contribute to the care of young born in the group (Mills 1990; Lewis and Pusey 1997). In facultative cooperative breeders such as the black-backed jackal *Canis mesomelas* and Arctic fox, the parents and nonbreeding helpers alike care for the young (Johnsingh 1982) although the number of helpers is small, and parents can successfully raise their young without helper assistance (Clutton-Brock 2006). Obligate cooperative breeders, such as the African wild dog and the meerkat *Suricata suricatta*, require assistance from nonbreeding helpers to successfully raise their offspring. In these species, nonbreeding helpers may even provide the majority of care, and their number typically exceeds the number of breeders (Lukas and Clutton-Brock 2012).

Weaning

The transition from a diet exclusively of milk to starting to obtain and ingest solid food is a crucial stage in the development of all mammalian young as it involves major changes in the behavior and physiology both of mothers and their offspring. An influential theory first proposed by Trivers (1974) essentially sees the weaning process as one of conflict between parents and young. It states that the optimal amount of parental investment in offspring is unequal for the two parties, such that offspring can be expected to try to obtain a greater amount of resources than their parents are willing to provide. However, the empirical evidence gathered since has refined this statement, leading to suggestions that the existence of such conflict might be overstated (Mock and Forbes 1992; Bateson 1994, 2014) and that the relationship between mother and offspring at weaning is more one of synchronization and cooperation than it is of conflict (Kölliker et al. 2005; Hinde et al. 2010; Cox and Hager 2016; Royle et al. 2016; Bánszegi et al. 2017a).

The young usually achieve considerable independence well before being completely weaned. They are able to digest solid food and to eliminate without maternal stimulation and can maintain an adequate body temperature without being brooded (Olmstead et al. 1979; Bateson 2014). Their locomotor abilities have also developed markedly (Peters 1983), increasing their ability to feed, hide, return to the nest, or defend themselves (Baerends-van Roon and Baerends 1979), probably boosting their confidence and leading them to increasingly explore unfamiliar environments (Romand and Ehret 1984). Hormonal changes of the mother during the lactational period may lead to a decline in her motivation to return to her young and in a reduction in her willingness to nurse, for example, by blocking access of the young to the nipples (Martin 1986; Bateson 2014), and to a reduction in responsiveness to their separation calls (Bánszegi et al. 2017a).

As weaning approaches, the content of the milk may change, generally increasing in fat and protein and decreasing in lactose as has been reported for several species, including humans (Neville et al. 1991; Verd et al. 2018). Information on this, however, is limited for *Carnivora*, with only a few studies of some seal species, probably stimulated by their sometimes extremely short nursing period. The change in milk composition during the weaning period is generally consistent with findings in other mammals although some decline in fat content has been found just before weaning (Bryden 1968; Riedman and Ortiz 1979; Carlini et al. 1994).

Postweaning Care

We continue the main theme of this chapter emphasizing the diversity of parental care among carnivores with a consideration of postweaning patterns leading to eventual independence of the young. In the pinnipeds, where information is available, the young rapidly achieve complete independence once they start a life at sea (Boness and Bowen 1996). However, in several other obligate carnivore taxa, most notably the *Felidae*, where the survival of the young will depend on their ability to hunt, learning the necessary skills may require the young to remain with their mother or other members of the group for a longer period. The degree of "teaching" such skills varies in carnivores. In pack-hunting species that often pursue large prey, the young can gain experience by merely joining the hunt, so the degree of teaching is relatively low. In contrast, young solitary hunters such as most *Felidae* have few opportunities to interact with live prey unless provided by adults, usually the mother; hence, the degree of teaching is high (Thornton and Raihani 2008).

Thus, among felids, behavior suggestive of teaching has been reported in tigers (Schaller 1967), cheetahs *Acinonyx jubatus* (Kruuk and Turner 1967), caracals *Caracal caracal* (Ewer 1969), black-footed cats *Felis nigripes,* and domestic cats (Leyhausen 1979; Caro 1980). Juvenile lions, tigers, and leopards begin to make hunting excursions with their mother at 12–15 months of age (Bekoff et al. 1984). Lionesses are reported to distinguish between serious hunts, with only adults participating, and training hunts, with juveniles following and watching adults in pursuit of prey (Schenkel 1966). Leyhausen (1979) has described in detail how female cats first only bring the prey and consume it in the presence of the young, then later allow them to interact with the already dead prey, and finally bring them live prey that is freed in their presence. The mother does not assist or interfere with the efforts of the young to kill the prey but brings it back if it escapes. Further suggestive evidence of teaching by bringing live prey to the young has also been reported in meerkats (Thornton and McAuliffe 2006), river otters *Lontra canadensis* (Liers 1951), the eastern spotted skunk *Spilogale putorius* (Thorne and Waggy 2017), and the dwarf mongoose *Helogale parvula* (Rood 1978), all of which hunt individually.

In contrast, there are no records of adult pack-hunting canids modifying their behavior in such a way as to promote learning when young are present at a hunt. The only anecdotal examples of teaching in canids are in bat-eared foxes and the red fox *Vulpes vulpes*, which forage individually for invertebrate and small vertebrate prey (Macdonald 1980; Nel 1999).

The importance of learning also depends on whether specialized techniques need to be used either because the prey is difficult to catch or might pose a danger to the hunter. Felids are known to often kill prey by precise bites to the nape, throat, or the snout (Kitchener et al. 2010), and meerkats kill potentially dangerous scorpions using a complex sequence of moves to avoid being stung (Thornton and Raihani 2008). In contrast, canids have less specialized canine teeth and bite the prey with more or less accuracy. As young canids can practice this relatively imprecise technique by joining the hunting pack, adults do not need to teach them and can feed them with regurgitated meat rather than by releasing live prey.

5 Conclusion

Given the diversity of carnivore lifestyles as illustrated by the various examples in this chapter, it is clear there is no typical pattern of carnivore parental care, not even within taxonomic *families*. From the view of general biology, notable in this respect is the lack of a clear correspondence between taxonomic groups and patterns of parental care, with the exception that while the *Canidae* have biparental care, often with other social support, in the *Felidae* – with the exception of the lion –, the three pinniped *families*, the *Ursidae,* and the *Mustelidae*, parental care is provided by the mother alone. Considering this diversity, caution must be taken not to inappropriately overgeneralize patterns of parental care from one particular species or taxonomic group, or from laboratory and highly domesticated or semidomesticated species, to other carnivores, or indeed to other mammals more broadly (see Macrì and Richter 2015 for a related warning regarding the use of a limited number of mammalian species and experimental paradigms in biomedical research). More studies are needed across a greater range of species and taxonomic groups and where possible in the natural conditions, or at least approximations of these, under which each species has evolved. The need for such studies is underlined by the important role of carnivores as apex predators in the food chain in many ecosystems and thus their key role in ecosystem management and conservation.

References

Andersen IL, Nævdal E, Bøe KE. Maternal investment, sibling competition, and offspring survival with increasing litter size and parity in pigs (*Sus scrofa*). Behav Ecol Sociobiol. 2011;65(6):1159–67. https://doi.org/10.1007/s00265-010-1128-4.

Baerends-van Roon JM, Baerends GP. The morphogenesis of the behaviour of the domestic cat, with a special emphasis on the development of prey-catching. Amsterdam: North-Holland Publishing Company; 1979.

Bánszegi O, Szenczi P, Urrutia A, Hudson R. Conflict or consensus? Synchronous change in mother–young vocal communication across weaning in the cat. Anim Behav. 2017a;130:233–40. https://doi.org/10.1016/j.anbehav.2017.06.025.

Bánszegi O, Jacinto E, Urrutia A, Szenczi P, Hudson R. Can but don't: olfactory discrimination between own and alien offspring in the domestic cat. Anim Cogn. 2017b;20(4):1–10. https://doi.org/10.1007/s10071-017-1100-z.

Bateson P. The dynamics of parent-offspring relationships in mammals. Trends Ecol Evol. 1994;9(10):399–403. https://doi.org/10.1016/0169-5347(94)90066-3.

Bateson P. Behavioural development in the cat. In: Turner DC, Bateson P, editors. The domestic cat: the biology of its behaviour. 3rd ed. Cambridge: Cambridge University Press; 2014. p. 11–26.

Bekoff M, Daniels TJ, Gittleman JL. Life history patterns and the comparative social ecology of carnivores. Annu Rev Ecol Syst. 1984;15(1):191–232. https://doi.org/10.1146/annurev.es.15.110184.001203.

Boness DJ, Bowen WD. The evolution of maternal care in pinnipeds: new findings raise questions about the evolution of maternal feeding strategies. Bioscience. 1996;46(9):645–54. https://doi.org/10.2307/1312894.

Bowen W. Behavioural ecology of pinniped neonates. In: Renouf D, editor. The behaviour of pinnipeds. Dordrecht: Springer; 1991. p. 66–127.
Bryden M. Lactation and suckling in relation to early growth of the southern elephant seal, *Mirounga leonina* (L.). Aust J Zool. 1968;16(5):739–48. https://doi.org/10.1071/ZO9680739.
Cameron MF, Siniff DB, Proffitt KM, Garrott RA. Site fidelity of Weddell seals: the effects of sex and age. Antarct Sci. 2007;19(2):149–55. https://doi.org/10.1017/S0954102007000223.
Carlini AR, Márquez MEI, Soave G, Vergani DF, de Ferrer PAR. Southern elephant seal, *Mirounga leonina*: composition of milk during lactation. Polar Biol. 1994;14(1):37–42. https://doi.org/10.1007/BF00240270.
Caro TM. Effects of the mother, object play, and adult experience on predation in cats. Behav Neural Biol. 1980;29(1):29–51. https://doi.org/10.1016/S0163-1047(80)92456-5.
Case TJ. Endothermy and parental care in the terrestrial vertebrates. Am Nat. 1978;112(987):861–74. https://doi.org/10.1086/283328.
Charrier I, Aubin T, Mathevon N. Mother–calf vocal communication in Atlantic walrus: a first field experimental study. Anim Cogn. 2010;13(3):471–82. https://doi.org/10.1007/s10071-009-0298-9.
Chilvers BL, Wilkinson IS. Philopatry and site fidelity of New Zealand Sea lions (*Phocarctos hookeri*). Wildl Res. 2008;35(5):463–70. https://doi.org/10.1071/WR07053.
Clutton-Brock TH. The evolution of parental care. Monographs in behavior and ecology. Princeton University Press: Princeton; 1991.
Clutton-Brock TH. Cooperative breeding in mammals. In: Kappeler PM, van Schaik CP, editors. Cooperation in primates and humans: mechanisms and evolution. Berlin/Heidelberg: Springer; 2006. p. 173–90.
Clutton-Brock TH. Cooperative breeding. Mammal societies. Chichester: Wiley; 2016. p. 557–604.
Clutton-Brock TH, Albon SD, Guinness FE. Fitness costs of gestation and lactation in wild mammals. Nature. 1989;337(6204):260–2. https://doi.org/10.1038/337260a0.
Cox C, Hager R. Mothers do not show increased offspring avoidance and elevated corticosterone levels during weaning conflict in rats. PLoS One. 2016;11(9):e0163195. https://doi.org/10.1371/journal.pone.0163195.
Derocher AE, Andersen M, Wiig Ø, Aars J, Hansen E, Biuw M. Sea ice and polar bear den ecology at Hopen Island, Svalbard. Mar Ecol Prog Ser. 2011;441:273–9.
Donohue MJ, Costa DP, Goebel ME, Baker JD. The ontogeny of metabolic rate and thermoregulatory capabilities of northern fur seal, *Callorhinus ursinus,* pups in air and water. J Exp Biol. 2000;203(6):1003–16. https://doi.org/10.1242/jeb.203.6.1003.
East ML, Burke T, Wilhelm K, Greig C, Hofer H. Sexual conflicts in spotted hyenas: male and female mating tactics and their reproductive outcome with respect to age, social status and tenure. Proc R Soc B Lond. 2003;270(1521):1247–54. https://doi.org/10.1098/rspb.2003.2363.
Eizirik E, Murphy WJ, Koepfli K-P, Johnson WE, Dragoo JW, Wayne RK, et al. Pattern and timing of diversification of the mammalian order Carnivora inferred from multiple nuclear gene sequences. Mol Phylogenet Evol. 2010;56(1):49–63. https://doi.org/10.1016/j.ympev.2010.01.033.
Estes JA. Enhydra lutris. Mamm Species. 1980;133:1–8. https://doi.org/10.2307/3503844.
Ewer RF. The "Instinct to Teach". Nature. 1969;222(5194):698. https://doi.org/10.1038/222698a0.
Faure U, Domokos C, Leriche A, Cristescu B. Brown bear den characteristics and selection in eastern Transylvania, Romania. J Mammal. 2020;101(4):1177–88. https://doi.org/10.1093/jmammal/gyaa047.
Federico V, Allainé D, Gaillard J-M, Cohas A. Evolutionary pathways to communal and cooperative breeding in carnivores. Am Nat. 2020;195(6):1037–55. https://doi.org/10.1086/708639.
Feldman HN. Maternal care and differences in the use of nests in the domestic cat. Anim Behav. 1993;45(1):13–23. https://doi.org/10.1006/anbe.1993.1002.
Fernández N, Palomares F. The selection of breeding dens by the endangered Iberian lynx (*Lynx pardinus*): implications for its conservation. Biol Conserv. 2000;94(1):51–61. https://doi.org/10.1016/S0006-3207(99)00164-0.

Fleming TH, Rauscher RJ. On the evolution of litter size in *Peromyscus leucopus*. Evolution. 1978;32(1):45–55. https://doi.org/10.2307/2407409.

Franco-Trecu V, García-Olazábal MD, Tassino B, Acevedo J. Parturition process in an amphibian mammal species: new evidences in South American fur seal (*Arctocephalus australis*). Anim Biol. 2016;66(1):21–9. https://doi.org/10.1163/15707563-00002484.

Garshelis DL. Variation in ursid life histories. In: Lindburg D, Baragona K, editors. Giant pandas. London: University of California Press; 2004. p. 53–73.

Gittleman JL, Thompson SD. Energy allocation in mammalian reproduction. Am Zool. 1988;28(3):863–75. https://doi.org/10.1093/icb/28.3.863.

Greene CE. Clinical microbiology and infectious diseases of the dog and cat. Philadelphia: WB Saunders Company; 1984.

Gross MR. The evolution of parental care. Q Rev Biol. 2005;80(1):37–45. https://doi.org/10.1086/431023.

Hamilton WD. The genetical evolution of social behaviour. II. J Theor Biol. 1964;7(1):17–52. https://doi.org/10.1016/0022-5193(64)90039-6.

Harcourt R. Factors affecting early mortality in the South American fur seal (*Arctocephalus australis*) in Peru: density-related effects and predation. J Zool. 1992;226(2):259–70. https://doi.org/10.1111/j.1469-7998.1992.tb03838.x.

Hart BL, Powell KL. Antibacterial properties of saliva: role in maternal periparturient grooming and in licking wounds. Physiol Behav. 1990;48(3):383–6. https://doi.org/10.1016/0031-9384(90)90332-X.

Hassanin A, Veron G, Ropiquet A, Jansen van Vuuren B, Lécu A, Goodman SM, et al. Evolutionary history of Carnivora (Mammalia, Laurasiatheria) inferred from mitochondrial genomes. PLoS One. 2021;16(2):e0240770. https://doi.org/10.1371/journal.pone.0240770.

Hepper PG. Long-term retention of kinship recognition established during infancy in the domestic dog. Behav Process. 1994;33(1–2):3–14. https://doi.org/10.1016/0376-6357(94)90056-6.

Hinde CA, Johnstone RA, Kilner RM. Parent-offspring conflict and coadaptation. Science. 2010;327(5971):1373–6. https://doi.org/10.1126/science.1186056.

Hindell MA, Little GJ. Longevity, fertility and philopatry of two female southern elephant seals (*Mirounga leonina*) at Macquarie Island. Mar Mam Sci. 1988;4(2):168–71. https://doi.org/10.1111/j.1748-7692.1988.tb00197.x.

Hoffman JI, Forcada J. Extreme natal philopatry in female Antarctic fur seals (*Arctocephalus gazella*). Mamm Biol. 2012;77(1):71–3. https://doi.org/10.1016/j.mambio.2011.09.002.

Holekamp KE, Boydston EE, Szykman M, Graham I, Nutt KJ, Birch S, et al. Vocal recognition in the spotted hyaena and its possible implications regarding the evolution of intelligence. Anim Behav. 1999;58(2):383–95. https://doi.org/10.1006/anbe.1999.1157.

Horrell I, Bennett J. Disruption of teat preferences and retardation of growth following cross-fostering of 1-week-old pigs. Anim Prod. 1981;33(01):99–106. https://doi.org/10.1017/S0003356100025253.

Hudson R, González-Mariscal G, Beyer C. Chin marking behavior, sexual receptivity, and pheromone emission in steroid-treated, ovariectomized rabbits. Horm Behav. 1990;24(1):1–13. https://doi.org/10.1016/0018-506X(90)90022-P.

Hudson R, Raihani G, González D, Bautista A, Distel H. Nipple preference and contests in suckling kittens of the domestic cat are unrelated to presumed nipple quality. Dev Psychobiol. 2009;51(4):322–32. https://doi.org/10.1002/dev.20371.

Hudson R, Rödel HG, Trejo Elizalde M, Arteaga L, Kennedy G, Smith B. Pattern of nipple use by puppies: a comparison of the dingo (*Canis dingo*) and the domestic dog (*Canis familiaris*). J Comp Psychol. 2016;130(3):269–77.

Insley SJ, Phillips AV, Charrier I. A review of social recognition in pinnipeds. Aquat Mamm. 2003;29(2):181–201. https://doi.org/10.1578/016754203101024149.

Jacinto E, Szenczi P, Hudson R, Bánszegi O. Olfactory discrimination between litter mates by mothers and alien adult cats: lump or split? Anim Cogn. 2019;22(1):61–9. https://doi.org/10.1007/s10071-018-1221-z.

Jackson CR, Power RJ, Groom RJ, Masenga EH, Mjingo EE, Fyumagwa RD, et al. Heading for the hills: risk avoidance drives den site selection in African wild dogs. PLoS One. 2014;9(6):e99686. https://doi.org/10.1371/journal.pone.0099686.

Johnsingh AJT. Reproductive and social behaviour of the Dhole, *Cuon alpinus* (Canidae). J Zool. 1982;198(4):443–63. https://doi.org/10.1111/jzo.1982.198.4.443.

Kauhala K, Helle E, Pietilä H. Time allocation of male and female raccoon dogs to pup rearing at the den. Acta Theriol. 1998;43:301–10. https://doi.org/10.4098/AT.arch.98-25.

Kitchener AC, Van Valkenburgh B, Yamaguchi N, Macdonald D, Loveridge A. Felid form and function. In: Macdonald DW, Loveridge A, editors. Biology and conservation of wild felids. Oxford: Oxford University Press; 2010. p. 83–106.

Klarenbeek M, Okkens A, Kooistra H, Mol J, Bevers M, Taverne M. Plasma oxytocin concentrations during late pregnancy and parturition in the dog. Theriogenology. 2007;68(8):1169–76. https://doi.org/10.1016/j.theriogenology.2007.08.017.

Kleiman DG, Malcolm JR. The evolution of male parental investment in mammals. In: Gubernick DJ, Klopfer PH, editors. Parental care in mammals. Boston: Springer; 1981. p. 347–87.

Koivula M, Koskela E, Mappes T, Oksanen TA. Cost of reproduction in the wild: manipulation of reproductive effort in the bank vole. Ecology. 2003;84(2):398–405. https://doi.org/10.1890/0012-9658(2003)084[0398:CORITW]2.0.CO;2.

Kölliker M, Brodie ED III, Moore AJ. The coadaptation of parental supply and offspring demand. Am Nat. 2005;166(4):506–16. https://doi.org/10.1086/491687.

König B, Riester J, Markl H. Maternal care in house mice (*Mus musculus*): II. The energy cost of lactation as a function of litter size. J Zool. 1988;216(2):195–210. https://doi.org/10.1111/j.1469-7998.1988.tb02425.x.

Kruuk H. The spotted hyena. A study of predation and social behavior. Chicago: Chicago University Press; 1973.

Kruuk H, Turner M. Comparative notes on predation by lion, leopard, cheetah and wild dog in the Serengeti area, East Africa. Mammalia. 1967;31(1):1–27. https://doi.org/10.1515/mamm.1967.31.1.1.

Laack LL, Tewes ME, Haines AM, Rappole JH. Reproductive life history of ocelots *Leopardus pardalis* in southern Texas. Acta Theriol. 2005;50(4):505–14. https://doi.org/10.1007/BF03192643.

Le Boeuf BJ, Whiting RJ, Gantt RF. Perinatal behavior of northern elephant seal females and their young. Behaviour. 1972;43(1/4):121–56. https://doi.org/10.1163/156853973X00508.

Lewis SE, Pusey AE. Factors influencing the occurrence of communal care in plural breeding mammals. In: Solomon G, French JA, editors. Cooperative breeding in mammals. New York: Cambridge University Press; 1997. p. 335–63.

Leyhausen P. Cat behaviour. The predatory and social behaviour of domestic and wild cats. New York: Taylor & Francis/Garland STPM Press; 1979.

Libal NS, Belant JL, Leopold BD, Wang G, Owen PA. Despotism and risk of infanticide influence grizzly bear den-site selection. PLoS One. 2011;6(9):e24133. https://doi.org/10.1371/journal.pone.0024133.

Liers EE. Notes on the river otter (*Lutra canadensis*). J Mammal. 1951;32(1):1–9. https://doi.org/10.2307/1375406.

Lindeberg H. Reproduction of the female ferret (*Mustela putorius furo*). Reprod Domest Anim. 2008;43(s2):150–6. https://doi.org/10.1111/j.1439-0531.2008.01155.x.

Lopate C. Management of pregnant and neonatal dogs, cats, and exotic pets. Oxford: Wiley; 2012.

Lukas D, Clutton-Brock TH. Cooperative breeding and monogamy in mammalian societies. Proc R Soc B Biol Sci. 2012;279(1736):2151–6. https://doi.org/10.1098/rspb.2011.2468.

Macdonald D. The red fox, *Vulpes vulpes*, as a predator upon earthworms, *Lumbricus terrestris*. Zeitschrift Tierpsychol. 1980;52(2):171–200. https://doi.org/10.1111/j.1439-0310.1980.tb00710.x.

Macdonald DW, Sillero-Zubiri C. The biology and conservation of wild canids. Oxford: Oxford University Press; 2004.

MacLeod KJ, Nielsen JF, Clutton-Brock TH. Factors predicting the frequency, likelihood and duration of allonursing in the cooperatively breeding meerkat. Anim Behav. 2013;86(5):1059–67. https://doi.org/10.1016/j.anbehav.2013.09.012.

Macrì S, Richter SH. The Snark was a Boojum – reloaded. Front Zool. 2015;12(1):S20. https://doi.org/10.1186/1742-9994-12-S1-S20.

Maehr DS, Land ED, Roof JC, McCown JW. Early maternal behavior in the Florida panther (*Felis concolor coryi*). Am Midl Nat. 1989;122(1):34–43. https://doi.org/10.2307/2425680.

Magoun AJ, Copeland JP. Characteristics of wolverine reproductive den sites. J Wildl Manag. 1998;62(4):1313–20. https://doi.org/10.2307/3801996.

Malcolm JR, Marten K. Natural selection and the communal rearing of pups in African wild dogs (*Lycaon pictus*). Behav Ecol Sociobiol. 1982;10(1):1–13. https://doi.org/10.1007/BF00296390.

Malmkvist J, Palme R. Periparturient nest building: implications for parturition, kit survival, maternal stress and behaviour in farmed mink (*Mustela vison*). Appl Anim Behav Sci. 2008;114(1–2):270–83. https://doi.org/10.1016/j.applanim.2008.01.018.

Manchi S, Swenson JE. Denning behaviour of Scandinavian brown bears *Ursus arctos*. Wildl Biol. 2005;11(2):123–32. 10.

Mappes T, Koskela E, Ylonen H. Reproductive costs and litter size in the bank vole. Proc R Soc B Lond. 1995;261(1360):19–24. https://doi.org/10.1098/rspb.1995.0111.

Martin P. An experimental study of weaning in the domestic cat. Behaviour. 1986;99(3/4):221–49. https://doi.org/10.1163/156853986X00568.

Matthews SM, Green DS, Higley JM, Rennie KM, Kelsey CM, Green RE. Reproductive den selection and its consequences for fisher neonates, a cavity-obligate mustelid. J Mammal. 2019;100(4):1305–16. https://doi.org/10.1093/jmammal/gyz069.

McNutt JW, Silk JB. Pup production, sex ratios, and survivorship in African wild dogs, *Lycaon pictus*. Behav Ecol Sociobiol. 2008;62(7):1061–7. https://doi.org/10.1007/s00265-007-0533-9.

Mills MGL. Kalahari hyaenas: comparative behavioural ecology of two species. 1st ed. London: Unwin Hyman; 1990.

Mittermeier RA, Wilson DE. Handbook of the mammals of the world: carnivores. Lynx Editions: Barcelona; 2009.

Miyazaki N. Caspian and Baikal seals: *Pusa caspica* and *Pusa sibirica*. In: Perrin WF, Würsig B, Thewissen JGM, editors. Encyclopedia of marine mammals. 2nd ed. London: Academic; 2009. p. 188–91.

Mock DW, Forbes LS. Parent-offspring conflict: a case of arrested development. Trends Ecol Evol. 1992;7(12):409–13. https://doi.org/10.1016/0169-5347(92)90022-4.

Montgomery TM, Pendleton EL, Smith JE. Physiological mechanisms mediating patterns of reproductive suppression and alloparental care in cooperatively breeding carnivores. Physiol Behav. 2018;193:167–78. https://doi.org/10.1016/j.physbeh.2017.11.006.

Moreira N. Reproduction in small female felids. In: Fowler ME, Cubas ZS, editors. Biology, medicine and surgery of south American wild animals. Ames: Iowa State University Press; 2001. p. 301–7.

Mustonen A-M, Nieminen P. Ecophysiology of overwintering in northern martes species. In: Aubry KB, Zielinski WJ, Raphael MG, Proulx G, Buskirk SW, editors. Biology and conservation of martens, sables, and fishers: a new synthesis. New York: Cornell University Press; 2012. p. 186–206.

Naaktgeboren C. Some aspects of parturition in wild and domestic Canidae. Int Zoo Yearb. 1968;8(1):8–13. https://doi.org/10.1111/j.1748-1090.1968.tb00420.x.

Nel JAJ. Social learning in canids: an ecological perspective. In: Box HO, Gibson KR, editors. Mammalian social learning: comparative and ecological perspectives. Cambridge: Cambridge University Press; 1999. p. 259–78.

Neville MC, Allen JC, Archer PC, Casey CE, Seacat J, Keller RP, et al. Studies in human lactation: milk volume and nutrient composition during weaning and lactogenesis. Am J Clin Nutr. 1991;54(1):81–92. https://doi.org/10.1093/ajcn/54.1.81.

Nielsen CK, Woolf A. Spatial organization of bobcats (*Lynx rufus*) in southern Illinois. Am Midl Nat. 2001;146(1):43–52. https://doi.org/10.1674/0003-0031(2001)146[0043:SOOBLR]2.0.CO;2.

Noonan MJ, Newman C, Buesching CD, Macdonald DW. Evolution and function of fossoriality in the Carnivora: implications for group-living. Front Ecol Evol. 2015;3(116) https://doi.org/10.3389/fevo.2015.00116.

Nowak RM. Walker's carnivores of the world. Baltimore: The Johns Hopkins University Press; 2005.

Oftedal OT, Gittleman JL. Patterns of energy output during reproduction in carnivores. In: Gittleman JL, editor. Carnivore behavior, ecology, and evolution. Boston: Springer; 1989. p. 355–78.

Oli MK, Jacobson HA, Leopold BD. Denning ecology of black bears in the White River National Wildlife Refuge, Arkansas. J Wildl Manag. 1997;61(3):700–6. https://doi.org/10.2307/3802177.

Olmstead CE, Villablanca JR, Torbiner M, Rhodes D. Development of thermoregulation in the kitten. Physiol Behav. 1979;23(3):489–95. https://doi.org/10.1016/0031-9384(79)90048-9.

Packard JM. Wolf behavior: reproductive, social, and intelligent. In: Mech LD, Boitani L, editors. Wolves: behavior, ecology, and conservation. Chicago: University of Chicago Press; 2010. p. 35–65.

Pal SK. Parental care in free-ranging dogs, *Canis familiaris*. Appl Anim Behav Sci. 2005;90(1):31–47. https://doi.org/10.1016/j.applanim.2004.08.002.

Pal SK, Ghosh B, Roy S. Agonistic behaviour of free-ranging dogs (*Canis familiaris*) in relation to season, sex and age. Appl Anim Behav Sci. 1998;59(4):331–48. https://doi.org/10.1016/S0168-1591(98)00108-7.

Périquet S, Mapendere C, Revilla E, Banda J, Macdonald DW, Loveridge AJ, et al. A potential role for interference competition with lions in den selection and attendance by spotted hyaenas. Mamm Biol. 2016;81(3):227–34. https://doi.org/10.1016/j.mambio.2015.10.005.

Peters SE. Postnatal development of gait behaviour and functional allometry in the domestic cat (*Felis catus*). J Zool. 1983;199(4):461–86. https://doi.org/10.1111/j.1469-7998.1983.tb05100.x.

Pfeifer S. Role of the nursing order in social development of mountain lion kittens. Dev Psychobiol. 1980;13(1):47–53. https://doi.org/10.1002/dev.420130108.

Pitcher BJ, Harcourt RG, Charrier I. Rapid onset of maternal vocal recognition in a colonially breeding mammal, the Australian sea lion. PLoS One. 2010;5(8):e12195. https://doi.org/10.1371/journal.pone.0012195.

Poglayen-Neuwall I. Copulatory behavior, gestation and parturition of the tayra. Zeitschrift Säugetierkunde. 1974;40:176–89.

Presley SJ. Eira barbara. Mamm Species. 2000;2000(636):1–6. https://doi.org/10.1644/1545-1410(2000)636<0001:EB>2.0.CO;2.

Raihani G, González D, Arteaga L, Hudson R. Olfactory guidance of nipple attachment and suckling in kittens of the domestic cat: inborn and learned responses. Dev Psychobiol. 2009;51(8):662–71. https://doi.org/10.1002/dev.20401.

Reichman OJ, Smith SC. Burrows and burrowing behavior by mammals. In: Genoways HH, editor. Current mammalogy. New York: Plenum Press; 1990. p. 197–244.

Richardson PRK. Aardwolf mating system: overt cuckoldry in an apparently monogamous mammal. S Afri J Sci. 1987;83(7):405–10.

Riedman M, Ortiz CL. Changes in milk composition during lactation in the northern elephant seal. Physiol Zool. 1979;52(2):240–9. https://doi.org/10.1086/physzool.52.2.30152567.

Rogers LL, McColley L, Dalton J, Stroner J, Hajicek D, Partin A, et al. Behavior in free-living American black bear dens: parturition, maternal care, and cub behavior. Animals. 2020;10(7):1123. https://doi.org/10.3390/ani10071123.

Roka B, Jha AK, Rai U. Breeding behaviour of red panda (*Ailurus fulgens fulgens*) in Padmaja Naidu Himalayan Zoological Park, Darjeeling, West Bengal. Mag Zoo Outreach Org. 2015. p. 10–13.

Romand R, Ehret G. Development of sound production in normal, isolated, and deafened kittens during the first postnatal months. Dev Psychobiol. 1984;17(6):629–49. https://doi.org/10.1002/dev.420170606.

Rood JP. Banded mongoose males guard young. Nature. 1974;248(5444):176. https://doi.org/10.1038/248176a0.

Rood JP. Dwarf mongoose helpers at the den. Z Tierpsychol. 1978;48(3):277–87. https://doi.org/10.1111/j.1439-0310.1978.tb00260.x.

Roper TJ. Badger *Meles meles* setts–architecture, internal environment and function. Mammal Rev. 1992;22(1):43–53. https://doi.org/10.1111/j.1365-2907.1992.tb00118.x.

Ross S, Kamnitzer R, Munkhtsog B, Harris S. Den-site selection is critical for Pallas's cats (*Otocolobus manul*). Can J Zool. 2010;88(9):905–13. https://doi.org/10.1139/Z10-056.

Royle NJ, Alonzo SH, Moore AJ. Co-evolution, conflict and complexity: what have we learned about the evolution of parental care behaviours? Curr Opin Behav Sci. 2016;12:30–6. https://doi.org/10.1016/j.cobeha.2016.08.004.

Ruggiero LF, Pearson E, Henry SE. Characteristics of American marten den sites in Wyoming. J Wildl Manag. 1998;62(2):663–73. https://doi.org/10.2307/3802342.

Ruprecht JS, Ausband DE, Mitchell MS, Garton EO, Zager P. Homesite attendance based on sex, breeding status, and number of helpers in gray wolf packs. J Mammal. 2012;93(4):1001–5. https://doi.org/10.1644/11-mamm-a-330.1.

Sauvé CC, Beauplet G, Hammill MO, Charrier I. Mother–pup vocal recognition in harbour seals: influence of maternal behaviour, pup voice and habitat sound properties. Anim Behav. 2015;105:109–20. https://doi.org/10.1016/j.anbehav.2015.04.011.

Sazatornil V, Rodríguez A, Klaczek M, Ahmadi M, Álvares F, Arthur S, et al. The role of human-related risk in breeding site selection by wolves. Biol Conserv. 2016;201:103–10. https://doi.org/10.1016/j.biocon.2016.06.022.

Schaller GB. The deer and the tiger. A study of wild life in India. Chicago: University of Chicago Press; 1967.

Schenkel R. Play, exploration and territoriality in the wild lion. Symp Zool Soc Lond. 1966;18:11–22.

Schmelz M, Duguid S, Bohn M, Völter CJ. Cooperative problem solving in giant otters (*Pteronura brasiliensis*) and Asian small-clawed otters (*Aonyx cinerea*). Anim Cogn. 2017;20(6):1107–14. https://doi.org/10.1007/s10071-017-1126-2.

Schneider TC, Kappeler PM. Gregarious sexual segregation: the unusual social organization of the Malagasy narrow-striped mongoose (*Mungotictis decemlineata*). Behav Ecol Sociobiol. 2016;70:913–26. https://doi.org/10.1007/s00265-016-2113-3.

Schoknecht PA. Growth and teat ownership in a litter of binturongs. Zoo Biol. 1984;3(3):273–7. https://doi.org/10.1002/zoo.1430030310.

Schulz TM, Bowen WD. The evolution of lactation strategies in pinnipeds: a phylogenetic analysis. Ecol Monogr. 2005;75(2):159–77. https://doi.org/10.1890/04-0319.

Seidensticker J. On the ecological separation between tigers and leopards. Biotropica. 1976;8(4):225–34. https://doi.org/10.2307/2989714.

Seidensticker JC, Hornocker MG, Wiles WV, Messick JP. Mountain lion social organization in the Idaho Primitive Area. Wildl Monogr. 1973;35:3–60.

Seryodkin IV, Kostyria AV, Goodrich JM, Miquelle DG, Smirnov EN, Kerley LL, et al. Denning ecology of brown bears and asiatic black bears in the Russian far east. Ursus. 2003;14(2):153–61.

Sieber OJ. Acoustic recognition between mother and cubs in raccoons (*Procyon lotor*). Behaviour. 1986;96(1/2):130–63. https://doi.org/10.1163/156853986X00252.

Smith TG, Stirling I. The breeding habitat of the ringed seal (*Phoca hispida*). The birth lair and associated structures. Can J Zool. 1975;53(9):1297–305. https://doi.org/10.1139/z75-155.

Smits CMM, Smith CAS, Slough BG. Physical characteristics of Arctic fox (*Alopex lagopus*) dens in Northern Yukon Territory, Canada. Arctic. 1988;41(1):12–6. https://doi.org/10.14430/arctic1687.

Squires JR, Decesare NJ, Kolbe JA, Ruggiero LF. Hierarchical den selection of Canada lynx in Western Montana. J Wildl Manag. 2008;72(7):1497–506. https://doi.org/10.2193/2007-396.

Sundell J. Reproduction of the least weasel in captivity: basic observations and the influence of food availability. Acta Theriol. 2003;48(1):59–72. https://doi.org/10.1007/BF03194266.

Sunquist ME. The social organization of tigers (*Panthera tigris*) in Royal Chitawan National Park, Nepal. Smithsonian Contrib Zool. Washington: Smithsonian Institution Press; 1981.

Szenczi P, Bánszegi O, Urrutia A, Faragó T, Hudson R. Mother-offspring recognition in the domestic cat: kittens recognize their own mother's call. Dev Psychobiol. 2016;58(5):568–77. https://doi.org/10.1002/dev.21402.

Szenczi P, Urrutia A, Hudson R, Bánszegi O. Are you my mummy? Long-term olfactory memory of mother's body odour by offspring in the domestic cat. Anim Cogn. 2022;25(1):21–26. https://doi.org/10.1007/s10071-021-01537-w.

Thompson PM, Miller D, Cooper R, Hammond PS. Changes in the distribution and activity of female harbour seals during the breeding season: implications for their lactation strategy and mating patterns. J Anim Ecol. 1994;63(1):24–30. https://doi.org/10.2307/5579.

Thorne ED, Waggy C. First reported observation of food provisioning to offspring by an Eastern spotted skunk, a small carnivore. Northeastern Nat. 2017;24(1) https://doi.org/10.1656/045.024.0108.

Thornton A, McAuliffe K. Teaching in wild meerkats. Science. 2006;313(5784):227–9. https://doi.org/10.1126/science.1128727.

Thornton A, Raihani NJ. The evolution of teaching. Anim Behav. 2008;75(6):1823–36. https://doi.org/10.1016/j.anbehav.2007.12.014.

Torriani MVG, Vannoni E, McElligott AG. Mother-young recognition in an ungulate hider species: a unidirectional process. Am Nat. 2006;168(3):412–20. https://doi.org/10.1086/506971.

Trillmich F. Mutual mother-pup recognition in Galápagos fur seals and sea lions: cues used and functional significance. Behaviour. 1981;78(1):21–42. https://doi.org/10.1163/156853981X00248.

Trimble M, Insley SJ. Mother–offspring reunion in the South American sea lion *Otaria flavescens* at Isla de Lobos (Uruguay): use of spatial, acoustic and olfactory cues. Ethol Ecol Evol. 2010;22(3):233–46. https://doi.org/10.1080/03949370.2010.502318.

Trivers RL. Parent-offspring conflict. Am Zool. 1974;14(1):249–64. https://doi.org/10.1093/icb/14.1.249.

Turner DC, Bateson P. The domestic cat: the biology of its behaviour. 3rd ed. Cambridge: Cambridge University Press; 2014.

Vaňková D, Bartoš L, Málek J. The role of vocalization in the communication between red deer hinds and calves. Ethology. 1997;103(10):795–808. https://doi.org/10.1111/j.1439-0310.1997.tb00121.x.

Verd S, Ginovart G, Calvo J, Ponce-Taylor J, Gaya A. Variation in the protein composition of human milk during extended lactation: a narrative review. Nutrients. 2018;10(8):1124.

White S, Briers RA, Bouyer Y, Odden J, Linnell JDC. Eurasian lynx natal den site and maternal home-range selection in multi-use landscapes of Norway. J Zool. 2015;297(2):87–98. https://doi.org/10.1111/jzo.12260.

Wolski TR, Houpt KA, Aronson R. The role of the senses in mare–foal recognition. Appl Anim Ethol. 1980;6(2):121–38. https://doi.org/10.1016/0304-3762(80)90064-4.

Woodroffe R, Vincent A. Mother's little helpers: patterns of male care in mammals. Trends Ecol Evol. 1994;9(8):294–7. https://doi.org/10.1016/0169-5347(94)90033-7.

Zhang Z, Swaisgood RR, Wu HUA, Li M, Yong Y, Hu J, et al. Factors predicting den use by maternal giant pandas. J Wildl Manag. 2007;71(8):2694–8. https://doi.org/10.2193/2006-504.

The Onset of Maternal Behavior in Sheep and Goats: Endocrine, Sensory, Neural, and Experiential Mechanisms

Frédéric Lévy

Abstract In sheep and goats, the onset of maternal behavior at parturition is characterized by a first phase called *maternal responsiveness* during which the mother is attracted to any newborn. In a second phase, called maternal selectivity, the mother establishes a selective bond with her young so that she only accepts it at suckling. After a description of the behavioral expression of both phases, this chapter reviews the physiological, sensory, and neural mechanisms involved. These two behavioral processes are synchronized with parturition by the vaginocervical stimulation induced by the expulsion of the newborn. Olfactory cues provided by the neonate are involved in maternal responsiveness and selectivity. Oxytocin supported by estrogens is the key factor for maternal responsiveness. The neural network involved in maternal responsiveness is mainly hypothalamic and is different from the circuitry involved in selectivity, which mainly concerns olfactory processing regions. Visual and auditory cues are necessary for offspring recognition at a distance. This multisensory recognition suggests that mothers form a mental image of their young. Maternal experience renders mothers more responsive to maternally relevant physiology and to young-related sensory inputs.

Keywords Oxytocin · Estrogens · Olfactory learning · Olfactory neurogenesis · Olfactory bulb · Vision · Audition · Attachment

1 Introduction

The expression of parental care in mammals varies widely across species and is closely related to the developmental stage of the young but also to the social structure of the species and their ecology. Sheep and goats are mostly seasonal breeders, and consequently, births occur during a short time period, providing the young with an optimal growth. Feral breeds of sheep and goats form social groups and

---

F. Lévy (✉)
CNRS, IFCE, INRAE, Université de Tours, PRC, F-37380, Nouzilly, France

G. González-Mariscal (ed.), *Patterns of Parental Behavior*, Advances in Neurobiology 27, https://doi.org/10.1007/978-3-030-97762-7_3

continuously move in search of food. The ability of the young to follow their mother quickly after birth is essential for their survival, especially in sheep, while in goats, the young stays hidden during the first days of life. Newborns possess fully developed sensory and motor systems as well as capacities for thermoregulation control to follow the mother. Thus, the demand of maternal behavior for precocial mammals (ungulates, pinnipeds, cetaceans) differs from altricial mammals (most rodents, canids, felids). Mothers of precocial young do not build a nest, and the mother-infant relationship is characterized by the maintenance of direct contact between both partners. Mothers of these species, therefore, develop a maternal care system in which a selective bond to one's own young is rapidly formed after parturition. Subsequently, mothers only allow their young to suck while rejecting any alien young that approaches the udder. This behavioral trait offers a unique opportunity to study the sensorial, physiological, and neural mechanisms involved in memorization and recognition of an individual. In addition, the onset of maternal behavior is strictly dependent on the process of parturition, while in most rodents, maternal care can be induced only by the presence of the young. Therefore, studying the onset of maternal behavior in these species is also appropriate to reveal the importance of parturition itself as the physiological event for synchronizing a multiplicity of neural mechanisms that coordinate maternal care. Another remarkable characteristic of sheep and goat maternal behavior is the primary importance of olfaction that is involved both in the attractiveness of any neonate at parturition and in the individual recognition of the young that allows the establishment of selective bond. Therefore, issues of the underlying bases of an individual odor and neural mechanisms responsible for learning this individual odor can be addressed.

This chapter reviews the current literature on the physiological, sensory, and neural mechanisms underlying maternal interest toward any neonate at parturition and the formation of a selective bond to one's own young. In a first section, expression of maternal behavior, i.e., maternal responsiveness and selectivity, is described. Immediate factors controlling maternal responsiveness (Sect. 3) and selectivity (Sect. 4) are then examined. In addition to these internal factors, Sect. 5 considers how external factors such as level of nutrition can affect the expression of maternal behavior. The final section concerns breed, temperament, and maternal experience as possible causes of variations in the quality and the quantity of the different items that composed maternal care.

2 Characteristics of the Onset of Maternal Behavior

Behavior During the Prepartum Period

In domestic and feral breeds of sheep and goats, the future mother shows a shift of behavior from gregariousness to isolation from the flock (Arnold and Dudzinski 1978; Lickliter 1984a; O'Brien 1984; Alexander et al. 1990a). This tendency to isolation is associated with a clear decrease in gregariousness just before parturition (Poindron et al. 1997, 2007a). For example, in bighorn sheep and European mouflon, females withdraw from their group about 2 weeks before parturition (Langbein et al. 1998) and domestic ewes show the lowest responses to social isolation 24 h before parturition (Soto et al. 2021). In Beetal goats, manifestation of isolation seeking begins 4–6 h before parturition (Das and Tomer 1997). In Saanen and Toggenburg domestic goats, the phenomenon of isolation and seeking a birth site ranged from 1.5 h to 8 h (Lickliter 1985), while in Murciano-Granadin goats, this isolation occurs mainly 1 h before parturition (Ramírez et al. 1995). Isolation is often accompanied with seeking a secluded or sheltered place to lamb. The shelter chosen varies according to the topography of the environment, the bush cover, and the presence of fences or cubicles in indoor conditions (Gonyou and Stookey 1984). Seeking shelter seems to depend on the size of the paddock: While in small paddocks Merino ewes lamb in shelter, there is no tendency for ewes to lamb close to the boundary fences in 12 ha paddocks (Stevens et al. 1981). Factors like food availability, protection against predators, or adverse climatic conditions (wind, rain) appear to influence the choice of the birthplace (Alexander et al. 1990a; Yamin et al. 1995; Pritchard et al. 2021). Isolation and lambing in a sheltered place also prevent the future mother from interventions of other ewes during parturition that could disturb the display of a proper maternal care and bonding. In indoor lambing pens, providing cubicles reduces mother-young separation and the stealing of lambs by other pre-parturient ewes (Gonyou and Stookey 1983). Thus, offering shelter to ewes during the lambing period allows an improvement of welfare and of lamb survival especially during inclement weather (Pritchard et al. 2021).

Pre-parturient behavior has been extensively characterized in ewes during the last hour prior to parturition. The typical behavioral traits are pawing, the ewe scraping the ground with her forefeet, which could be reminiscent to nest building (Echeverri et al. 1992; Arnold and Morgan 1975; Alexander et al. 1988) and ground sniffing, increasing after amniotic fluids are discharged, due to attraction to these fluids (Lévy et al. 1983). Circling and flanking that are relatively rare outside the parturition period increase during the last hour before parturition (Echeverri et al. 1992). Some behavioral sequences also are characteristics of the pre-parturient ewe. For example, ground sniffing frequently precedes walking and pawing the ground, and circling leads to pawing. These sequences are thought to be related to the search for an appropriate birth site (Echeverri et al. 1992). Both non-agonistic, such as sniffing conspecific's vulva, and agonistic behaviors, such as butting and pushing, occur at high levels as lambing approaches, especially in intensive farm management where animal density can be elevated (Das and Tomer 1997; Lickliter 1985; Ramírez et al. 1995; Echeverri et al. 1992). These agonistic behaviors contribute to isolation during lambing and consequently facilitate the formation of a bond with the neonate. During the hours preceding parturition, ewes and goats emit frequent vocalizations of low amplitude and frequency characterized by low-pitched bleats produced with the mouth closed (Lickliter 1985; Ramírez et al. 1995; Arnold and Morgan 1975; Sèbe et al. 2007).

Behavior at Parturition and Maternal Responsiveness

In sheep, there is no consistent peak of births throughout the 24 h cycle (Arnold and Morgan 1975; Alexander et al. 1993). However, some studies report a slight tendency for lambing at particular times of day rather than overnight (Dwyer and Lawrence 1998; Cloete et al. 2002; Fahmy et al. 1997). Differences could be due to the breed but also to environmental factors such as time of feeding (Arnold and Morgan 1975; Alexander 1988). In goats, the picture is quite different since most of births occur during daytime (Das and Tomer 1997; Lickliter 1985; Allan et al. 1991). During the late phase of labor, the amniotic sac ruptures. The majority of ewes and goats sniff and lick the ground where amniotic fluids spill and give birth at this place. Length of labor, defined as the interval between the onset of visible contractions and the expulsion of the newborn, is variable according to the studies. However, most ewes have a labor of less than 1 h with very few more than 2 h (Arnold and Morgan 1975; Regueiro et al. 2021; Alexander et al. 1990b; Dwyer et al. 2001; Dwyer et al. 1996). A similar interval duration is reported in goats (Das and Tomer 1997; Ramírez et al. 1995; Sambraus and Wittmann 1989) although some studies indicate a longer parturition length (>130 min; Kasikci 2018). Multiple parturitions, birth weight, breed, and maternal experience affect parturition length (Cloete et al. 2002; Regueiro et al. 2021; Cloete et al. 1998; Otal et al. 2010; Martinez et al. 2009; Dwyer and Lawrence 1999a). The majority of the females are lying down during at least part of the labor process. Fetuses usually present in an anterior position with the head resting on the extended forelegs.

Immediately after birth, mothers stand up and start avidly licking (Fig. 1a), nudging the neonate, usually beginning with the head and continuing eventually onto the body with a particular attention to the anogenital region (Lickliter 1985; Alexander 1988). Licking behavior might be accompanied with pawing the newborn lamb to incite it to stand (Fig. 1b). Mothers lick their newborn almost continuously for the first 90 min and intermittently for the following 2 h (Lickliter 1985; Dwyer and Lawrence 1998; Sambraus and Wittmann 1989). The amount of time spent grooming depends on breed and multiple parturitions, twinborn lambs receiving less overall grooming attention than single lambs (Dwyer and Lawrence 1998; O'Connor et al. 1992). Licking behavior toward the neonate is mediated through attractiveness of amniotic fluid and is restricted to the time of parturition as amniotic fluid is aversive outside the parturition period (Lévy et al. 1983). Grooming stimulates activity of the lamb and prevents heat loss by drying it and is important for the development of maternal responsiveness and maternal selectivity (Poindron et al. 2010; Otal et al. 2009). Occasionally, when the neonate urinates, goats display flehmen behavior in which they arch the head upward and, with an open jaw, draw the urine into the vomeronasal organ, a secondary olfactory organ. While grooming the newborn, both ewes and goats emit low-pitched bleats (Lickliter 1985; Sèbe et al. 2007; Sambraus and Wittmann 1989; Dwyer et al. 1998; Shillito 1972). These vocalizations are almost emitted during the immediate postpartum period, and in sheep, their rate is affected by breed and experience (Dwyer et al. 1998). They have a

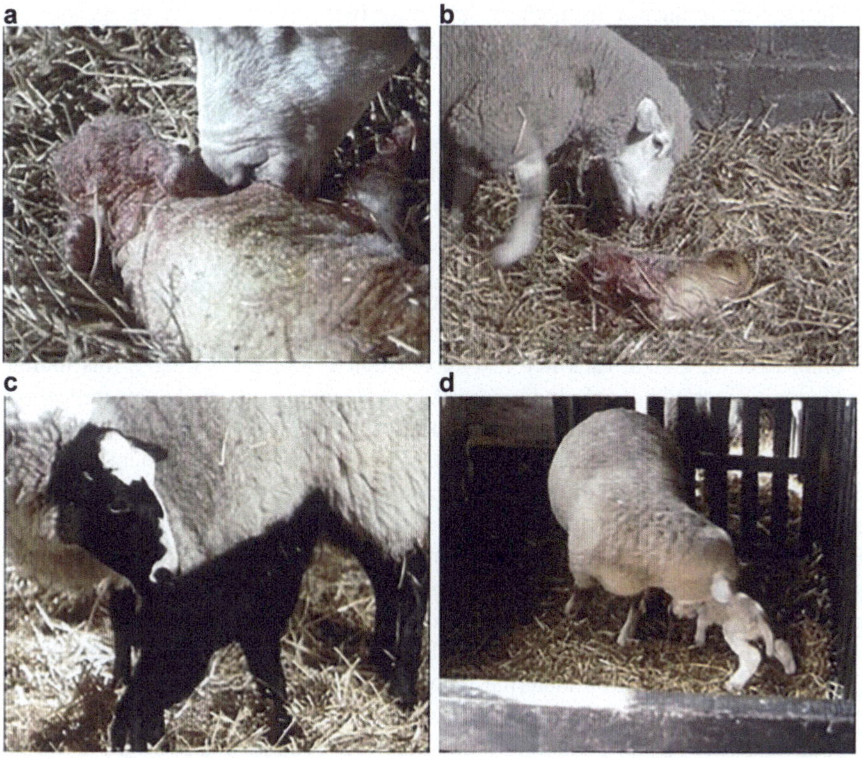

Fig. 1 Onset of maternal behavior in sheep. (**a**) Immediately after parturition, a mother licks her lamb. (**b**) A mother paws the lamb to stimulate it for standing. (**c**) A mother adopts a parallel-inverse position with their young for suckling. (**d**) A mother rejects an alien lamb approaching the udder for suckling

calming effect and also play a crucial role in the development of the mother-young bond as 2-day-old lambs are able to recognize low-pitched bleats of their own mother (Sèbe et al. 2010). The neonate is usually able to stand within the first 15–20 min. The mother helps the newborn in searching the udder by arching the back and bending the hind legs, thus facilitating its access and by gently pushing it with her muzzle and grooming the anogenital region (Fig. 1c). Suckling is generally achieved during the second hour after birth, which allows ingestion of colostrum, of primary importance for lamb survival, and attachment to the mother (Nowak et al. 2007).

The placenta is usually expelled after about 3 h in sheep (Arnold and Morgan 1975) and 2–3 h in goats (Lickliter 1985; Otal et al. 2010). Surprisingly, does and ewes, as herbivorous animals, temporarily change their feeding preference and ingest their placenta. However, partial ingestion occurs in the majority of cases (Alexander et al. 1990b; Otal et al. 2010; Gonzalez-Stagnaro and Madrid-Bury 2004). The proportion of animals displaying placentophagia depends on litter size with a higher frequency in mothers delivering twins, or triplets in sheep, than in mothers with singletons, and on breed (Alexander et al. 1990b; Otal et al. 2010; Ramirez et al. 1995). Placentophagia is thought to provide beneficial effects to mothers, such as analgesia enhancement, additional food supply, and promoting lactogenesis (Mota-Rojas et al. 2020). Placentophagia is also assumed to minimize the risk of predation by removing olfactory attractant of the afterbirth (Bridges 2015). In this respect, one may wonder the reason why complete placentophagia is less widespread in these ungulate species.

While after some hours sheep mothers leave the birth site followed by their lambs (follower species), the doe leaves her kids for several hours, and the kids hide for 4–7 days (hider species). Mothers visit them for nursing only a few times per day (Poindron et al. 2007a; Lickliter 1984b; O'Brien 1983). Interestingly, this hiding behavior is reported not only in feral goats but also in domesticated goats indicating it has not been lost by domestication (Lickliter 1984b).

Maternal responsiveness constitutes the first phase of the development of maternal behavior at parturition during which maternal care is direct toward any newborn and exchange with an alien newborn is easily performed (Poindron et al. 1980). Behaviors indicative of maternal responsiveness are licking, emission of low-pitched bleats, suckling and absence of agonistic behavior, high-pitched bleats, and suckling refusal. Criteria used to define maternal responsiveness may vary between studies, to the use of one criterion such as licking (Krehbiel et al. 1987) and suckling (Poindron et al. 1980) to a maternal index composed of maternal acceptance and rejection behaviors (Perrin et al. 2007).

Full maternal responsiveness in both sheep and goats is related to the event of parturition and fades after some hours in the absence of the newborn (Lévy et al. 1991; Collias 1956; Klopfer et al. 1964; Ramírez et al. 1996). In sheep, if the newborn lamb is removed at birth, before any contact with the mother, the majority of ewes are unable to display maternal care when reunited after 12 h of separation (Poindron et al. 1979). In goats, the length of separation is even shorter since a separation of 1 h (Klopfer et al. 1964; Ramírez et al. 1996) brings about the extinction of the maternal response in all the mothers. The difference in separation length between sheep and goats could be due to the elimination of birth fluids by cleaning the kid during the separation period that renders it less attractive to the does. Thus, maximal maternal responsiveness is limited to a short period. Herscher et al. (1963) define it as a period in which "the sensibility of the mother is highest on facing certain particular stimuli." The term of sensitive period was then used, which is related to the sensitivity to any neonate and means that the contact with the neonate at a particular period will result in lasting effects on the maternal behavior (Poindron et al. 2007b). This characteristic emphasizes the importance of parturition as a key event for triggering maternal responsiveness.

Individual Recognition of the Young: Maternal Selectivity

During the phase of maternal responsiveness, while ewes and does are interacting with their neonate, sniff them, lick them, suckle them, and vocally communicate with them, they learn the sensory identity. This learning leads to the establishment of individual recognition so that after a few hours, mothers are able to recognize their own offspring and accept them at suckling while they reject any suckling attempt by any other young (Fig. 1d). This discriminative process is called maternal selectivity (Poindron et al. 2007b; Lévy et al. 2004). If the offspring is removed, a high level of locomotor activity and the emission of distress bleats also characterizes maternal selectivity (Poindron et al. 1994). At reunion, the motivation to approach the young is greater than in the case of an alien young (Shillito and Alexander 1975). Similarly, when put in a two-choice test, mothers approach specifically their lamb and stay close to it in comparison to an alien young (Shillito and Alexander 1975; Ferreira et al. 2000; Keller et al. 2003). Thus, maternal selectivity has been considered as a model of attachment since it fulfills the criteria defined by Ainsworth and Bell (1970) studying mother-infant relationships in humans (Nowak et al. 2011).

Two types of recognition of the young can be characterized in both sheep and goats. A proximal recognition (<0.25 m) is based on learning of the olfactory signature of the young and leads to the exclusive acceptance of the familiar young at suckling. Mothers, made anosmic before parturition, show no signs of discrimination and accept any young at suckling (Baldwin and Shillito 1974; Bouissou 1968; Morgan et al. 1975; Klopfer and Gamble 1966; Lévy et al. 1995a) even if the newborns differ in the color of their coat (Romeyer et al. 1994a). In addition, a distal recognition allows mothers to identify their young from a distance, and this involves visual and auditory cues (Keller et al. 2003; Terrazas et al. 1999).

Proximal and distal recognition also differ in their dynamics of establishment. Sheep and goat mothers are able to perform selective nursing (proximal recognition) within 2 h of contact with their young after parturition (Romeyer et al. 1994a; Poindron and Le Neindre 1980), most ewes being selective even after 30 min of contact after birth (Keller et al. 2003). Discrimination at a distance with the help of both visual and auditory cues is functional slightly later since mothers are able to show a preference for their lambs, only after 6 h of contact in ewes and after 4 h in does (Keller et al. 2003; Poindron et al. 2003).

The fact that the establishment of proximal and distal recognition is closely related in time could indicate that proximal recognition represents a first step critical for the later development of distal recognition. However, both recognitions develop independently from each other. Anosmic ewes can discriminate their own lamb from a distance even though they are not able to discriminate their own lamb at a short distance and are willing to nurse any lamb (Ferreira et al. 2000). Therefore, no intermodal sensory processing is involved in the complete development of maternal discrimination, at least during the early phase of mother-young interaction. Nonetheless, at 1 month of lactation, anosmic ewes display discriminative behavior at suckling with more rejection of the alien lamb suggesting that audition and vision can compensate for the absence of olfaction allowing proximal recognition (Ferreira et al. 2000). Similarly, Alexander and Stevens (Alexander and Shillito 1981) reported that washing the familiar lamb at 5 days postpartum disrupts proximal recognition while it has no behavioral consequence when performed at 1 month postpartum. Likewise, anosmia performed at 2 weeks postpartum does not prevent recognition at suckling as when anosmia is performed before parturition (Poindron 1976). Hence, some compensatory mechanisms could exist between the different sensory modalities so that the predominance of olfaction for proximal recognition progressively changes to a more multisensory control as lactation advances.

3 Immediate Factors Controlling Maternal Responsiveness

The close synchrony between parturition and the onset of maternal responsiveness indicates that physiological changes associated with the development of pregnancy and parturition could be involved in the control of the behavior. In fact, these physiological changes, mainly endocrine and neuroendocrine, induce a maximal state of responsiveness to the sensory cues emanating from young which thereafter maintain maternal responsiveness (Rosenblatt 1994).

Physiological Changes During Pregnancy and Parturition

As in many other mammals like rodents and lagomorphs, an increase in the ratio of estradiol to progesterone occurs at parturition in sheep and goats. In sheep, 2 to 4 days before parturition, the concentration of plasma progesterone falls while estradiol rises one day before parturition to reach a peak at parturition and returns to basal levels within the first 4 h after parturition (Challis and Linzell 1971; Chamley et al. 1973; Shipka and Ford 1991). While similar changes have been observed in goats, there is no large acute prepartum increase in estradiol concentrations of the great magnitude known for sheep (Currie et al. 1988). Elevated plasma concentrations of glucocorticoids, especially cortisol, measured around parturition, have been reported in sheep (Strott et al. 1974) and goats (Hydbring et al. 1999).

The change of estradiol and progesterone steroid balance induces a rise in concentrations of the pituitary hormone prolactin, which occurs 1 day before parturition in sheep (Chamley et al. 1973), during the last 3 days in goats (Currie et al. 1988). These prolactin concentrations are maintained at a high level after parturition because of suckling stimulation (Chamley et al. 1973; Davis et al. 1971).

Contrary to the other hormones, the rise in plasma oxytocin (OT) concentration, released from the posterior pituitary, is strictly contemporaneous with the actual phase of expulsion and is caused by the vaginocervical stimulation. In both species, OT concentrations show significant elevation only at birth and up to 15 min postpartum due to the Ferguson reflex (Currie et al. 1988; Kendrick et al. 1991a). Changes in OT concentrations also occur during parturition at the central level in various brain regions. OT release increases in the cerebrospinal fluid, the olfactory bulb (OB), the bed nucleus of the stria terminalis (BNST), the septum, the medial preoptic area (MPOA), the paraventricular nucleus of the hypothalamus (PVN), and the substantia nigra (Da Costa et al. 1996; Kendrick et al. 1986, 1988a, b, 1992a; Lévy et al. 1992). OT increased release is associated with the increase of OT immunoreactivity and mRNA level in the OB, the BNST, the MPOA, the supraoptic nucleus (SON), and PVN (Broad et al. 1993). Expression of OT receptor mRNA also enhances in these brain regions, and this enhancement is promoted by the increase in estradiol release (Broad et al. 1999a). Thus, at the time of parturition, there is a dramatic increase in OT synthesis, storage, and release, and this tremendous activation is amplified by an increase in the number of OT receptors.

Parturition is also the time of an increased release of several neurotransmitters. For instance, the ascending noradrenergic system shows a general pattern of activation that parallels the OT system since increase in noradrenaline release is observed in the same brain regions (Kendrick et al. 1986, 1988a, 1992a; Lévy et al. 1993). Release of GABA and glutamate also increases in the OB, the BNST, the MPOA, and the PVN (Da Costa et al. 1996; Kendrick et al. 1986, 1988a, 1992a, b, 1997a, b).

Thus, parturition is accompanied with simultaneous activations occurring in different brain areas not directly connected. This suggests that the physiological regulation of maternal motivation is multi-determined. Nevertheless, these studies correlate physiological events with not only the emergence of maternal responsiveness but also with the onset of parturition and lactation. Therefore, only some of these physiological events could be specific for the induction of maternal behavior.

Hormonal Factors Controlling Maternal Responsiveness

In sheep, a positive effect of estradiol on maternal responsiveness has been reported on several occasions although results have not always been consistent. Earliest studies show that in nonpregnant ewes, a short-term treatment (1 week) or a long-term treatment (4 weeks) of progesterone following estradiol at high doses are effective but only in 50% of the ewes exposed to lambs for 1 h (Le Neindre et al. 1979). This treatment also induces abnormal behavior such as permanent estrus and male sexual behavior. Further investigations have revealed evidence for facilitation of maternal responsiveness by steroids. Spontaneous acceptance of a lamb occurs only when estrogen concentrations are high, at estrus, and at the very end of pregnancy (Poindron and Le Neindre 1980). The period during which parturient ewes are positively responsive to lambs, the sensitive period, can be extended for 4 h to 24 h by estradiol injection at a supraphysiological doses (20 mg; Poindron et al. 1979). A similar dose of estradiol benzoate also induces maternal responsiveness in nongestant ewes. However, this treatment is pharmacologic since it induces abnormal sexual behavior. In studies performed using more physiologically appropriate doses

(200 μg of estradiol) and longer period of treatment (6 weeks), full maternal responsiveness is never observed in any of the tested ewes (Kendrick et al. 1992b; Kendrick and Keverne 1991). In goats, attempts to stimulate maternal responsiveness with steroid treatment that induces lactation have failed so far (Rosenblatt and Siegel 1981). Thus, contrary to rodents and lagomorphs, in sheep and even less in goats, steroids are not sufficient to induce the full repertoire of maternal responsiveness, and as we will see, they have priming effect allowing other factors to trigger the behavior.

The synchrony between the process of expulsion of the fetus and the onset of maternal responsiveness has led to suspect the involvement of vaginocervical stimulation in activating the behavior. This hypothesis is now supported by numerous evidences in both non-parturient and parturient sheep and goats. Maternal responsiveness is induced in nonpregnant ewes primed with progesterone and estradiol after administration of 5 min of vaginocervical stimulation (Keverne et al. 1983). The dramatic effects of vaginocervical stimulation have been confirmed in two other breeds of sheep receiving low doses of estradiol and even in ewes at estrus (Kendrick et al. 1991b, 1992b; Poindron et al. 1988). This stimulation not only increases the number of maternal ewes but also induces licking, maternal bleats, and acceptance at the udder and reduces aggressive behavior toward the young. Ewes also display an attraction toward amniotic fluids that is a characteristic trait of maternal ewes at parturition. A facilitating action of vaginocervical stimulation has been also investigated in parturient ewes. Vaginocervical stimulation re-induces maternal responses 1 h (Keverne et al. 1983) or 2 h (Lévy et al. 2010) and even 24 h after parturition (Kendrick et al. 1991b). In contrast, if stimulation of the genital tract is prevented at the time of parturition by peridural anesthesia, maternal behavior is disrupted, and attraction to amniotic fluids is lost (Krehbiel et al. 1987; Lévy et al. 1990a). Similarly, in goats, vaginocervical stimulation is a key factor for the induction of maternal responsiveness. Blocking vaginocervical stimulation by peridural anesthesia at the first signs of parturition impairs the display of maternal care (Poindron et al. 2007a), and an artificial vaginocervical stimulation applied at 2 h postpartum induces again maternal responsiveness (Romeyer et al. 1994b).

The neuroendocrine mechanisms by which vaginocervical stimulation elicits maternal responsiveness implicate the release of OT in various brain regions. Indeed, the involvement of OT was first supported by studies reporting that a massive release of OT occurs in the brain, similar to that in the peripheral circulation, at parturition and in cycling ewes after an experimental vaginocervical stimulation (Kendrick et al. 1986, 1988b; Lévy et al. 1992). The actual demonstration of a direct link between activation of the OT system and maternal responsiveness is revealed by the display of maternal responses induced by intracerebroventricular injection of OT in non-gestant ewes (Kendrick et al. 1987; Keverne and Kendrick 1991). Interestingly, OT, like vaginocervical stimulation, was ineffective when given without estradiol priming. OT involvement was also reported in parturient ewes. Blocking vaginocervical stimulation at parturition with peridural anesthesia inhibits central release of OT and maternal responsiveness. Maternal care and attraction to amniotic fluids are restored by intracerebroventricular injection of OT (Lévy et al. 1992).

Thus, OT is a physiological signal of primary importance to synchronize the onset of maternal responsiveness with the appearance of the young. This signal is further amplified by other neuropeptide systems such as opiates. Intracerebroventricular infusions of morphine potentiate cerebrospinal fluid concentrations of OT after vaginocervical stimulation and induce maternal responsiveness similar to that of the parturient ewe (Keverne and Kendrick 1991). The facilitating action of opiates have been also evidenced in parturient ewes whose maternal responsiveness is blocked by an injection of naltrexone, an opioid antagonist (Caba et al. 1995). Whether other neuroendocrine systems, like arginine-vasopressin system or prolactin, are involved directly or indirectly through the OT system remains an open question.

Sensory Factors Controlling Maternal Responsiveness

The onset phase of maternal responsiveness controlled by neuroendocrine systems induces a maximal state of responsiveness to the sensory cues emanating from the young, which thereafter maintain maternal responsiveness (Rosenblatt and Lehrman 1963). In sheep, mothers lose their maternal responsiveness if the newborn is removed at parturition (Poindron et al. 1980).

A series of studies have demonstrated that olfactory cues provided by the newborn are of primary importance for the maintenance of maternal responsiveness beyond the sensitive period. Deprivation of olfactory cues by placing the lamb in an airtight transparent box from birth to 8 h after parturition decreases maternal responsiveness tested after this period of separation with the lamb. When only lamb odors can be perceived by the mother, the proportion of ewes is greater than when only visual cues are provided (Poindron et al. 1988). It appears that goats differ from sheep, as in goats suppressing the sensory cues provided by licking and suckling impairs the maintenance of maternal responsiveness after 4 h of treatment while in sheep it does not even following 12 h of treatment (Bordi et al. 1994; Romeyer et al. 1993a). This impairment could result from a loss of olfactory perception due to the privation of licking the kid. By licking, olfactory molecules provided by the newborn can be transferred via the tongue to the vomeronasal organ. If an effect of licking privation was confirmed, the vomeronasal system could be involved in the regulation of maternal responsiveness in the goat, contrary to sheep (Lévy et al. 1995a).

What could be the type of olfactory cues meaningful for the mother to maintain her responsiveness to the young? Early studies indicate that newly born kids (Gubernick et al. 1979) or lambs (Poindron et al. 1980) are more readily accepted by parturient females than day-old ones. One of the elements differentiating both types of young is the presence of amniotic fluids on their coat, which are very attractive for the mother at parturition (Lévy et al. 1983; Vince et al. 1985). This olfactory attraction is necessary for the maintenance of maternal responsiveness by attracting the mother to her newborn. Washing the newborn with either soap or water reduces licking behavior, and it prevents udder acceptance and increases aggressive behavior (Levy and Poindron 1987). Amniotic fluids can also facilitate the adoption of 1-day-old lambs whose coats are treated with amniotic fluids (Lévy and Poindron 1984; Basiouni and Gonyou 1988). Amniotic fluids from the tested mother or an alien mother have similar effects on maternal acceptance, suggesting that amniotic fluids contain olfactory cues responsible for a general attractiveness (Arnould et al. 1991). These compounds appear to differ from those eliciting repulsion in non-gestant ewes (Uriarte et al. 2012). A chemical characterization has identified volatile organic compounds of amniotic fluids, and it would be interesting to test whether some of these compounds are involved in the attractiveness of amniotic fluids (Viviers et al. 2015). Thus, amniotic fluids are "a potent organizer of maternal behavior" in focusing the attention of the mother from the licking of her body to the licking of the neonate, as was stated in the cat by Schneirla et al. (1963).

Another experimental strategy used to investigate the role of olfaction in the maintenance of maternal responsiveness is to observe the effects of prepartum anosmia. In sheep, disruption of the olfactory epithelium by irrigation of the nasal cavities with zinc sulfate induces a reduction in the time spent licking and suckling the newborn, and in the number of maternal bleats, and an increase in the number of protest bleats (Lévy et al. 1995a). On the contrary, lesion of the vomeronasal organ causes little disturbances in the onset of maternal care. In goats, anosmia has no consequence on the display of maternal responsiveness (Romeyer et al. 1994a; Hernandez et al. 2002). However, the effects of anosmia were not investigated in females without a previous maternal experience, and it could be possible that this experience compensate for the loss of olfactory perception. Whether a difference exists in the role of the main olfactory system between both species remains an open question.

Brain Regions Involved in Maternal Responsiveness

Hormonal changes occurring at parturition render the mother responsive to lamb cues, mainly olfactory, so that maternal responsiveness can emerge. Hormones act on a primary neural circuitry of which some elements have been identified in sheep.

One approach for defining brain regions involved in maternal responsiveness has been the use of the so-called expression of immediate early genes (c-fos) or the protein Fos as markers of neuronal activation. Increase in c-fos expression concomitant to the expression of maternal responsiveness has been observed in extensive neural circuitry including various limbic (BNST, lateral septum, medial amygdala, hippocampus) and hypothalamic areas (MPOA, PVN, SON, ventromedial nucleus; Da Costa et al. 1997; Keller et al. 2004a).

However, this approach is correlational and does not reveal the functional involvement of these brain structures. To answer this question, reversible inactivation of some of these structures, using infusion of an anesthetic, was undertaken before parturition and during the first 2 h postpartum (Perrin et al. 2007). MPOA inactivation significantly impairs the whole expression of maternal behavior as a massive decrease in licking behavior, nursing, and maternal vocalizations is reported (Fig. 2). Interestingly, inactivation of the septum or the diagonal band of Broca has no behavioral consequences. In addition, when maternal interest is challenged by a separation/reunion lamb test, mothers with MPOA inactivation exhibit little reaction after separation of their lambs and do not show any motivation to reunite with them. These findings allow considering the MPOA as a key structure for the control of maternal responsiveness as was reported in various studies in the rat (Numan and Insel 2003; Olazabal et al. 2013). Surprisingly, whereas ewes with BNST inactivation express normal maternal behavior at parturition, they are less motivated to join their lambs, suggesting that this structure regulates approach behavior to the young as it has been demonstrated in the rat (Numan and Numan 1996). The MPOA is also important for the maintenance of maternal responsiveness beyond parturition. Deficits in responses to the separation/reunion lamb test are

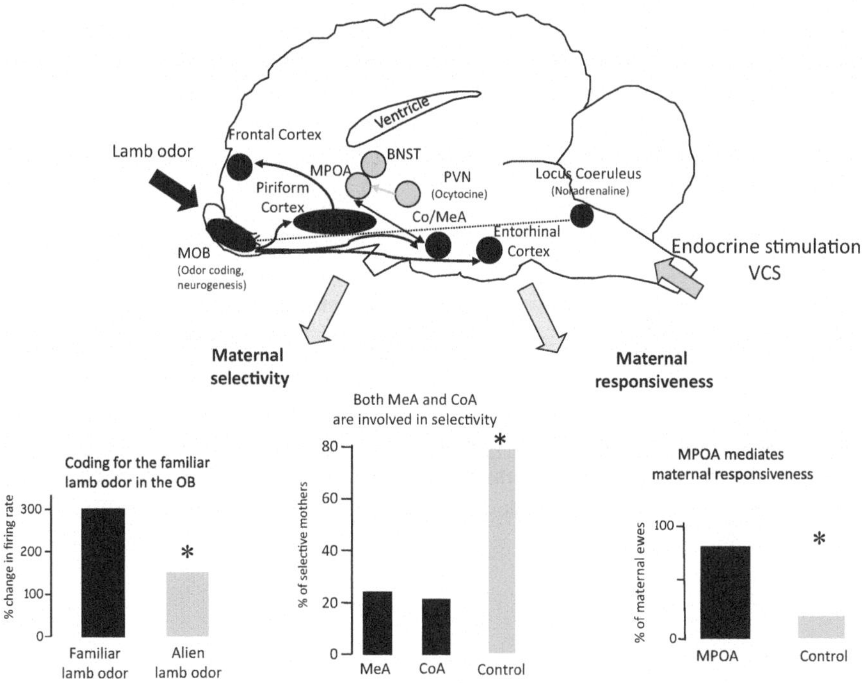

Fig. 2 A neural model of the regulation of maternal responsiveness and selectivity at parturition in sheep. The neural network represents a set of brain structures involved in maternal responsiveness (*grey*) and in selectivity (*dark*). The functional role of some of these structures has been tested (see graphs below). BNST: bed nucleus of the stria terminalis; CoA: cortical amygdala; MeA: medial amygdala; MOB: main olfactory bulb; MPOA: medial preoptic area; PVN: paraventricular nucleus; VCS vaginocervical stimulation. (Adapted from Lévy and Keller 2009)

found in maternal ewes infused at 2 h postpartum with an anesthetic for a 12-h period (Perrin et al. 2007). This result is in accordance with a previous study showing an enhanced Fos expression in the MPOA at 7 h postpartum following the reintroduction of the lamb after a short separation period (Keller et al. 2005).

The PVN is the main source of OT release in the brain, and at parturition, there is an extensive activation of the whole OT system of the PVN (Da Costa et al. 1996; Broad et al. 1993, 1999b). As a matter of fact, OT infusion into the PVN in nonpregnant ewes receiving a steroid pretreatment induces the full maternal repertoire (Da Costa et al. 1996). However, since enhancement of OT release also occurs at parturition in the MPOA, the BNST, and the OB, these brain regions could be part of the neural network involved in maternal responsiveness (Kendrick and Keverne 1992). This hypothesis was tested in steroid-treated non-gestant ewes infused with OT in the MPOA or in the OB. These treatments do not induce the full repertoire of maternal behavior but only reduce aggression toward lambs, and no effects on acceptance behaviors are promoted (Kendrick 2000). Part of the effects of OT in inducing maternal responsiveness could be mediated through its modulation of neurotransmitter releases (Lévy et al. 1995b; Kendrick et al. 1997a). For instance, noradrenaline is released together with OT during parturition in the MPOA, the BNST, the PVN, and the OB (Kendrick et al. 1992a, 1997a; Lévy et al. 1995b). Moreover, retrodialysis infusions of OT increase noradrenaline release in the MPOA (Kendrick et al. 1992a) and in the OB (Lévy et al. 1995b). Therefore, it is possible that OT acts on presynaptic noradrenergic terminals, and in this way, a restricted pattern of potentiated noradrenaline release could occur at sites controlling maternal responsiveness.

4 Immediate Factors Controlling Maternal Selectivity

Sheep and goat mothers learn and memorize the identity of their young using olfaction, vision, and audition. As revealed in previous sections, mothers first learn olfactory cues provided by the young that permit recognition at a short distance and acceptance at suckling. A few hours later when the exclusive bond is formed, mothers learn the young's visual appearance and voice that are more salient cues for recognition at a distance.

Olfactory Cues

Like many young of ungulates, lambs and kids adopt a parallel inverse position when suckling, allowing the mothers to sniff the trunk and the anogenital region during nursing (Fig. 1c). This olfactory investigation is of primary importance for identification of the young. When ewes or does are experimentally rendered anosmic, they accept at suckling alien young as well as their own although sniffing of the anogenital region persists (Poindron et al. 2007b; Lévy et al. 2004). Interestingly, in sheep, only the main olfactory system is involved in young identification since lesions of the accessory olfactory system does not prevent maternal selectivity at suckling (Lévy et al. 1995a). Whether in goats the accessory olfactory system plays a role in selectivity remains to be clarified.

The issue of the chemical nature and the source of odors responsible for the olfactory signature of the young are still not elucidated. Olfactory cues responsible for the identity of the young are probably not very volatile since mothers are not able to recognize their lamb at a distance greater than 25 cm (Keller et al. 2003; Poindron et al. 2003; Alexander 1978). Wool from various body regions could provide cues to recognition since sheep mothers are very interested in wool taken from the rump, trunk, or head-neck of their own lamb (Alexander and Stevens 1982). A chemical analysis of the cranial wool of lamb identifies 133 volatile organic compounds that could be implicated in the olfactory identity of the lamb (Burger et al. 2011). Unfortunately, alien lambs dressed in jackets sprayed with a mixture of these compounds were rejected at suckling. The anal odor could also serve a specific release to suckling (Alexander et al. 1983a). The absence of interest for lamb feces suggests that sebaceous glands exterior to the anus could be part of the source of odor. Odors associated with amniotic fluids seem not be determinant for the olfactory recognition of the lamb. Ewes that already are selective at suckling reject an alien lamb coated with amniotic fluid coming from their own lamb (Porter et al. 1994). However, one study reports a relatively high proportion of selective mothers just after parturition suggesting that amniotic fluid could bear some individual chemosensory information (Keller et al. 2003).

What are the processes involved in the formation of the individual odor of the lamb? Evidence for a genetic contribution to olfactory signatures has been demonstrated in rodents and humans (Restrepo et al. 2006). Consequently, a degree of genetic relatedness of individuals and a similarity of their odor types is predictable. In sheep, mothers have been tested for their ability to detect olfactory resemblance between their twins. Mothers more readily accept their familiar lamb than to its twin isolated at birth. However, they treat more positively their separated twin than unfamiliar alien lambs (Porter et al. 1991). In a follow-up experiment, it was evidenced that familiar and isolated monozygotic twins are not differentiated by their mothers while this is not the case for the corresponding dizygotic twins (Romeyer et al. 1993b). It thus seems that the odor of monozygotic twins is more akin than those of dizygotic twins, supporting the hypothesis that olfactory signatures are genetically influenced, as reported in rodents and humans.

The environment also influences individually distinctive odor signatures. A possible role of acquired maternal labels, for example, through the maternal diet has been reported in the spiny mouse (Porter and Doane 1978). These diet-specific odors could be deposited onto the young as their mothers lick them or they could be transferred to the litter through milk. The existence of acquired maternal labels has also been proposed in goats (Gubernick 1981). However, recognition of the familiar kid can occur without labeling: Mother goats are selective to their own kids placed after birth into a wire meshed cage that prevents their mother from labeling through licking and/or suckling (Romeyer et al. 1993a). In addition, such maternal labels are not necessary for the rejection of alien lambs since mother goats reject both labeled and unlabeled alien lambs. Similar conclusions can be drawn from studies in sheep. Mothers develop a selective bond with their lambs, even when direct physical contact is prevented, provided that they have access to the lamb's odor (Romeyer et al. 1993b). Mothers also reject alien young regardless of whether those lambs have been isolated (presumably unlabeled) after birth or housed with their own mother (presumably labeled; Porter et al. 1991) or fed by an alien mother (presumably labeled through milk; Lévy et al. 1991). However, maternal labels although unnecessary can be incorporated into the lamb odor. Ewes are able to discriminate between their monozygotic twins after having full contact with one of those lambs (presumably labeled) for 4 h whereas not when physical contact with one of the twins has been prevented during the exposure period (Romeyer et al. 1993b). Thus, it seems that the lamb odor results from chemical by-products of bodily processes and acquired olfactory cues especially those transferred from the mother and those resulting from the activity of the bodily microbiota.

Visual and Auditory Cues

Maternal recognition of lamb or kid at great distance is dependent on visual and/or acoustic cues and is established a few hours after parturition (Keller et al. 2003; Terrazas et al. 1999; Poindron et al. 2003). Recognition of the lamb based only on visual cues has been explored in a few studies, but the salient cues involved are not determined yet. Early studies report that the appearance of the lamb's face is of primary importance while changes in the appearance of the rump or the front legs do not disturb lamb recognition (Alexander and Shillito 1977). The significance of faces is further demonstrated by the ability of ewes to discriminate photographs of faces between unfamiliar lambs (Ferreira et al. 2004) and between familiar and unfamiliar lambs (Kendrick et al. 1996). Peripheral features of the faces could be of significance since removing the peripheral, but not internal, features abolishes the recognition between faces of adult sheep (Peirce et al. 2000). Lamb discrimination is only accomplished on photographs of 3-week-old lambs, suggesting that these facial features are not salient before this age (Kendrick et al. 1996). The ability of goats to discriminate the face of their familiar kid is not documented, while goats can discriminate members of their group using head's cues (Keil et al. 2012) and can form categories based on the visual appearance of symbols and generalize across new symbols (Meyer et al. 2012). In addition, kids use pelage color to recognize their mothers in a large group-choice situation (Ruiz-Miranda 1993). Because kids can be distinguished according to the color of their coat, mother goats might be able to recognize the face of their young earlier than sheep mothers.

Vocal discrimination is effective sooner than visual discrimination since ewes display a preference for vocalizations of their lambs at 24 h of age (Sèbe et al. 2007). This early ability is facilitated by a high vocal activity from both the mother and its young for the first hours after parturition (Sèbe et al. 2007; Dwyer et al. 1998; Poindron et al. 1980) providing a favorable context to learn the vocal identity of the lamb. In addition, an association between the vocal activity of the dyad and nursing is established within 6 h postpartum, shaping very early mother-young communication (Sebe et al. 2008). Another factor facilitating the early development of vocal discrimination resides in the fact that lambs' calls are individually distinctive based on the temporal and the frequency distribution of the call (Sèbe et al. 2018). However, it remains to be determined which parameters are relevant for lamb's recognition by the mother. The response to lambs' calls is maximal when motivation for feeding is high, suggesting that vocal recognition is not only efficient for the maintenance of mother-young contact but also ensures a high probability of preferential maternal nursing (Sebe et al. 2008). In hider species, like goats, mothers memorize the location where their offspring are hidden and call to initiate nursing. Therefore, the requirement for mothers to recognize their young vocally is less critical. However, evidence has been provided for offspring vocalization individuality even during the hiding phase, at 1 week postpartum, based on frequency distribution and duration of the bleats. This individual acoustic signature probably serves as a support for the recognition of the kid evidenced at 2 days after birth (Briefer and McElligott 2011; Terrazas et al. 2003). Thus, it appears that vocal discrimination plays an important role in the early development of the bond with the young. It can be effective at much greater distances than allowed by visual recognition and when mother and young cannot see each other. However, although auditory cues are sufficient to allow identification soon after birth, addition of visual cues increases the efficiency of offspring recognition.

Brain Regions Involved in Selectivity

Fos studies reported in section "Brain regions involved in maternal responsiveness" also reveal that olfactory memory processes associated with selectivity require a neural network different from the one involved in maternal responsiveness (Fig. 2). Olfactory processing regions, the OB, the piriform cortex, the frontal medial cortex, and the orbitofrontal cortex, are mainly activated in mothers that are undergoing olfactory formation (Da Costa et al. 1997; Keller et al. 2004a). In addition, the cortical nucleus of the amygdala, the entorhinal cortex, and the hippocampal formation show increased Fos labeling in anosmic (and consequently not selective) mothers in comparison to intact ones (Keller et al. 2004a).

The Olfactory Bulb

The OB, the first relay olfactory information processing, was extensively studied with a combination of different methodological approaches. Using electrophysiological recordings of the mitral cells, the main cells of the OB that send olfactory signals to various olfactory cortical structures, parturition was revealed as a key event that modifies the processing of lamb odors (Kendrick et al. 1992c). During pregnancy, lamb or amniotic fluid odors preferentially activate none of these cells, and instead, the majority of them respond to food odors. After birth, the number of cells that respond preferentially to lamb odors markedly increases, indicating that the change in salience of lamb odors that occurs at parturition is mediated by a shift in olfactory cell responsiveness. In addition, while the majority of mitral cells do not differentiate between lamb odors, a proportion of the cells do respond preferentially to the odor of the ewe's own lamb (Fig. 2). Thus, a coding for the familiar lamb odor is set up at the level of the first relay of odor processing when the recognition of lamb odors is a behavioral priority.

Measures of the release of glutamate, the neurotransmitter of mitral cells, and gamma-aminobutyric acid (GABA), a modulator of the activity of mitral cells, by in vivo microdialysis are consistent with the changed firing activity of these neurons. Although lamb odors do not induce release of these neurotransmitters prepartum, after parturition, own lamb odors selectively increase glutamate and GABA release (Kendrick et al. 1992c). Functional studies support these correlations. Infusion of a GABA receptor antagonist in the OB prevents lamb recognition after the selective bonding had been formed (Kendrick 1994).

Centrifugal pathways of the OB, like noradrenergic inputs, play a role in establishing changes within the OB, which are responsible for the formation of a lamb olfactory memory. During the learning phase of lamb odor, an increase in noradrenaline release occurs in the OB (Lévy et al. 1993). Lesions of noradrenergic projections to the OB or direct infusions of a ß-adrenergic antagonist reduce the number of selective ewes without affecting maternal responsiveness and odor perception (Lévy et al. 1990b; Pissonnier et al. 1985). Noradrenergic activation stimulates nitric oxide release, which in turn potentiates glutamate release. Blockade of the neuronal enzyme nitric oxide synthase prevents both the potentiation of glutamate release and formation of the olfactory memory (Kendrick et al. 1997b). OT release within the OB (see section "Brain regions involved in maternal responsiveness") at parturition may also facilitate lamb odor memory because it increases noradrenaline concentration in this brain region (Lévy et al. 1995b). Further evidence that OT can modulate olfactory processing directly in the OB comes from a study showing a change of mitral cell activity by microiontophoretic applications of OT (Yu et al. 1996). Thus, OT would facilitate maternal responsiveness, by mediating olfactory attraction to amniotic fluids (Lévy et al. 1990a), and selectivity, by facilitating olfactory memory formation induced by noradrenaline release.

In addition to these changes, the OB retains another form of brain plasticity that is a lifelong continuous generation of neurons originating from neural stem cells in the sub-ventricular zone located on the wall of the lateral ventricles (Fig. 3a). These cells produce transient amplifying cells, which divide to produce neuroblasts. The neuroblasts migrate along the rostral migratory stream, and after reaching the OB, the majority of them mature into interneurons, which regulate the activity of the mitral cells (Fig. 3b; Lledo et al. 2006; Lepousez et al. 2015). This neuroplasticity could provide an additional mechanism through which olfaction can contribute to

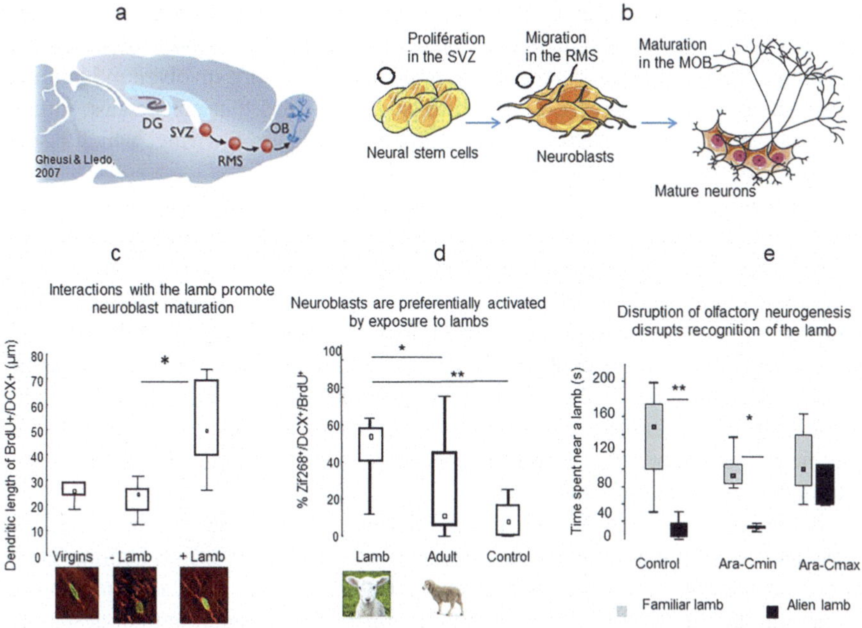

Fig. 3 Olfactory neurogenesis and maternal behavior. (**a**) Continuing neurogenesis mainly occurs within the sub-ventricular zone (SVZ) and the dentate gyrus (DG). The SVZ gives rise to neuroblasts that migrate through the rostral migratory stream (RMS) and differentiate into mature neurons in the olfactory bulb (OB; Gheusi and Lledo 2007). (**b**) The main steps of olfactory neurogenesis include the proliferation, migration, differentiation, and integration of newborn neurons. (**c**) Photographs showing 3-month-old neuroblasts and their dendrites labeled by the bromodeoxyuridine (BrdU) and doublecortin (DCX) (adapted from Brus et al. 2014). (**d**) Zif268 is a marker of neural activity (adapted from Corona et al. 2017). (**e**) Infusion of the antimitotic Ara-C into the ventricles leads to a reduction <70% (AraC-min) or >70% (Ara-Cmax) in the number of olfactory neuroblasts (adapted from Corona et al. 2018)

the enhancement of lamb olfactory learning. A first evidence to support this hypothesis comes from experiments showing an alteration of olfactory neurogenesis during the establishment of maternal behavior. A decrease in cell proliferation in the sub-ventricular zone of the ventricle is observed in mothers that remain with their lambs for the first 2 postpartum days when compared to mothers separated from them (Brus et al. 2010). In addition, the number of neuroblasts, which migrate to the OB, also decreases in mothers interacting with their lambs. However, maturation of the remaining neuroblasts, revealed by their dendritic length and their number of nodes, is enhanced (Fig. 3c; Brus et al. 2014). Interactions with young and associated olfactory learning, rather than parturition, are responsible for these modifications because they are prevented by separating mothers from their lambs at parturition. Numerous studies report that olfactory experience sculpts newborn neurons (Lazarini and Lledo 2011). In the context of motherhood, olfactory exposure to pups induces changes in structural synaptic plasticity of newly born olfactory neurons in mice (Belnoue et al. 2016; Kopel et al. 2012). The decrease in the number of neuroblasts would reduce cell competition and consequently increase their maturation, facilitating their integration in the neural network involved in learning. In support of this hypothesis, a modeling study indicates that an increase in cell proliferation causes a reduction in the amount of synaptic rewiring, which is not beneficial for learning (Butz et al. 2008).

A further step in the demonstration of the importance of olfactory neurogenesis in the context of maternal behavior consists in testing if increased activation of newborn neurons after exposure to lambs is of behavioral relevance. By pairing neurogenic markers with markers of neural activity, activation of olfactory newborn neurons have been compared between mothers exposed either to lambs or to an adult conspecific (Corona et al. 2017). Exposure to lambs increases the percentage of activated neuroblasts in the OB compared to exposure to an unfamiliar ewe, indicating that the preferential activation is not achieved for any social odors but is specific to lamb odors (Fig. 3d). Hence, newborn neurons of the OB could participate in the processing of lamb odors.

To further understand the functional relevance of olfactory neurogenesis, the consequences of a disruption of neurogenesis on recognition of the familiar lamb were assessed during the early postpartum period (Corona et al. 2018). Infusion of the antimitotic drug Ara-C into the ventricles during the second month of gestation, leading to a 70% reduction in olfactory neurogenesis, disrupts discrimination of the familiar lamb. During selectivity tests, mothers with reduced olfactory neurogenesis do not display aggressive behavior toward alien lambs. In addition, when ewes are given the choice between familiar and unfamiliar anesthetized lambs so that only olfactory cues are available, mothers with a 70% reduction in neurogenesis are not able to discriminate their own lamb from an alien lamb (Fig. 3e). These results indicate that adult-born olfactory neurons are to some extent involved in lamb olfactory memory. The mechanisms controlling the integration of olfactory adult-born neurons during the formation of the lamb memory are not elucidated. The increased activity of noradrenergic afferents to the OB at parturition could be a potential candidate as it was demonstrated in mice that olfactory perceptual learning requires both the presence of adult-born cells and noradrenaline (Moreno et al. 2012).

Beyond the Olfactory Bulb

Plastic changes that occur in the OB represent the first changes in the processing of a lamb odor memory. Based on studies showing Fos activation during formation of such memory, the involvement of some secondary olfactory processing regions was investigated by silencing these regions using infusion of an anesthetic.

Cortices that receive extensive olfactory projections do not seem to be required for the formation of olfactory lamb memory. Infusion of tetracaine within the piriform cortex for 4 h postpartum only reduces rejection of an unfamiliar lamb from the same breed but not from a different one suggesting the involvement of this olfactory cortex in the fine discrimination of similar lamb odors. However, the assumption that lambs of the same breed share common olfactory signatures remains to be elucidated (Broad et al. 1999b). It would be of interest to inactivate more specifically the posterior part of the pyriform cortex known to be engaged in olfactory memorization (Mouly et al. 2001; Mouly and Gervais 2002). The medial frontal cortex that shows increased c-fos expression at parturition when ewes are exposed to lambs (Da Costa et al. 1997) was also reversibly inactivated in parturient ewes. The treatment decreases aggressive motor responses (head butts) and protest vocalizations directed toward strange lambs without interfering with acceptance behaviors toward lambs. These rejection behaviors toward strange lambs appear 1 h after the termination of tetracaine infusions, indicating that despite inactivation of the medial frontal cortex, an olfactory memory for the familiar lamb has been formed (Broad et al. 2002).

A functional role for both the cortical and the medial nuclei of the amygdala in the formation of lamb olfactory memory, which receives olfactory input from the OB, has been revealed using infusion of lidocaine for the first 8 h postpartum (Keller et al. 2004b). Inactivation of either the cortical or the medial nucleus results in acceptance at suckling of the alien lamb, a lack of aggressive behavior, and a decrease in the percentage of selective ewes for the first 8 h postpartum (Fig. 2). The loss of selectivity cannot be the consequence of an impairment of maternal responsiveness because ewes display the full repertoire of maternal acceptance toward their lambs at parturition. The termination of lidocaine infusions in nonselective mothers does not restore selectivity, and lidocaine infusions in selective mothers at 2 days postpartum do not impair selectivity, indicating that the loss of selectivity cannot be explained by an impairment of memory retrieval. These results support the view that both nuclei form a hub in the network controlling olfactory lamb recognition memory.

The basal forebrain cholinergic system innervating cortical and limbic areas is well known for its contribution to attention processes and memory function (Knox 2016). This brain region is also essential for the formation of olfactory recognition of the lamb. At parturition and during the first hours postpartum, the cholinergic system is activated, especially in the OB (Kendrick et al. 1986; Lévy et al. 1993). Moreover, the blockade of central muscarinic receptors by injections of scopolamine at parturition impairs olfactory recognition of the familiar lamb tested at 4 h postpartum without interfering with maternal responsiveness (Lévy et al. 1997). Central muscarinic receptors are also implicated in post-acquisition processes of lamb odor recognition since injection of scopolamine after the establishment of selectivity impairs retention of this memory (Ferreira et al. 1999). The extended lesion of the basal forebrain cholinergic system (higher than 75%), using a specific cholinergic immunotoxin, ME20.4IgG-saporin, induces a severe impairment of olfactory lamb recognition within 4 h following birth (Ferreira et al. 2001). No deficit of maternal responsiveness was reported, indicating that the impairment of recognition could not result from disturbance of lamb nursing, an olfactory deficit, or a loss of aggressiveness toward the alien lamb. However, which of the projections sites of the cholinergic system is involved remains to be determined. A functional interaction between noradrenergic and cholinergic projections to the OB can be suggested since increased release of both neurotransmitters occurs in the OB at parturition. In addition, impairment of the noradrenergic afferents to the OB does not prevent individual recognition of the lamb in all animals (Lévy et al. 1990b). The hypothesis of a synergistic action between both systems would be interesting to test.

The neural network controlling olfactory recognition of the lamb undoubtedly includes more brain structures. For instance, the entorhinal cortex may be important since it shows increased Fos activation during the formation of olfactory lamb memory (Da Costa et al. 1997; Keller et al. 2004a), and it is critical in olfactory recognition memory of conspecifics in rodents (Petrulis and Eichenbaum 2003; Bannerman et al. 2002). In addition, increased Fos activation in the orbitofrontal cortex in selective mothers is consistent with electrophysiological studies showing cells responding to odors, in particular to biological meaning cues of conspecifics (Onoda et al. 1984).

A hypothetical pathway supporting lamb olfactory memory could be the following: Olfactory inputs from the OB to both cortical and medial nuclei of the amygdala may, in turn, be relayed to the entorhinal cortex and also to the orbitofrontal cortex since tract-tracing studies indicate such connections in sheep (Meurisse et al. 2009).

5 Environmental Factors Regulating the Onset of Maternal Behavior: Nutrition Levels

The previous sections have described the endocrine and neuroendocrine factors as well as the sensory cues that determine the onset of maternal responsiveness and selectivity. In addition to these internal factors, external factors such as changes in the environment affect the expression of maternal behavior. For example, the low availability of food in winter or in semiarid zones during pregnancy is not only associated with a reduction in colostrum and milk production, but it also impairs lamb survival by affecting maternal care and neonate behavior at birth (Dwyer et al. 2003). Undernutrition during the last 6 weeks of pregnancy results in mothers of twins deserting the birth site in comparison with well-fed ones (Putu et al. 1988), and ewes on low-pasture allowance are less likely to stay with their litter (Everett-Hincks et al. 2005). A simple reason could be that undernourished mothers are likely to be more attracted to food than to their lambs and move quickly from the birth site to graze (Nowak 1996). The poor maternal care can be also due to lamb weakness and inability to follow the mother. However, some studies indicate that maternal nutrition also affects the expression of maternal behavior. A moderate reduction in nutritional intake during pregnancy induces decreased grooming of the lamb at parturition, increased aggressiveness, and low scores of maternal attachment in Scottish Blackface mothers (Dwyer et al. 2003). Similarly, feed-restricted Ghezel ewes during late pregnancy are slower to groom their lambs and reject more frequently their lamb at suckling than ewes offered ad libitum intake (Ahmadzadeh et al. 2020). A plausible cause of the effects of undernutrition on maternal care may lay in steroid changes in late gestation induced by food restriction. Both studies report a higher progesterone plasmatic concentration in ewes offered a restricted regime. A reduced estradiol to progesterone ratio is also observed in Dwyer et al.'s study (Dwyer et al. 2003). Other factors such as a prolonged parturition and a higher incidence of incorrect lamb presentation at delivery could also affect the expression of maternal care (Dwyer et al. 2003). However, other studies report no effect of ewe nutrition on maternal behavior (Everett-Hincks et al. 2005; Corner et al. 2010), and some methodological differences can explain this discrepancy. Indeed, in Dwyer et al. (2003) and Ahmadzadeh et al.'s studies (2020), maternal behavior was observed without human intervention. In Everett-Hincks et al. (2005) and Corner et al.'s studies (2010) maternal behavior was quantified during tagging procedure, and the response of the mother to her lambs was assessed while the lambs were being restrained by a human. In addition, Everett-Hincks et al. (2005) and Corner et al. (2010) used mixed-age ewes, whereas Dwyer et al. (2003) used only primiparous ewes, and it is possible that mothers without any previous experience are more sensitive to the effects of undernutrition.

In goats, the effect of undernutrition on maternal care is poorly documented. One study shows that 70% of the nutritional requirements for maintenance and fetal growth given during the last 3 months of gestation induce no impairment of licking, maternal bleating, and nursing at parturition but impair nonolfactory recognition of the own kid (Terrazas et al. 2009). Surprisingly, in goats maintained under extensive semiarid conditions, supplementation with concentrate during late gestation does not result in improvement of maternal behavior (Luna-Orozco et al. 2015) while feeding supplemental maize during the last days of pregnancy improves maternal care at birth and maternal selectivity (Ramírez-Vera et al. 2012). Clearly, more investigations are required to determine whether a more severe restriction of nutrition or a situation that requires the does to expend more energy to feed would impair maternal care. In addition, both in sheep and goats, deficiencies in trace elements and other nutrients resulting from undernutrition may affect maternal behavior, and this possibility has not been explored yet.

6 Personality Traits and Experiential Factors Regulating the Onset of Maternal Behavior

Although the expression of maternal behavior is quite similar in both sheep and goats, significant variations are observed in the quality and the quantity of the different items that constitute maternal care (Dwyer 2008a). These differences are found between females of different breeds and also within the same breed, according to the temperament of the animal or to its maternal experience.

Breed Differences

There are over 200 breeds of domestic sheep created to serve multiple production purposes (i.e., meat, milk, wool), and these different breeds are adapted to different environments (mountains, hills, lowlands) and to different weather conditions. These multiple selections and adaptations to various habitats have resulted in variations in social behavior and particularly in the expression of maternal behavior (Dwyer and Lawrence 1999b). Indeed, many aspects of maternal behavior at parturition vary between breeds. First, differences are reported in the shelter-seeking behavior (Pritchard et al. 2021; Whateley et al. 1974), pawing the ground, and walking in circles during labor (Arnold and Morgan 1975). Breed differences are also reported in behaviors expressed at parturition toward lambs. Licking behavior, emission of maternal bleats, acceptance at udder, rejection behaviors toward the lambs, and time spent at the birth site are the main behavioral items affected by breeds (Dwyer and Lawrence 1998, 2000a, Alexander et al. 1983b, 1990b; Dwyer et al. 1996; Poindron et al. 1984a; Simitzis et al. 2016). These differences are mainly detectable in primiparous ewes since mothers rapidly adopt the appropriate behavior toward the lamb within the first hours after parturition, and minor differences between breeds are observed 3 h after parturition (Poindron et al. 1984b). Not only the expression of maternal behavior at parturition, but also preference for the own lamb, differs between breeds. For instance, Suffolk mothers have a longer latency to approach their own lambs and spend less time with it than Scottish Blackface mothers when given a choice between their own and an alien lamb in a Y-maze (Dwyer 2008b).

Why do some breeds show inferior maternal behavior? The reasons are not understood. Dwyer and Lawrence (1998) reporting a poorer quality of maternal behavior in a lowland breed selected for growth (Suffolk) in comparison with a less-selected hill breed of sheep (Blackface) suggest that "an increase in selection for superior production characteristics had led to a decrease in the expression of behaviors associated with survivability under extensive conditions." However, some of the maternal behaviors could directly be under genetic control. The maternal behavior score at lambing, based on the proximity of the mother to her lamb as it is tagged and assessed outdoors, shows some heritability albeit low (Lambe et al. 2001; Plush et al. 2011; Everett-Hincks and Cullen 2009). Scoring maternal behavior outdoors in extensive farming systems could introduce factors of variations such as the social and grazing environments. When an indoor behavior test assessing attraction and reactivity to separation from the litter at 24 h postpartum is performed, high estimated heritabilities are found for call bleats and maternal bleats. Genetic correlations are moderate for proximity to the litter in the presence of a human (Hazard et al. 2020). It is therefore conceivable to include some of the heritable behavioral traits in sheep breeding to improve maternal attachment.

Temperament

Temperament can be defined as a set of individual differences in behavior that occur in early life, and that is relatively stable across situations and over the course of time. Some evidence indicates that traits of temperament can influence the quality of maternal behavior. Merino mothers selected for their calm temperament spent more time grooming their lambs and bleat more frequently to their lambs than ewes selected for their nervous temperament (Bickell et al. 2010) (Murphy et al. cited in Dwyer 1998). Ewes selected for their superior mothering ability display less emotive reactions in an open-field test than unselected ewes (Kilgour and Szantar-Coddington 1995). A positive correlation is reported between agitation score in social isolation and improved maternal score (Plush et al. 2011). Similarly, ewes showing the highest levels of social reactivity and plasma cortisol responses to isolation and reunion with congeners spend more time licking their lambs, display less refusal at suckling, and vocalize more when separated from their lambs (Coulon et al. 2014). The temperament or emotional reactivity could in part determine maternal styles defined as a consistent group of maternal behavior items expressed by an individual. In sheep, a majority of Suffolk mothers expresses a maternally rejecting style, whereas most Blackface mothers show a high level of maternal care and are infrequently rejecting mothers. In addition, both behavioral styles are consistent across parities (Dwyer and Lawrence 2000b). Thus, selection for these traits of temperament by improving mother ability could indirectly enhance lamb survival (Plush et al. 2011; Brown et al. 2016).

Maternal Experience

Although the onset of maternal behavior is deeply driven by physiological factors that ensure a proper maternal care, previous maternal experience is a requisite for optimal expression of maternal behavior at parturition in most mammals, at least in ungulates, rabbits, and primates (Gonzalez-Mariscal and Poindron 2002; Fleming and Li 2002). In sheep, inexperienced mothers display various behavioral disturbances: decreased licking, emission of maternal bleats, udder acceptance, and increased withdrawal from the lamb and aggressive behavior at least within the first hours after parturition (O'Connor et al. 1992; Poindron and Le Neindre 1980; Poindron et al. 1984b; Dwyer and Lawrence 2000b; Lv et al. 2016; Ekiz et al. 2007). These deficits in maternal care of inexperienced females differ according to the breed. Suffolk primiparous ewes show higher rejection of the lamb in comparison to Blackface ewes (Dwyer and Lawrence 2000b). In the Romanov breed, primiparous mothers display more licking behavior toward the lamb than Préalpes and Ile-de-France breeds (Poindron et al. 1984b). These deficits result in higher mortality of lambs from primiparous ewes (Putu et al. 1986). However, the behavioral disturbances observed in primiparous mothers rapidly disappear within the first hours postpartum, indicating that maternal behavior is built up as the mother interacts with her young (Poindron et al. 1984b). In goats, behaviors of isolation seeking and intolerance to conspecifics before parturition are less observed in primiparous goats that are more likely to follow the flock (Das and Tomer 1997; Lickliter 1984b; Malfatti et al. 1991). Unfortunately, apart from studies describing maternal behavior of primiparous ewes (Otal et al. 2010; Martinez et al. 2009), there is no comprehensive comparative study of maternal behavior at parturition between primiparous and multiparous goats to assess the importance of maternal experience in this species.

The disturbances of maternal responsiveness expressed by primiparous ewes could affect the establishment of olfactory recognition of the lamb. In Kendrick's review (Kendrick 1994), unpublished observations indicate a slower latency to bond by 2 h postpartum in primiparous ewes. However, another study reports that when primiparous ewes showing a lack of maternal responsiveness are excluded, the remaining mothers are as efficient in their olfactory recognition at 2 h postpartum as are their multiparous counterparts (Keller et al. 2003). It seems, therefore, that once inexperienced mothers establish the contact with their lamb, the olfactory learning process is immediately functional, and maternal experience does not influence this process. The fact that in the latter study ewes were kept individually in close contact with their lamb supports this hypothesis. In contrast to olfactory recognition, previous maternal experience facilitates the development of visual and auditory recognition. Multiparous mothers are able to recognize their lamb using nonolfactory cues at 6 h, whereas primiparous mothers do so only after 24 h of mother-young contact (Keller et al. 2003). The facilitating effect of maternal experience on nonolfactory recognition was further confirmed in another breed of sheep (Gonzalez et al. 2015). Even a single maternal experience is sufficient since biparous mothers achieve this recognition at 8 h postpartum (Keller et al. 2003). The differential effect of maternal experience between olfactory and nonolfactory recognition of the lamb confirms the relative independence of the two systems of recognition (see section "Sensory factors controlling maternal responsiveness"; Ferreira et al. 2000). Visual and auditory cues could represent a multisensory process more complex than learning a single (olfactory) sensory cue, and that requires maternal experience.

Differences in the sensory, physiological, and neurobiological mechanisms involved in maternal responsiveness and maternal selectivity between primiparous and multiparous females have been identified that may be responsible for the behavioral differences. For instance, the importance of sensory cues varies with experience. Depriving ewes of amniotic fluid by washing the neonate induces a reduction of licking in inexperienced mothers, and most of them fail to accept their lamb while experienced mothers establish an appropriate maternal behavior (Lévy and Poindron 1987). Amniotic fluids are necessary for primiparous females, while experienced mothers compensate for the absence of amniotic fluids.

The action of the physiological factors is greatly influenced by maternal experience. In inexperienced non-gestant ewes, neither steroid priming nor vaginocervical stimulation induces acceptance of the lamb (Le Neindre et al. 1979; Keverne and Kendrick 1991). In addition, estrogen-primed nulliparous ewes are unresponsive to other artificial means of inducing maternal responsiveness, either with OT or opiates (Keverne and Kendrick 1991). It is possible that maternal experience enhances the brain's sensitivity to hormonal factors involved in maternal responsiveness. For instance, multiparous mothers show a higher density of estrogen receptors-alpha in the PVN and the MPOA, key structures for the induction of maternal responsiveness (Meurisse et al. 2005). The increase in estrogen receptors could account for the higher capacity of experienced mothers to respond to estrogen administration. Furthermore, maternal experience also enhances expression of OT receptor mRNA in the PVN. Increased release of OT reported in the OB of multiparous mothers following parturition could enhance the attractiveness of amniotic fluids or facilitate the olfactory memory for lambs (Lévy et al. 1995b). The increased release of OT could be the consequence of a larger release of neurotransmitters reported in multiparous females. For instance, more noradrenaline and acetylcholine are measured in the OB of experienced than inexperienced mothers (Lévy et al. 1993; Keverne et al. 1993; Lévy et al. 1995b), and infusion of OT induces an increase of noradrenaline concentrations in multiparous but not in primiparous mothers (Lévy et al. 1995b). These differences in transmitter release disappear 6 h after parturition, revealing a neural maturation of the OB as interactions with the lamb progresses.

The neural network involved in maternal responsiveness could undergo maturation induced by maternal experience. For instance, whereas inactivation of the MPOA at parturition results in a loss of interest toward the lamb in primiparous mothers, it causes few impairments in multiparous mothers (Perrin et al. 2007). Thus, maternal experience modifies the neural network involved in maternal behavior so that other brain structures could compensate for MPOA inactivation.

7 Conclusion

Sheep and goats offer the unique possibility to investigate the mechanisms involved in maternal care associated with a selective attachment to the young, as rodent mothers do not show selective interaction with their litter. These two facets, which lead to maternal attachment, are synchronized with parturition by uterine stimulation induced by the expulsion of the neonate. However, the mechanisms governing both processes differ, especially the two neural networks currently identified (Fig. 2). Whereas hypothalamic nuclei (the PVN, the MPOA, and the BNST) are involved in maternal responsiveness, olfactory processing regions (the OB, the cortical and medial nuclei of the amygdala, and the frontal cortices) and the noradrenergic and cholinergic systems constitute the circuitry involved in selectivity. A hypothetical neural model can be proposed to explain how both neural networks come into play based on anatomical studies showing interconnections between the two (Meurisse et al. 2009; Lévy et al. 1999). At parturition, vaginocervical stimulation induces activation of the parvocellular OT neurons in the PVN (Fig. 2). This activation coordinates release of OT in a number of brain regions and specifically in the MPOA/BNST and the OB. OT could act on the OB to promote attraction to amniotic fluids and consequently licking the neonate. Brain regions involved in the performance of other acceptance behaviors (nursing, bleating) are currently unknown. Once nursing occurs, it activates the MPOA/BNST through the activation of the OT system in the PVN. Vaginocervical stimulation also induces activation of both the cholinergic system of the basal forebrain and the noradrenergic system of the locus coeruleus, both of which project to the OB. The noradrenergic activation induces neural changes within the OB, which results in an enhanced response to the familiar lamb odor favoring its discrimination. The changes in mitral cell activity induced by the familiar lamb odor activate both the cortical and medial nuclei of the amygdala. Thus, these nuclei would respond more strongly to odor from the familiar lamb than from any alien lamb. In turn, this olfactory network stimulated by the familiar odor would activate the MPOA/BNST/PVN network to promote maternal acceptance. This would result in a tighter coupling between both networks such that after a few hours postpartum, familiar lamb odors systematically evoke maternal acceptance. On the other hand, stimulation of the OB by alien lamb odor would result in an inhibition or a lack of activation of the MPOA/BNST/PVN network. Concomitantly, unfamiliar lamb odor would activate brain regions that regulate rejection behavior, like the medial frontal cortex of which inactivation impairs the aggressive motor rejection response (Broad et al. 2002). Undoubtedly, other brain regions involved in both aspects of maternal behavior are required to complete this model.

To date, no study has identified the sensory cues relevant for recognition of the familiar lamb. The chemical nature of odors responsible for attractiveness of the neonate could be singular compounds, such as dodecyl propionate, reported to promote anogenital licking in the mother rat (Brouette-Lahlou et al. 1991). Cues coding for individuality could be more complex, constituted by a mixture of compounds probably localized in the wool and others produced by skin glands. Unveiling the identity of these odors would be of great help to sheep farmers as they could induce attraction to a newborn in non-maternal mothers or adoption of lambs by mothers with dead lambs. Similarly, features of the faces that support the visual recognition of the familiar lamb and acoustic parameters involved in the acoustic signature are currently unknown.

Beyond the identification of these sensory cues, the multisensory recognition of the young raises the issue of their interaction. In the rat mother, a synergistic effect has been evidenced between olfactory and acoustic cues. Orientation to ultrasonic calls of pups out of the nest is facilitated by pup odors (Farrell and Alberts 2002). Pup calls are responsible for the initiation of anogenital licking by the mother, which is initiated by pup preputial secretions (Brouette-Lahlou et al. 1999). In the sea lion, acoustic and olfactory cues act synergistically and with different function to achieve pup recognition (Wierucka et al. 2018). In sheep mothers, anosmic ewes unable to recognize the odor of their lamb still accept them at suckling, suggesting a compensatory mechanism by auditory and visual cues (see section "Individual recognition of the young: maternal selectivity"). At the brain level, interactions between different sensory cues have been evidenced. Exposure to pup odors changes responses of the auditory cortex to ultrasound calls in parturient mothers (Cohen et al. 2011). In sheep mothers, bleats from the familiar lamb induce similar levels of activation as exposure to pictures of lambs' faces in the temporal cortex (Kendrick et al. 2001). Thus, ewes may form and use mental multisensory images of their lamb, thus strengthening their mutual bond. How and where in the brain this image is formed offers a wealth of possibilities for fascinating research.

Although the mechanisms governing maternal responsiveness and selectivity, especially the effects of vaginocervical stimulation, seem to be similar between sheep and goats, many questions remain answered concerning the neuroendocrine control of maternal behavior in goats (Poindron et al. 2007a). For instance, it appears that a treatment of estradiol and progesterone does not induce maternal behavior in nonpregnant goats (Rosenblatt and Siegel 1981). This result is surprising when considering the importance of steroids for the onset of maternal behavior in mammals (Gonzalez-Mariscal and Poindron 2002). In addition, our understanding of the neural mechanisms that control maternal behavior in goats remains poor. It appears that the vomeronasal organ and the associated olfactory brain regions could play a role, contrary to what occurs in sheep. Furthermore, since goats hide their kids during the first week of lactation, they could have developed spatial memory skills to join them for suckling. Consequently, it is possible that the hippocampus, a key brain structure involved in spatial memory, is activated. This behavioral characteristic could result in a neural network involved in maternal behavior different from the one existing in sheep.

References

Ahmadzadeh L, Hosseinkhani A, Taghizadeh A, Ghasemi-Panahi B, Hamidian G. Effect of late gestational feed restriction and glucogenic precursor on behaviour and performance of Ghezel ewes and their offspring. Appl Anim Behav Sci. 2020;231:105030.

Ainsworth MD, Bell SM. Attachment, exploration, and separation: illustrated by the behavior of one-year-olds in a strange situation. Child Dev. 1970;41(1):49–67.

Alexander G. Odour, and the recognition of lambs by Merino ewes. Appl Anim Ethol. 1978;4:153–8.

Alexander G. What makes a good mother?: components and comparative aspects of maternal behaviour in ungulates. Proc Aust Soc Anim Prod. 1988;17:25–41.

Alexander G, Shillito EE. Importance of visual clues from various body regions in maternal recognition of the young in Merino sheep (Ovis aries). Appl Anim Ethol. 1977;3:137–143.

Alexander G, Shillito EE. Recognition of washed lambs by merino ewes. Appl Anim Ethol. 1981;7:77–86.

Alexander G, Stevens D. Odour cues to maternal recognition of lambs: an investigation of some possible sources. Appl Anim Ethol. 1982;9(2):165–75.

Alexander G, Stevens D, Bradley LR. Washing lambs and confinement as aids to fostering. Appl Anim Ethol. 1983a;10:251–61.

Alexander G, Stevens D, Kilgour R, de Langen H, Mottershead BE, Lynch JJ. Separation of ewes from twin lambs: incidence in several sheep breeds. Appl Anim Ethol. 1983b;10:301–17.

Alexander G, Stevens D, Bradley LR. Maternal behaviour in ewes following caesarian section. Appl Anim Ethol. 1988;19:273–7.

Alexander G, Stevens D, Bradley LR. Distribution of field birth-sites of lambing ewes. Aust J Exp Agric. 1990a;30:759–67.

Alexander G, Stevens D, Bradley LR, Barwick SA. Maternal-behavior in Border Leicester, Glen Vale (Border Leicester derived) and Merino sheep. Aust J Exp Agric. 1990b;30(1):27–38.

Alexander G, Stevens D, Baker P, Bradley LR. The timing of birth in grazing Merino sheep. Aust J Exp Agric. 1993;33(5):557–60.

Allan CJ, Holst PJ, Hinch GN. Behaviour of parturient Australian bush goats. I. Doe behaviour and kid vigour. Appl Anim Behav Sci. 1991;32:55–64.

Arnold GW, Dudzinski ML. Maternal behaviour. In: Arnold GW, Dudzinski ML, editors. Ethology of free-ranging domestic animals. Amsterdam: Elsevier Scientific Publishing Company; 1978. p. 137–65.

Arnold GW, Morgan PD. Behaviour of the ewe and lamb at lambing and its relationship to lamb mortality. Appl Anim Ethol. 1975;2(1):25–46.

Arnould C, Piketty V, Lévy F. Behavior of ewes at parturition toward amniotic fluids from sheep, cows and goats. Appl Anim Behav Sci. 1991;32(2-3):191–6.

Baldwin BA, Shillito EE. The effects of ablation of the olfactory bulbs on parturition and maternal behaviour in Soay sheep. Anim Behav. 1974;22(1):220–3.

Bannerman DM, Lemaire M, Yee BK, Iversen SD, Oswald CJ, Good MA, et al. Selective cytotoxic lesions of the retrohippocampal region produce a mild deficit in social recognition memory. Exp Brain Res. 2002;142(3):395–401.

Basiouni GF, Gonyou HW. Use of birth fluids and cervical stimulation in lamb fostering. J Anim Sci. 1988;66(4):872–9.

Belnoue L, Malvaut S, Ladeveze E, Abrous DN, Koehl M. Plasticity in the olfactory bulb of the maternal mouse is prevented by gestational stress. Sci Rep. 2016;6:37615.

Bickell SL, Nowak R, Poindron P, Ferguson D, Blache D. Maternal behaviour at parturition in outdoor conditions differs only moderately between single-bearing ewes selected for their calm or nervous temperament. Anim Prod Sci. 2010;50(7):675–82.

Bordi A, De Rosa G, Napolitano F, Litterio M, Marino V, Rubino R. Postpartum development of the mother-young relationship in goats. Appl Anim Behav Sci. 1994;42:145–52.

Bouissou MF. Effet de l'ablation des bulbes olfactifs sur la reconnaissance du jeune par sa mère chez les Ovins. Revue de Comportement Animal. 1968;2:77–83.

Bridges RS. Neuroendocrine regulation of maternal behavior. Front Neuroendocrinol. 2015;36:178–96.

Briefer E, McElligott AG. Mutual mother–offspring vocal recognition in an ungulate hider species (Capra hircus). Anim Cogn. 2011;14(4):585–98.

Broad KD, Kendrick KM, Sirinathsinghji DJ, Keverne EB. Changes in oxytocin immunoreactivity and mRNA expression in the sheep brain during pregnancy, parturition and lactation and in response to oestrogen and progesterone. J Neuroendocrinol. 1993;5(4):435–44.

Broad KD, Lévy F, Evans G, Kimura T, Keverne EB, Kendrick KM. Previous maternal experience potentiates the effect of parturition on oxytocin receptor mRNA expression in the paraventricular nucleus. Eur J Neurosci. 1999a;11(10):3725–37.

Broad KD, Da Costa AP, Keverne EB, Kendrick KM, editors. The role of the pyriform cortex in olfactory memory. Society for Neuroscience Abstract; 1999b.

Broad KD, Hinton MR, Keverne EB, Kendrick KM. Involvement of the medial prefrontal cortex in mediating behavioural responses to odour cues rather than olfactory recognition memory. Neuroscience. 2002;114(3):715–29.

Brouette-Lahlou I, Amouroux R, Chastrett I, Cosnier J, Stoffelsma J, Vernet-Maury E. Dodecyl propionate, the attractant from rat pups preputial gland, characterization and identification. J Chem Ecol. 1991;17:1343–54.

Brouette-Lahlou I, Godinot F, Vernet-Maury E. The mother rat's vomeronasal organ is involved in detection of dodecyl propionate, the pup's preputial gland pheromone. Physiol Behav. 1999;66(3):427–36.

Brown DJ, Fogarty NM, Iker CL, Ferguson DM, Blache D, Gaunt GM. Genetic evaluation of maternal behaviour and temperament in Australian sheep. Anim Prod Sci. 2016;56(4):767–74.

Brus M, Meurisse M, Franceschini I, Keller M, Lévy F. Evidence for cell proliferation in the sheep brain and its down-regulation by parturition and interactions with the young. Horm Behav. 2010;58(5):737–46.

Brus M, Meurisse M, Keller M, Lévy F. Interactions with the young down-regulate adult olfactory neurogenesis and enhance the maturation of olfactory neuroblasts in sheep mothers. Front Behav Neurosci. 2014;8.

Burger BV, Viviers MZ, Le Roux NJ, Morris J, Bekker JPI, Le Roux M. Olfactory cue mediated neonatal recognition in sheep, Ovis aries. J Chem Ecol. 2011;37(10):1150–63.

Butz M, Teuchert-Noodt G, Grafen K, van Ooyen A. Inverse relationship between adult hippocampal cell proliferation and synaptic rewiring in the dentate gyrus. Hippocampus. 2008;18(9):879–98.

Caba M, Poindron P, Krehbiel D, Lévy F, Romeyer A, Vénier G. Naltrexone delays the onset of maternal behavior in primiparous parturient ewes. Pharmacol Biochem Behav. 1995;52(4):743–8.

Challis JRG, Linzell JL. The concentration of total unconjugated oestrogens in the plasma of pregnant goats. Reprod Fertil Dev. 1971;26:401–4.

Chamley WA, Buckmaster J, Cereni ME, Cumming IA, Goding JR, Obst JM, et al. Changes in the levels of progesterone, corticosteroids, estrone, oestradiol-17ß, luteinizing hormone and prolactin in the peripheral plasma of the ewe during late pregnancy and at parturition. Biol Reprod. 1973;9:30–5.

Cloete SWP, Scholtz AJ, Ten Hoope JM, Lombard PJA, Franken MC. Ease of birth relation to pelvic dimensions, litter weight and conformation of sheep. Small Rumin Res. 1998;31(1):51–60.

Cloete SWP, Scholtz AJ, Taljaard R. Lambing behaviour of Merino ewes from lines subjected to divergent selection for multiple rearing ability from the same base population. S Afr J Anim Sci. 2002;32(1):57–65.

Cohen L, Rothschild G, Mizrahi A. Multisensory integration of natural odors and sounds in the auditory cortex. Neuron. 2011;72(2):357–69.

Collias NE. The analysis of socialization in sheep and goats. Ecology. 1956;37(2):228–39.

Corner RA, Kenyon PR, Stafford KJ, West DM, Morris ST, Oliver MH. The effects of pasture availability for twin- and triplet-bearing ewes in mid and late pregnancy on ewe and lamb behaviour 12 to 24 h after birth. Animal. 2010;4(1):108–15.

Corona R, Meurisse M, Cornilleau F, Moussu C, Keller M, Lévy F. Exposure to young preferentially activates adult-born neurons in the main olfactory bulb of sheep mothers. Brain Struct Funct. 2017;222(3):1219–29.

Corona R, Meurisse M, Cornilleau F, Moussu C, Keller M, Lévy F. Disruption of adult olfactory neurogenesis induces deficits in maternal behavior in sheep. Behav Brain Res. 2018;347:124–31.

Coulon M, Lévy F, Ravel C, Nowak R, Boissy A. Mild effects of gestational stress and social reactivity on the onset of mother-young interactions and bonding in sheep. Stress. 2014;17(6):460–70.

Currie WB, Gorewit RC, Michel FJ. Endocrine changes, with special reference on estradiol 17-beta, prolactin and oxytocin before and during labour and delivery in goats. J Reprod Fertil. 1988;82:299–308.

Da Costa AP, Guevara-Guzman RG, Ohkura S, Goode JA, Kendrick KM. The role of oxytocin release in the paraventricular nucleus in the control of maternal behaviour in the sheep. J Neuroendocrinol. 1996;8(3):163–77.

Da Costa AP, Broad KD, Kendrick KM. Olfactory memory and maternal behaviour-induced changes in c-fos and zif/268 mRNA expression in the sheep brain. Brain Res Mol Brain Res. 1997;46(1-2):63–76.

Das N, Tomer OS. Time pattern on parturition sequences in Beetal goats and crosses: comparison between primiparous and multiparous does. Small Rumin Res. 1997;26(1-2):157–61.

Davis SL, Reichert LE, Niswender LE. Serum levels of prolactin as measured by radioimmunoassay. Biol Reprod. 1971;4:145–7.

Dwyer CM. Individual variation in the expression of maternal behaviour: a review of the neuroendocrine mechanisms in the sheep. J Neuroendocrinol. 2008a;20(4):526–34.

Dwyer CM. Genetic and physiological determinants of maternal behavior and lamb survival: implications for low-input sheep management1,2. J Anim Sci. 2008b;86(suppl_14):E246–E58.
Dwyer CM, Lawrence AB. Variability in the expression of maternal behaviour in primiparous sheep: effects of genotype and litter size. Appl Anim Behav Sci. 1998;58(3-4):311–30.
Dwyer CM, Lawrence AB. Does the behaviour of the neonate influence the expression of maternal behaviour in sheep? Behaviour. 1999a;136:367–89.
Dwyer CM, Lawrence AB. Ewe-ewe and ewe-lamb behaviour in a hill and a lowland breed of sheep: a study using embryo transfer. Appl Anim Behav Sci. 1999b;61(4):319–34.
Dwyer CM, Lawrence AB. Effects of maternal genotype and behaviour on the behavioural development of their offspring in sheep. Behaviour. 2000a;137(12):1629–54.
Dwyer CM, Lawrence AB. Maternal behaviour in domestic sheep (Ovis aries): constancy and change with maternal experience. Behaviour. 2000b;137(10):1391–413.
Dwyer CM, Lawrence AB, Brown HE, Simm G. Effect of ewe and lamb genotype on gestation length, lambing ease and neonatal behaviour of lambs. Reprod Fertil Dev. 1996;8(8):1123–9.
Dwyer CM, McLean KA, Deans LA, Chirnside J, Calvert SK, Lawrence AB. Vocalisations between mother and young in sheep: effects of breed and maternal experience. Appl Anim Behav Sci. 1998;58(1-2):105–19.
Dwyer CM, Lawrence AB, Bishop SC. The effects of selection for lean tissue content on maternal and neonatal lamb behaviours in Scottish Blackface sheep. Anim Sci. 2001;72:555–71.
Dwyer CM, Lawrence AB, Bishop SC, Lewis M. Ewe–lamb bonding behaviours at birth are affected by maternal undernutrition in pregnancy. Br J Nutr. 2003;89(1):123–36.
Echeverri AC, Gonyou HW, Ghent AW. Preparturient behavior of confined ewes: time budgets, frequencies, spatial distribution and sequential analysis. Appl Anim Behav Sci. 1992;34(4):329–44.
Ekiz B, Kocak O, Ozcan M, Yilmaz A. Effects of parity and litter size on maternal behaviour in Kivircik ewes. Acta Agriculturae Scandinavica Section A-Animal Science. 2007;57(2):81–8.
Everett-Hincks JM, Cullen NG. Genetic parameters for ewe rearing performance1. J Anim Sci. 2009;87(9):2753–8.
Everett-Hincks JM, Blair HT, Stafford KJ, Lopez-Villalobos N, Kenyon PR, Morris ST. The effect of pasture allowance fed to twin- and triplet-bearing ewes in late pregnancy on ewe and lamb behaviour and performance to weaning. Livest Prod Sci. 2005;97(2–3):253–66.
Fahmy MH, Robert S, Castonguay F. Ewe and lamb behaviour at parturition in prolific and nonprolific sheep. Can J Anim Sci. 1997;77(1):9–15.
Farrell WJ, Alberts JR. Maternal responsiveness to infant rat ultrasonic vocalizations during the maternal behavior cycle and following steroid and experiential induction regimens. J Comp Psychol. 2002;116:286–96.
Ferreira G, Gervais R, Durkin TP, Lévy F. Postacquisition scopolamine treatments reveal the time course for the formation of lamb odor recognition memory in parturient ewes. Behav Neurosci. 1999;113(1):136–42.
Ferreira G, Terrazas A, Poindron P, Nowak R, Orgeur P, Lévy F. Learning of olfactory cues is not necessary for early lamb recognition by the mother. Physiol Behav. 2000;69(4–5):405–12.
Ferreira G, Meurisse M, Gervais R, Ravel N, Lévy F. Extensive immunolesions of basal forebrain cholinergic system impair offspring recognition in sheep. Neuroscience. 2001;106(1):103–15.
Ferreira G, Keller M, Saint-Dizier H, Perrin G, Levy F. Transfer between views of conspecific faces at different ages or in different orientations by sheep. Behav Process. 2004;67(3):491–499. https://doi.org/10.1016/j.beproc.2004.08.005
Fleming AS, Li M. Psychobiology of maternal behavior and its early determinants in nonhuman mammals. In: Bornstein MH, editor. Handbook of parenting. 2nd ed. Mahwah: Lawrence Erlbaum Associates; 2002. p. 61–97.
Gheusi G, Lledo PM. Control of early events in olfactory processing by adult neurogenesis. Chem Senses. 2007;32(4):397–409.
Gonyou HW, Stookey JM. Use of lambing cubicles and the behavior of ewes at parturition. J Anim Sci. 1983;56(4):787–91.

Gonyou HW, Stookey JM. Behavior of parturient ewes in group-lambing pens with and without cubicles. Appl Anim Behav Sci. 1984;14(2):163–71.

Gonzalez EGY, Cuellar A, Hernandez H, Nandayapa E, Alvarez L, Tortora J, et al. Maternal experience in Romanov sheep impairs mother-lamb recognition during the first 24 hours postpartum. J Vet Behav Clin Appl Res. 2015;10(1):66–72.

Gonzalez-Mariscal G, Poindron P. Parental care in mammals: immediate internal and sensory factors of control. Hormones, brain and behavior. I. Elsevier Science; 2002. p. 215–98.

Gonzalez-Stagnaro C, Madrid-Bury N. Parturition in native goats. Revista Cientifica-Facultad De Ciencias Veterinarias. 2004;14(2):124–32.

Gubernick DJ. Mechanisms of maternal "labelling" in goats. Anim Behav. 1981;29:305–6.

Gubernick DJ, Corbeau Jones K, Klopfer PH. Maternal imprinting in goats? Anim Behav. 1979;27:314–5.

Hazard D, Mace T, Kempeneers A, Delval E, Foulquie D, Bouix J, et al. Genetic parameters estimates for ewes' behavioural reactivity towards their litter after lambing. J Anim Breed Genet. 2020;137(4):374–83.

Hernandez H, Serafin N, Terrazas A, Marnet P, Kann G, Delgadillo J, et al. Maternal olfaction differentially modulates oxytocin and prolactin release during suckling in goats. Horm Behav. 2002;42(2):232–44.

Herscher L, Richmond JB, Moore AU. Maternal behavior in sheep and goats. In: Rheingold HL, editor. Maternal behavior in mammals. New York: Wiley; 1963. p. 203–32.

Hydbring E, Madej A, MacDonald E, Drugge B, Berglund B, Olsson K. Hormonal changes during parturition in heifers and goats are related to the phases and severity of labour. J Endocrinol. 1999;160:75–85.

Kasikci D. Maternal behaviors of Saanen × Hair Goat crosses at partum and early postpartum doe-kid relationships. Appl Ecol Environ Res. 2018;16(6):7791–8.

Keil NM, Imfeld-Mueller S, Aschwanden J, Wechsler B. Are head cues necessary for goats (Capra hircus) in recognising group members? Anim Cogn. 2012;15(5):913–21.

Keller M, Meurisse M, Poindron P, Nowak R, Ferreira G, Shayit M, et al. Maternal experience influences the establishment of visual/auditory, but not olfactory recognition of the newborn lamb by ewes at parturition. Dev Psychobiol. 2003;43(3):167–76.

Keller M, Meurisse M, Lévy F. Mapping the neural substrates involved in maternal responsiveness and lamb olfactory memory in parturient ewes using Fos imaging. Behav Neurosci. 2004a;118(6):1274–84.

Keller M, Perrin G, Meurisse M, Ferreira G, Lévy F. Cortical and medial amygdala are both involved in the formation of olfactory offspring memory in sheep. Eur J Neurosci. 2004b;20(12):3433–41.

Keller M, Meurisse M, Lévy F. Mapping of brain networks involved in consolidation of lamb recognition memory. Neuroscience. 2005;133(2):359–69.

Kendrick KM. Neurobiological correlates of visual and olfactory recognition in sheep. Behav Process. 1994;33:89–112.

Kendrick KM. Oxytocin, motherhood and bonding. Exp Physiol. 2000;85:111–24.

Kendrick KM, Keverne EB. Importance of progesterone and estrogen priming for the induction of maternal behavior by vaginocervical stimulation in sheep: effects of maternal experience. Physiol Behav. 1991;49(4):745–50.

Kendrick KM, Keverne EB. Control of synthesis and release of oxytocin in the sheep brain. Ann N Y Acad Sci. 1992;652:102–21.

Kendrick KM, Keverne EB, Baldwin BA, Sharman DF. Cerebrospinal fluid levels of acetylcholinesterase, monoamines and oxytocin during labour, parturition, vaginocervical stimulation, lamb separation and suckling in sheep. Neuroendocrinology. 1986;44:149–56.

Kendrick KM, Keverne EB, Baldwin BA. Intracerebroventricular oxytocin stimulates maternal behaviour in the sheep. Neuroendocrinology. 1987;46:56–61.

Kendrick KM, Keverne EB, Chapman C, Baldwin BA. Intracranial dialysis measurement of oxytocin, monoamine and uric acid release from the olfactory bulb and substantia nigra of sheep during parturition, suckling, separation from lambs and eating. Brain Res. 1988a;439(1-2):1–10.

Kendrick KM, Keverne EB, Shapman C, Baldwin BA. Microdialysis measurement of oxytocin, aspartate, gama-aminobutyric acid and glutamate release from the olfactory bulb of the sheep during vaginocervical stimulation. Brain Res. 1988b;442:171–4.

Kendrick KM, Keverne EB, Hinton MR, Goode JA. Cerebrospinal fluid and plasma concentrations of oxytocin and vasopressin during parturition and vaginocervical stimulation in the sheep. Brain Res Bull. 1991a;26(5):803–7.

Kendrick KM, Lévy F, Keverne EB. Importance of vaginocervical stimulation for the formation of maternal bonding in primiparous and multiparous parturient ewes. Physiol Behav. 1991b;50(3):595–600.

Kendrick KM, Keverne EB, Hinton MR, Goode JA. Oxytocin, amino acid and monoamine release in the region of the medial preoptic area and bed nucleus of the stria terminalis of the sheep during parturition and suckling. Brain Res. 1992a;569(2):199–209.

Kendrick KM, da Costa AP, Hinton MR, Keverne EB. A simple method for fostering lambs using anoestrus ewes with artificially induced lactation and maternal behaviour. Appl Anim Behav Sci. 1992b;34:345–57.

Kendrick KM, Lévy F, Keverne EB. Changes in the sensory processing of olfactory signals induced by birth in sheep. Science. 1992c;256(5058):833–6.

Kendrick KM, Atkins K, Hinton MR, Heavens P, Keverne B. Are faces special for sheep? Evidence from facial and object discrimination learning tests showing effects of inversion and social familiarity. Behav Process. 1996;38:19–35.

Kendrick KM, Da Costa APC, Broad KD, Ohkura S, Guevara R, Lévy F, et al. Neural control of maternal behaviour and olfactory recognition of offspring. Brain Res Bull. 1997a;44(4):383–95.

Kendrick KM, Guevara-Guzman R, Zorrilla J, Hinton MR, Broad KD, Mimmack M, et al. Formation of olfactory memories mediated by nitric oxide. Nature. 1997b;388(6643):670–4.

Kendrick KM, Leigh A, Peirce J. Behavioural and neural correlates of mental imagery in sheep using face recognition paradigms. Anim Welf. 2001;10:S89–S101.

Keverne EB, Kendrick KM. Morphine and corticotrophin-releasing factor potentiate maternal acceptance in multiparous ewes after vaginocervical stimulation. Brain Res. 1991;540(1-2):55–62.

Keverne EB, Lévy F, Poindron P, Lindsay DR. Vaginal stimulation: an important determinant of maternal bonding in sheep. Science. 1983;219(4580):81–3.

Keverne EB, Lévy F, Guevara-Guzman R, Kendrick KM. Influence of birth and maternal experience on olfactory bulb neurotransmitter release. Neuroscience. 1993;56(3):557–65.

Kilgour RJ, Szantar-Coddington MR. Arena behaviour of ewes selected for superior mothering ability differs from that of unselected ewes. Anim Reprod Sci. 1995;37(2):133–41.

Klopfer PH, Gamble J. Maternal imprinting in goats: the role of chemical senses. Zeitschrift Fur Tierpsychologie. 1966;23(5):588–92.

Klopfer PH, Adams DK, Klopfer MS. Maternal imprinting in goats. Proc Natl Acad Sci U S A. 1964;52:911–4.

Knox D. The role of basal forebrain cholinergic neurons in fear and extinction memory. Neurobiol Learn Mem. 2016;133:39–52.

Kopel H, Schechtman E, Groysman M, Mizrahi A. Enhanced synaptic integration of adult-born neurons in the olfactory bulb of lactating mothers. J Neurosci. 2012;32(22):7519–27.

Krehbiel D, Poindron P, Lévy F, Prud'Homme MJ. Peridural anesthesia disturbs maternal behavior in primiparous and multiparous parturient ewes. Physiol Behav. 1987;40(4):463–72.

Lambe NR, Conington J, Bishop SC, Waterhouse A, Simm G. A genetic analysis of maternal behaviour score in Scottish Blackface sheep. Anim Sci. 2001;72(2):415–25.

Langbein J, Scheibe KM, Eichhorn K. Investigations on periparturient behaviour in free-ranging mouflon sheep (Ovis orientalis musimon). J Zool. 1998;244:553–61.

Lazarini F, Lledo PM. Is adult neurogenesis essential for olfaction? Trends Neurosci. 2011;34(1):20–30.

Le Neindre P, Poindron P, Delouis C. Hormonal induction of maternal behavior in non-pregnant ewes. Physiol Behav. 1979;22(4):731–4.

Lepousez G, Nissant A, Lledo PM. Adult neurogenesis and the future of the rejuvenating brain circuits. Neuron. 2015;86(2):387–401.

Lévy F, Keller M. Olfactory mediation of maternal behavior in selected mammalian species. Behav Brain Res. 2009;200(2):336–45.

Lévy F, Poindron P. Influence of amniotic fluids in the manifestation of maternal-behavior in parturient ewe. Biol Behav. 1984;9(3):271–8.

Levy F, Poindron P. The importance of amniotic fluids for the establishment of maternal-behavior in experienced and inexperienced ewes. Anim Behav. 1987;35:1188–92.

Lévy F, Poindron P. The importance of amniotic fluids for the establishment of maternal behavior in experienced and non-experienced ewes. Anim Behav. 1987;35:1188–92.

Lévy F, Poindron P, Le Neindre P. Attraction and repulsion by amniotic fluids and their olfactory control in the ewe around parturition. Physiol Behav. 1983;31:687–92.

Lévy F, Keverne EB, Piketty V, Poindron P. Physiological determinism of olfactory attraction for amniotic fluids in sheep. In: MacDonald DW, Müller-Schwarze D, Natynczuck SE, editors. Chemical signals in vertebrates, vol. 5. Oxford: Oxford University Press; 1990a. p. 162–5.

Lévy F, Gervais R, Kindermann U, Orgeur P, Piketty V. Importance of beta-noradrenergic receptors in the olfactory bulb of sheep for recognition of lambs. Behav Neurosci. 1990b;104:464–9.

Lévy F, Gervais R, Kindermann U, Litterio M, Poindron P, Porter R. Effects of early post-partum separation on maintenance of maternal responsiveness and selectivity in parturient ewes. Appl Anim Behav Sci. 1991;31(1-2):101–10.

Lévy F, Kendrick KM, Keverne EB, Piketty V, Poindron P. Intracerebral oxytocin is important for the onset of maternal behavior in inexperienced ewes delivered under peridural anesthesia. Behav Neurosci. 1992;106(2):427–32.

Lévy F, Guevara-Guzman R, Hinton MR, Kendrick KM, Keverne EB. Effects of parturition and maternal experience on noradrenaline and acetylcholine release in the olfactory bulb of sheep. Behav Neurosci. 1993;107:662–8.

Lévy F, Locatelli A, Piketty V, Tillet Y, Poindron P. Involvement of the main but not the accessory olfactory system in maternal-behavior of primiparous and multiparous ewes. Physiol Behav. 1995a;57(1):97–104.

Lévy F, Kendrick KM, Goode JA, Guevara-Guzman R, Keverne EB. Oxytocin and vasopressin release in the olfactory bulb of parturient ewes: changes with maternal experience and effects on acetylcholine, γ-aminobutyric acid, glutamate and noradrenaline release. Brain Res. 1995b;669(2):197–206.

Lévy F, Richard P, Meurisse M, Ravel N. Scopolamine impairs the ability of parturient ewes to learn to recognise their lambs. Psychopharmacology. 1997;129(1):85–90.

Lévy F, Meurisse M, Ferreira G, Thibault J, Tillet Y. Afferents to the rostral olfactory bulb in sheep with special emphasis on the cholinergic, noradrenergic and serotonergic connections. J Chem Neuroanat. 1999;16(4):245–63.

Lévy F, Keller A, Poindron P. Olfactory regulation of maternal behavior in mammals. Horm Behav. 2004;46(3):284–302.

Lévy F, Keller M, Cornilleau F, Moussu C, Ferreira G. Vaginocervical stimulation of ewes induces the rapid formation of a new bond with an alien young without interfering with a previous bond. Dev Psychobiol. 2010;52(6):537–44.

Lickliter RE. Mother-infant spatial relationships in domestic goats. Appl Anim Behav Sci. 1984a;13:93–100.

Lickliter RE. Hiding behaviour in domestic goat kids. Appl Anim Behav Sci. 1984b;12:245–51.

Lickliter RE. Behavior associated with parturition in the domestic goat. Appl Anim Behav Sci. 1985;13(4):335–45.

Lledo PM, Alonso M, Grubb MS. Adult neurogenesis and functional plasticity in neuronal circuits. Nat Rev Neurosci. 2006;7(3):179–93.

Luna-Orozco JR, Meza-Herrera CA, Contreras-Villarreal V, Hernandez-Macias N, Angel-Garcia O, Carrillo E, et al. Effects of supplementation during late gestation on goat performance and behavior under rangeland conditions. J Anim Sci. 2015;93(8):4153–60.
Lv S-J, Yang Y, Li F-K. Parity and litter size effects on maternal behavior of Small Tail Han sheep in China. Anim Sci J. 2016;87(3):361–9.
Malfatti A, Lucamoni A, Debenedetlt A. Behavior associated with parturition in goats. Appl Anim Behav Sci. 1991;30(1-2):191.
Martinez M, Otal J, Ramirez A, Hevia ML, Quiles A. Variability in the behavior of kids born of primiparous goats during the first hour after parturition: effect of the type of parturition, sex, duration of birth, and maternal behavior. J Anim Sci. 2009;87(5):1772–7.
Meurisse M, Gonzalez A, Delsol G, Caba D, Lévy F, Poindron P. Estradiol receptor-alpha expression in hypothalamic and limbic regions of ewes is influenced by physiological state and maternal experience. Horm Behav. 2005;48(1):34–43.
Meurisse M, Chaillou E, Levy F. Afferent and efferent connections of the cortical and medial nuclei of the amygdala in sheep. J Chem Neuroanat. 2009;37(2):87–97.
Meyer S, Nurnberg G, Puppe B, Langbein J. The cognitive capabilities of farm animals: categorisation learning in dwarf goats (Capra hircus). Anim Cogn. 2012;15(4):567–76.
Moreno MM, Bath K, Kuczewski N, Sacquet J, Didier A, Mandairon N. Action of the noradrenergic system on adult-born cells is required for olfactory learning in mice. J Neurosci. 2012;32(11):3748–58.
Morgan PD, Boundy CAP, Arnold GW, Lindsay DR. The roles played by the senses of the ewe in the location and recognition of lambs. Appl Anim Ethol. 1975;1:139–50.
Mota-Rojas D, Orihuela A, Strappini A, Villanueva-Garcia D, Napolitano F, Mora-Medina P, et al. Consumption of maternal placenta in humans and nonhuman mammals: beneficial and adverse effects. Animals. 2020;10(12):16.
Mouly AM, Gervais R. Polysynaptic potentiation at different levels of rat olfactory pathways following learning. Learn Mem. 2002;9(2):66–75.
Mouly AM, Fort A, Ben-Boutayab N, Gervais R. Olfactory learning induces differential long-lasting changes in rat central olfactory pathways. Neuroscience. 2001;102(1):11–21.
Nowak R. Neonatal survival: contributions from behavioural studies in sheep. Appl Anim Behav Sci. 1996;49(1):61–72.
Nowak R, Keller M, Val-Laillet D, Lévy F. Perinatal visceral events and brain mechanisms involved in the development of mother-young bonding in sheep. Horm Behav. 2007;52(1):92–8.
Nowak R, Keller M, Lévy F. Mother–young relationships in sheep: a model for a multidisciplinary approach of the study of attachment in mammals. J Neuroendocrinol. 2011;23(11):1042–53.
Numan M, Insel TR. In: Ball GF, Balthazart J, Nelson RJ, editors. The neurobiology of parental behavior. Springer; 2003. p. 417.
Numan M, Numan MJ. A lesion and neuroanatomical tract-tracing analysis of the role of the bed nucleus of the stria terminalis in retrieval behavior and other aspects of maternal responsiveness in rats. Dev Psychobiol. 1996;29(1):23–51.
O'Brien PH. Feral goat parturition and lying-out sites: spatial, physical and meteorological characteristics. Appl Anim Behav Sci. 1983;10:325–39.
O'Brien PH. Leavers and stayers: maternal post-partum strategies in feral goats. Appl Anim Behav Sci. 1984;12:233–43.
O'Connor CE, Lawrence AB, Wood-Gush DGM. Influence of litter size and parity on maternal behaviour at parturition on Scottish Blackface sheep. Appl Anim Behav Sci. 1992;33:345–55.
Olazabal DE, Pereira M, Agrati D, Ferreira A, Fleming AS, Gonzalez-Mariscal G, et al. Flexibility and adaptation of the neural substrate that supports maternal behavior in mammals. Neurosci Biobehav Rev. 2013;37(8):1875–92.
Onoda N, Imamura K, Obata E, Iino M. Response selectivity of neocortical neurons to specific odors in the rabbit. J Neurophysiol. 1984;52(4):638–52.
Otal J, Lévy F, Cornilleau F, Moussu C, Keller M, Poindron P. Preventing physical interactions between parturient ewes and their neonate differentially impairs the development of mater-

nal responsiveness and selectivity depending on maternal experience. Appl Anim Behav Sci. 2009;120(3-4):140–9.

Otal J, Martinez M, Quiles A, Hevia ML, Ramirez A. Effect of litter size and sex on the birth weight of newborn kids and in the behaviour of primiparous goats before, during and after the parturition. Can J Anim Sci. 2010;90(4):483–90.

Peirce JW, Leigh AE, Kendrick KM. Configurational coding, familiarity and the right hemisphere advantage for face recognition in sheep. Neuropsychol. 2000;38(4):475–483.

Perrin G, Meurisse M, Lévy F. Inactivation of the medial preoptic area or the bed nucleus of the stria tenninalis differentially disrupts maternal behavior in sheep. Horm Behav. 2007;52(4):461–73.

Petrulis A, Eichenbaum H. The perirhinal-entorhinal cortex, but not the hippocampus, is critical for expression of individual recognition in the context of the Coolidge effect. Neuroscience. 2003;122(3):599–607.

Pissonnier D, Thiéry JC, Fabre-Nys C, Poindron P, Keverne EB. The importance of olfactory bulb noradrenalin for maternal recognition in sheep. Physiol Behav. 1985;35(3):361–3.

Plush KJ, Hebart ML, Brien FD, Hynd PI. The genetics of temperament in Merino sheep and relationships with lamb survival. Appl Anim Behav Sci. 2011;134(3-4):130–5.

Poindron P. Mother-young relationships in intact or anosmic ewes at the time of suckling. Biol Behav. 1976;2:161–77.

Poindron P, Le Neindre P. Endocrine and sensory regulation of maternal behavior in the ewe. In: Rosenblatt JS, Hinde RA, Beer C, editors. Advances in the study of behavior. 11. New York: Academic; 1980. p. 75–119.

Poindron P, Martin GB, Hooley RD. Effects of lambing induction on the sensitive period for the establishment of maternal behaviour in sheep. Physiol Behav. 1979;23(6):1081–7.

Poindron P, Le Neindre P, Raksanyi I, Trillat G, Orgeur P. Importance of the characteristics of the young in the manifestation and establishment of maternal behaviour in sheep. Reprod Nutr Dev. 1980;20(3B):817–26.

Poindron P, Leneindre P, Lévy F, Keverne EB. The physiological mechanisms of the acceptance of the newborn among sheep. Biol Behav. 1984a;9(1):65–88.

Poindron P, Raksanyi I, Orgeur P, Le Neindre P. Comparaison du comportement maternel en bergerie à la parturition chez des brebis primipares ou multipares de race Romanov, Préalpes de Sud et Ile-de-France. Génétique, Sélection, Evolution. 1984b;16(4):503–22.

Poindron P, Lévy F, Krehbiel D. Genital, olfactory, and endocrine interactions in the development of maternal behaviour in the parturient ewe. Psychoneuroendocrinology. 1988;13(1-2):99–125.

Poindron P, Caba M, Gomora Arrati P, Krehbiel D, Beyer C. Responses of maternal and non-maternal ewes to social and mother-young separation. Behav Process. 1994;31(1):97–110.

Poindron P, Soto R, Romeyer A. Decrease of response to social separation in preparturient ewes. Behav Process. 1997;40(1):45–51.

Poindron P, Gilling G, Hernandez H, Serafin N, Terrazas A. Early recognition of newborn goat kids by their mother: I. Nonolfactory discrimination. Dev Psychobiol. 2003;43(2):82–9.

Poindron P, Terrazas A, de Oca MD, Serafin N, Hernandez H. Sensory and physiological determinants of maternal behavior in the goat (Capra hircus). Horm Behav. 2007a;52(1):99–105.

Poindron P, Lévy F, Keller M. Maternal responsiveness and maternal selectivity in domestic sheep and goats: the two facets of maternal attachment. Dev Psychobiol. 2007b;49(1):54–70.

Poindron P, Otal J, Ferreira G, Keller M, Guesdon V, Nowak R, et al. Amniotic fluid is important for the maintenance of maternal responsiveness and the establishment of maternal selectivity in sheep. Animal. 2010;4(12):2057–64.

Porter RH, Doane HM. Studies of maternal behavior in spiny mice (Acomis cahirinus). Zeitschrift Fur Tierpsychologie. 1978;47:225–35.

Porter RH, Lévy F, Poindron P, Litterio M, Schaal B, Beyer C. Individual olfactory signatures as major determinants of early maternal discrimination in sheep. Dev Psychobiol. 1991;24(3):151–8.

Porter RH, Romeyer A, Lévy F, Krehbiel D, Nowak R. Investigations of the nature of lambs individual odor signatures. Behav Process. 1994;31(2-3):301–8.

Pritchard CE, Williams AP, Davies P, Jones D, Smith AR. Spatial behaviour of sheep during the neonatal period: preliminary study on the influence of shelter. Animal. 2021;15(7):100252.
Putu IG, Poindron P, Lindsay DR. A high level of nutrition during late pregnancy improves subsequent maternal behaviour of merino ewes. Proc Aust Soc Anim Prod. 1988;17:294–7.
Putu IG, Poindron P, Oldham CM, Gray SJ, Ballard M. Lamb desertion in primiparous and multiparous Merino ewes induced to lamb with dexamethasone. Proc Aust Soc Anim Prod. 1986;16:315–318.
Ramírez A, Quiles A, Hevia M, Sotillo F. Behavior of the Murciano-Granadina goat in the hour before parturition. Appl Anim Behav Sci. 1995;44:29–35.
Ramirez A, Quiles A, Hevia M, Sotillo F. Observations on the birth of goats. Can J Anim Sci. 1995;75(1):165–7.
Ramírez A, Quiles A, Hevia ML, Sotillo F, Ramírez MC. Effects of immediate and early postpartum separation on maintenance of maternal responsiveness in parturient multiparous goats. Appl Anim Behav Sci. 1996;48:215–24.
Ramírez-Vera S, Terrazas A, Delgadillo JA, Flores JA, Serafín N, Vielma J, et al. Inclusion of maize in the grazing diet of goats during the last 12 days of gestation reinforces the expression of maternal behaviour and selectivity during the sensitive period. Livest Sci. 2012;148(1):52–9.
Regueiro M, López-Mazz C, Jorge-Smeding E, Baldi F, Banchero G. Duration of phase II of labour negatively affects maternal behaviour and lamb viability in wool-type primiparous ewes under extensive rearing. Appl Anim Behav Sci. 2021;234:105207.
Restrepo D, Lin W, Salcedo E, Yamazaki K, Beauchamp G. Odor types and MHC peptides: complementary chemosignals of MHC haplotype? Trends Neurosci. 2006;29(11):604–9.
Romeyer A, Porter RH, Lévy F, Nowak R, Orgeur P, Poindron P. Maternal labelling is not necessary for the establishment of discrimination between kids by recently parturient goats. Anim Behav. 1993a;46(4):705–12.
Romeyer A, Porter RH, Poindron P, Orgeur P, Chesné P, Poulain N. Recognition of dizygotic and monozygotic twin lambs by ewes. Behaviour. 1993b;127:119–39.
Romeyer A, Poindron P, Orgeur P. Olfaction mediates the establishment of selective bonding in goats. Physiol Behav. 1994a;56(4):693–700.
Romeyer A, Poindron P, Porter RH, Lévy F, Orgeur P. Establishment of maternal bonding and its mediation by vaginocervical stimulation in goats. Physiol Behav. 1994b;55(2):395–400.
Rosenblatt JS. Psychobiology of maternal behavior: contribution to the clinical understanding of maternal behavior among humans. Acta Paediatr Suppl. 1994;397:3–8.
Rosenblatt JS, Lehrman DS. Maternal behavior in the laboratory rat. In: Rheingold HL, editor. Maternal behavior in mammals. New York: Wiley; 1963. p. 8–57.
Rosenblatt JS, Siegel HI. Factors governing the onset and maintenance of maternal behavior among nonprimate mammals. In: Gubernick DJ, Klopfer PH, editors. Parental care in mammals. New York: Plenum Press; 1981. p. 13–76.
Ruiz-Miranda CR. Use of pelage pigmentation in the recognition of mothers in a group by 2- to 4-month-old domestic goat kids. Appl Anim Behav Sci. 1993;36(4):317–26.
Sambraus HH, Wittmann M. Observations of the birth and suckling behavior of goats. Tierarztliche Praxis. 1989;17(4):359–65.
Schneirla TC, Rosenblatt JS, Tobach E. Maternal behavior in the cat. In: Rheingold HL, editor. Maternal behavior in mammals. New York: Wiley; 1963. p. 122–68.
Sèbe F, Nowak R, Poindron P, Aubin T. Establishment of vocal communication and discrimination between ewes and their lamb in the first two days after parturition. Dev Psychobiol. 2007;49(4):375–86.
Sebe F, Aubin T, Boue A, Poindron P. Mother-young vocal communication and acoustic recognition promote preferential nursing in sheep. J Exp Biol. 2008;211(Pt 22):3554–62.
Sèbe F, Duboscq J, Aubin T, Ligout S, Poindron P. Early vocal recognition of mother by lambs: contribution of low- and high-frequency vocalizations. Anim Behav. 2010;79(5):1055–66.
Sèbe F, Poindron P, Ligout S, Sèbe O, Aubin T. Amplitude modulation is a major marker of individual signature in lamb bleats. Bioacoustics. 2018;27(4):359–75.

Shillito EE. Vocalization in sheep. J Physiol London. 1972;226(2):P45.

Shillito EE, Alexander G. Mutual recognition amongst ewes and lambs of four breeds of sheep. Appl Anim Ethol. 1975;1:151–65.

Shipka MP, Ford SP. Relationship of circulating estrogen and progesterone concentrations during late pregnancy and the onset phase of maternal behavior in the ewe. Appl Anim Behav Sci. 1991;31:91–9.

Simitzis P, Galani K, Vasiliou P, Koutsouli P, Bizelis I. Effect of breed and litter size on the display of maternal perinatal and offspring postnatal behavior in dairy sheep. J Vet Behav. 2016;13:10–8.

Soto R, Terrazas A, Poindron P, González-Mariscal G. Regulation of maternal behavior, social isolation responses, and postpartum estrus by steroid hormones and vaginocervical stimulation in sheep. Horm Behav. 2021;136:105061.

Stevens D, Alexander G, Lynch JJ. Do Merino ewes seek isolation or shelter at lambing? Appl Anim Ethol. 1981;7:149–55.

Strott CA, Sundel H, Stahlman MT. Maternal and fetal plasma progesterone, cortisol, testosterone and 17beta-estradiol in preparturient sheep: response to fetal ACTH infusion. Endocrinology. 1974;95(5):1327–39.

Terrazas A, Ferreira G, Lévy F, Nowak R, Serafin N, Orgeur P, et al. Do ewes recognize their lambs within the first day postpartum without the help of olfactory cues? Behav Process. 1999;47(1):19–29.

Terrazas A, Serafin N, Hernandez H, Nowak R, Poindron P. Vocal recognition of 2-day-old goat kids by their mother: evidence of individual acoustic signatures in neonates. 2003.

Terrazas A, Robledo V, Serafín N, Soto R, Hernández H, Poindron P. Differential effects of undernutrition during pregnancy on the behaviour of does and their kids at parturition and on the establishment of mutual recognition. Animal. 2009;3(2):294–306.

Uriarte N, Agrati D, Banchero G, González A, Zuluaga MJ, Cawen E, et al. Different chemical fractions of fetal fluids account for their attractiveness at parturition and their repulsiveness during late-gestation in the ewe. Physiol Behav. 2012;107(1):45–9.

Vince MA, Lynch JJ, Mottershead B, Green G, Elwin R. Sensory factors involved in immediately postnatal ewe/lamb bonding. Behaviour. 1985;94:60–84.

Viviers MZ, Bekker JP, Burger BV, le Roux NJ, Morris J, le Roux M. Characterization of amniotic fluid of Dohne Merino ewes (Ovis aries) and its possible role in neonatal recognition. Z Naturforsch C J Biosci. 2015;70(5-6):115–28.

Whateley J, Kilgour R, Dalton DC. Behaviour of hill country sheep breeds during farming routines. Proc N Z Soc Anim Prod. 1974:28–36.

Wierucka K, Pitcher BJ, Harcourt R, Charrier I. Multimodal mother-offspring recognition: the relative importance of sensory cues in a colonial mammal. Anim Behav. 2018;146:135–42.

Yamin M, Payne G, Blackshaw JK. The time of birth and the choice of birth sites by Booroola Merino ewes and Angora goats. Appl Anim Behav Sci. 1995;45(1-2):89–96.

Yu GZ, Kaba H, Okutani F, Takahashi S, Higuchi T, Seto K. The action of oxytocin originating in the hypothalamic paraventricular nucleus on mitral and granule cells in the rat main olfactory bulb. Neuroscience. 1996;72(4):1073–82.

Parental Behaviour in Sows

Jinhyeon Yun and Olli Peltoniemi

Abstract Parental behaviour in sows begins with prepartum nest-building behaviour, which mainly consists of rooting, pawing, and foraging, to achieve a structure for farrowing and to nurse and protect the offspring. A hormonal background is linked to the onset and cessation of nest-building behaviour. The ambient farrowing environment that can adequately address the needs for nest-building behaviour thus plays an important role in good parental behaviour of the sows during and after farrowing. In addition to these environmental factors, here we also discuss other well-known intrinsic factors, such as heredity, parental experience, and litter size, affecting parental behaviour of the sows for successful farrowing and lactating performance.

Keywords Nest-building · Parturition process · Nursing · Maternal instinct · Savaging behaviour · Farrowing environment

1 Nest-Building Behaviour

Background

In the wild, the pig builds a nest for its litter prior to parturition. This behaviour is intrinsic and present in both wild breeds (European wild boar) and domestic breeds (Jensen 1986). This suggests that this behavioural need has maintained over the 10,000 years of domestication.

In nature, a pregnant sow tends to spend her time in a small group of four to six sows. At the end of pregnancy, the sow typically seeks an isolated site to build a

J. Yun (✉)
Department of Animal Science, College of Agriculture and Life Sciences, Chonnam National University, Gwangju, South Korea
e-mail: pilot9939@jnu.ac.kr

O. Peltoniemi
Department of Production Animal Medicine, Faculty of Veterinary Medicine, University of Helsinki, Helsinki, Finland

G. González-Mariscal (ed.), *Patterns of Parental Behavior*, Advances in Neurobiology 27, https://doi.org/10.1007/978-3-030-97762-7_4

nest. During the 24 h preceding parturition, the sow isolates herself and begins expressing nest-building behaviour, which is maximally expressed between 4 and 12 h prior to the birth of the first piglet (Algers and Uvnäs-Moberg 2007; Jensen 1986).

Hormonal Control of Nest-Building Behaviour

The hormonal factors that trigger nest-building behaviour include a rise in prolactin levels (Castrén et al. 1994; Fig. 1) induced by a decrease in progesterone and an increase in prostaglandin $F_{2\alpha}$ (reviewed by Algers and Uvnäs-Moberg, 2007). Additionally, a steep increase of oxytocin secretion in an episodic manner (Oliviero et al. 2008a), together with the sensation of a ready-built nest underneath the udder, appears to stop nest-building behaviour prior to the birth of the first piglet (Algers and Uvnäs-Moberg 2007). Prolonged nest-building behaviour, maintained during parturition, may be considered abnormal, and it appears as indicative of problems in the hormonal process of parturition (Algers and Uvnäs-Moberg 2007).

Environmental Requirements

Under natural or semi-natural conditions, the sow digs the ground with its forelegs and snout and builds the nest with nesting materials, such as branches, straw, and leaves, collected from the nearby environment (Jensen 1986; Mayer et al. 2002). In

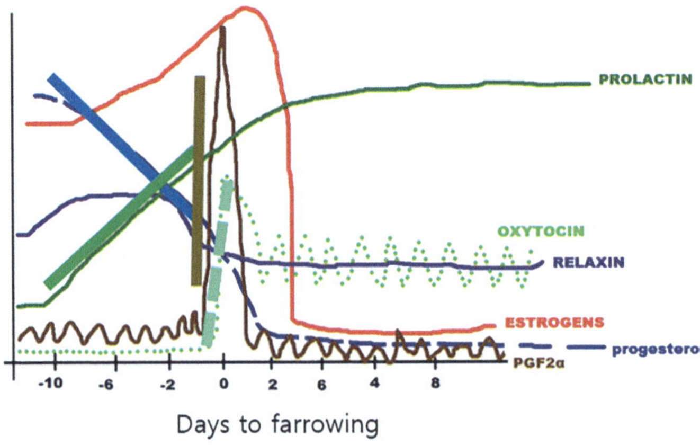

Fig. 1 Hormonal triggers of nest-building behaviour in a sow include a decrease in progesterone (bold blue bar) and an increase in prolactin and prostaglandin $F_{2\alpha}$ (bold green and brown bars). Cessation of nest building is brought about by increased oxytocin activity and the sensation of a ready-built nest against the udder. (Reviewed by Algers and Uvnäs-Moberg 2007)

agreement, nest-building behaviour in domestic sows can be stimulated by exogenous environmental factors (Jensen 1993; Algers and Uvnäs-Moberg 2007; Yun et al. 2014a). The size or radius of the nest varies depending on an individual's experience (Jensen 1986; Mayer et al. 2002) and on the surrounding environment (Mayer et al. 2002).

In the modern pig industry, keeping sows in small groups of four to six individuals towards the end of pregnancy may not be practically feasible. However, an increasing group size is likely to cause stress, which may accumulate so that adverse effects on health and pregnancy are encountered (Peltoniemi et al. 2016; Yun et al. 2019). Therefore, optimizing sow management to prevent stress appears advantageous to the pig industry (Peltoniemi et al. 2016, 2021). Feeding management practices, such as electric feeding systems that provide an undisturbed, individual method for feeding sows, appear to be useful in reducing stress responses commonly seen in conjunction with the feeding time of group-housed sows prior to their isolation into individual farrowing pens.

Isolation, once parturition is approaching, may not be possible in a commercial piggery, yet considering the need for isolation is imperative (Yun et al. 2019). Similar opportunities to build a nest with plentiful and varied nesting materials, as in nature, may not be feasible in most piggeries. Yet providing domestic sows with abundant nesting material and an adequately sized pen of more than six square metres, with an appealing piglet nest, are among the fundamental criteria that a hyperprolific sow requires for successful farrowing (Yun et al. 2015, 2019; Fig. 2). Nevertheless, the ongoing increase in litter size presents the industry with a challenge regarding farrowing duration.

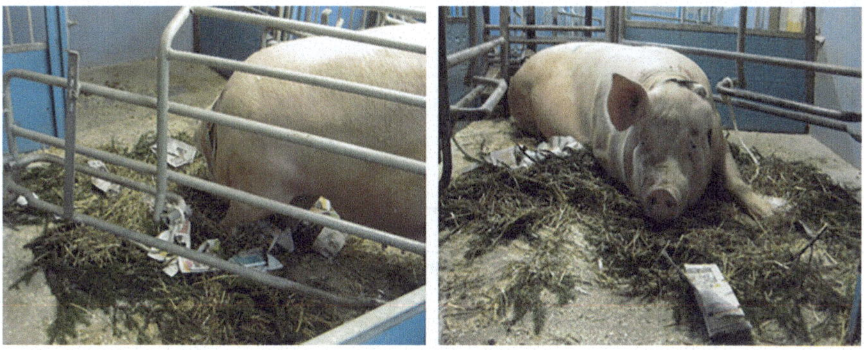

Fig. 2 A hybrid sow (Finnish Yorkshire × Finnish Landrace) builds a nest prior to parturition in a modern loose housing system with a provision of nesting materials

Effects of Modern Housing in Confinement on Nest-Building Behaviour

Confining sows in crates during farrowing is a common practice worldwide. In a typical farrowing unit, the sow is only allowed limited space for movement, and a nest-building substrate is also often absent or very limited (Vestergaard and Hansen 1984; Cronin et al. 1996; Edwards and Fraser 1997; Jensen et al. 1997; Thodberg et al. 1999; Gu et al. 2011; Yun et al. 2019). Restricting sow movement using a cage has been argued to reduce neonatal piglet mortality (Hansen and Curtis 1980; Hales et al. 2014). However, nest-building behaviour as triggered by endogenous hormonal activity during the prepartum period cannot be properly expressed by caged sows (Yun et al. 2014a; Fig. 3). If no nest-building substrate is available in the pen, confined sows typically express prolonged and unsuccessful nest-building behaviour (Damm et al. 2000, 2003; Yun et al. 2014a, 2015, 2019). This cannot be alleviated even by providing nesting materials if the crate inhibits the sow's movement (Jarvis et al. 2002; Yun and Valros 2015). The induced stress reportedly reduces oxytocin concentrations in the sow during the parturition process. Furthermore, behavioural abnormalities and cortisol concentrations remain at high levels after parturition under such conditions (Lawrence et al. 1994; Jarvis et al. 2002; Oliviero et al. 2008a). In fact, a prolonged farrowing process was reportedly one of the complications caused by confinement stress (Oliviero et al. 2010).

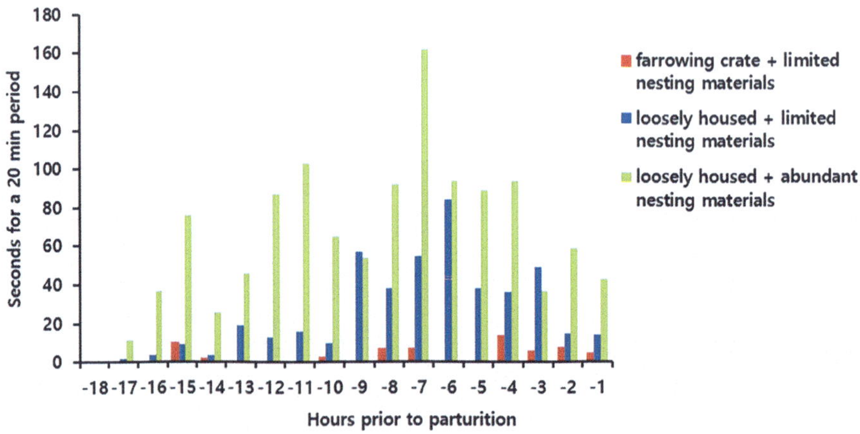

Fig. 3 Duration of nest-building behaviour of prepartum sows across 20-min periods per hour, from 18 h prior until birth of the first piglet (Modified from Yun et al. 2014a). The sows were housed in farrowing crates (0.80 × 2.10 M) (red colour in the graph) or loosely housed within a trapezoid shape (0.80 × 2.10 × 1.70 M), with a bucketful of sawdust on the ground (blue colour in the graph) or with two sawdust buckets, a shredded newspaper, three chopped straw buckets, seven tree branches, and three natural sisal ropes 50 cm in length on the ground (green colour in the graph)

Consequences of Lacking Nest-Building Behaviour

Lacking nest-building behaviour may lead to increases in circulating cortisol and ACTH concentrations (Jarvis et al. 1997), which indicate a physiological stress response. Addressing whether the need for nest-building is intrinsic, Gustafsson et al. (1999) found that domestic sows were able to build nests that were identical to wild boar nests. This was the case even after several previous farrowing experiences in confined crates without bedding. This innate behaviour is therefore a clear indication of impending parturition, and the need for its expression becomes manifest regardless of housing conditions or the availability of bedding material.

A prolonged farrowing process will subject piglets to asphyxia during parturition, and the litter will therefore be less vital at birth (Herpin et al. 2001). In contrast, Yun et al. (2013) found that providing larger space and abundant nest-building materials before parturition tended to increase sow plasma oxytocin concentrations (25 vs. 18 pg/ml in sows with abundant nesting materials vs. sows with crates, respectively). Abundant nesting materials also increased piglet serum IgG and IgM concentrations during early lactation (15 vs. 10 mg/ml (IgG) and 0.9 vs. 0.7 mg/ml (IgM) in sows with abundant nesting materials vs. with sows in crates, respectively; Yun et al. 2014b). Allowing for expression of nesting behaviour to take place may reduce farrowing duration and thereby allow for more vital piglets (Jensen 1986; Islas-Fabila et al. 2018). Colostrum intake will also be improved due to a shorter time interval from the start of farrowing to first suckling (Manjarin et al. 2018). A faster and uncomplicated farrowing also reduces pain and inflammation in the sow (Björkman et al. 2017; Kaiser et al. 2018). Allowing the sow to farrow free and providing a substrate (straw, sawdust, paper) 1–2 days before farrowing supports nest-building behaviour of the sow which, in turn, may significantly reduce farrowing duration and stillbirth rate (Oliviero et al. 2008b; Gu et al. 2011).

2 Farrowing

At the end of nest-building behaviour, approximately 2–4 h prior to the birth of the first piglet, the sow is usually laying down in lateral recumbence (Algers and Uvnäs-Moberg 2007). The abdominal muscles show more regular straining, and this period coincides with increased oxytocin activity, known to trigger uterine contractions. For educational purposes, parturition in the pig is divided into three phases, while farrowing duration in the scientific literature commonly refers to the second phase only (lasting from the birth of the first piglet to the birth of the last piglet; Fig. 4, data based on Peltoniemi et al. 2019).

The first stage of farrowing is characterized by cervical dilation. This is a complex biochemical process, which involves cytokines, prostaglandins, peptide (relaxin) and steroid hormones (Taverne and Noakes 2009). This process is well coordinated with myometrial contractions, which are an outcome of the increased

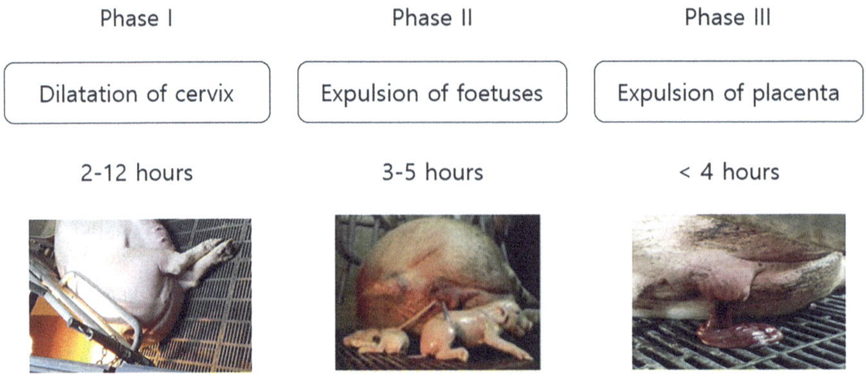

Fig. 4 Phases of farrowing in the pig. (Data refers to Peltoniemi et al. 2016, 2019)

oxytocin activity resulting from changes in the steroid environment described above (decreased progesterone and increased oestrogen). The second stage involves further straining of the abdominal muscles, rupture of the allantochorionic sac, and expulsion of the foetuses. In the contemporary sow, the average duration of this stage appears to be around 4–8 h with a 20-min interval between the foetuses being born (van Dijk et al. 2005; Oliviero et al. 2008b; Gu et al. 2011; Peltoniemi et al. 2019). The birth intervals are primarily influenced by oxytocin secretion, which is significantly increased during this stage, peaking after the birth of each individual foetus (Castrén et al. 1993; Oliviero et al. 2008a). During the third stage of farrowing, characterized by expulsion of foetal membranes, uterine contractions persist in a peristaltic manner, high in frequency but lower in amplitude compared with the second stage. This stage usually takes no longer than 4 h (Taverne and Noakes 2009). In the pig, foetal membranes are usually expelled in three to four parts, with an overlap of foetuses and membranes indicating some delay in the parturition process (Björkman et al. 2017).

3 Lactation

Nursing Behaviour

During parturition, colostrum secretions are continuously available for the litter. Across lactation, however, nursing events become coordinated by the sow (Rushen and Fraser 1989). She begins nursing by exposing her udder towards the piglets, adopting a side-lying posture and making a grunting sound to encourage piglet gathering. Nursing typically consists of three phases, including the pre-massage, milk let-down, and the post-massage periods, while milk let-down is not always included. At the end of the pre-massage period, sows use an increased grunting rate to signal the time for nutritive sucking to the piglets after a time lapse of approximately 20–25 s (Rushen and Fraser 1989). Piglets can only obtain milk from the sows at the milk let-down period. Thereafter, the piglets continue to proceed with the udder massage, which is performed during the post-massage period. This post-udder massage has been suggested as a means to acquire sufficient milk ejection from the sow at the subsequent nursing (Algers and Jensen 1985; Yun et al. 2013).

Aggressive Behaviour

Young sows, more typically than old sows, may show aggressive behaviour instead of maternal care, which is usually expressed towards piglets in the postpartum period. Sometimes, sows may even intentionally abuse their piglets to the point of injuring or killing them. Such "savaging" behaviour has been described as an important cause of neonatal piglet mortality, contributing to approximately 25% of total piglet mortality (Ahlström et al. 2002). The incidence of savaging behaviour is reportedly in the order of 7–12% (van der Steen et al. 1988). Savaging is more common in sows (gilts) during their first parity as they are more likely to be nervous following piglet birth and due to piglet movements (Johnson and McGlone 2011). However, sows that show aggressive savaging behaviour towards their piglets at their first farrowing are much more likely to be aggressive again at consecutive farrowings (Harris et al. 2003). Farrowing crates that prevent interactions between the sows and their offspring can also reportedly worsen this cannibalistic condition compared with loose housing systems (Jarvis et al. 2004). Furthermore, crate sows that are restless during the peripartum period may be more likely to become cannibalistic sows. Savaging behaviour is considered so harmful that sedatives have been prescribed in the pig industry to calm down sows showing such behaviour.

Is Maternal Behaviour Inherited?

Maternal care and behaviour have often been measured in terms of maternal responsiveness of the dam to stress signals, such as screaming by the neonatal offspring, which in the pig can be brought about, e.g., by separating piglets from their sows. Hellbrügge et al. (2008) observed, across lactation days 1–21, that the heritability of aggressive behaviour was much higher (0.32) compared to other behavioural traits such as a response to a playback of a piglet's stress call or a response to an unknown voice stimulus. In an earlier study, van der Steen et al. (1988) found that in addition to a considerable component of heritability, low birth weight predisposed sows to aggressive behaviour. Gene mapping exercises to investigate a more exact pattern of heritability linked to savaging or infanticide indicate certain candidate genes for this behaviour in chromosomes SSC2, SSC10, and SSCX (Quilter et al. 2007).

In an interesting study comparing sow lines from the 1970s to modern, highly prolific sow lines, Canario et al. (2014) found that the maternal behaviour of sows, along with the activity of neonatal piglets, has changed considerably over past generations as breeding has advanced. They found that reactivity in modern sows was associated with a higher risk of piglet death and that piglets from modern sow lines were less active after birth and also suffered from more complications such as respiratory distress (Canario et al. 2014).

Cross-Fostering in the Pig

Interestingly, sows may accept piglets from other litters to nurse and care for after parturition is complete. We successfully transferred 1-day-old European wild boar piglets to a first parity Yorkshire × Landrace hybrid sow after which the sow cared well for the piglets for weeks to come, and they were weaned successfully at the end of a lengthy lactation of 8 weeks (Peltoniemi et al. unpublished). In fact, cross-fostering in nature may appear within a nucleus group of a few sows. However, cross-fostering needs to occur within a day or so after birth to be successful. Price et al. (1994) cross-fostered piglets aged 1, 2, 4, or 7 days, with considerably lower rates of successful suckling events registered for piglets 2 days of age or older compared to those that were cross-fostered during their first day of life. Neonatal piglets 1 day of age quickly integrated into their new sow litter environment, containing a sow's original, own piglets in addition to the newcomers (Price et al. 1994). In a commercial setting of piglet production, cross-fostering is actually commonly practised to cope with the surplus of piglets in large litters, which would otherwise be left without a teat if left in the original litter they were born into (Peltoniemi et al. 2019).

4 Conclusions

Parental behaviour in sows mainly consists of prepartum nest-building behaviour, farrowing process, and lactational nurturing behaviour. Benefits of prepartum nest-building behaviour associated with hormonal support for the farrowing process and mammary development have been well documented. The ambient farrowing environment that can adequately address the needs for nest-building behaviour thus appears to play an important role in good parental behaviour of the sows during the parturition and lactation periods. In the meantime, since parental behaviour of the sows including nest-building is most significantly affected by litter size, it appears to need additional approaches with regard to degrees of adequate expression as well as environmental provision of sow behavioural needs in large litters. With respect to prepartum nest-building behaviour of the hyperprolific sow, future research should focus on its causal relationships with increased litter size, extended gestation periods, and enlarged body shape, and also their potential impacts on parturition and lactation performance.

References

Ahlström S, Jarvis S, Lawrence A. Savaging gilts are more restless and more responsive to piglets during the expulsive phase of parturition. Appl Anim Behav Sci. 2002;76:83–91.

Algers B, Jensen P. Communication during suckling in the domestic pig. Effects of continuous noise. Appl Anim Behav Sci. 1985;14:49–61.

Algers B, Uvnäs-Moberg K. Maternal behavior in pigs. Horm Behav. 2007; https://doi.org/10.1016/j.yhbeh.2007.03.022.

Björkman S, Oliviero C, Rajala-Schultz PJ, Soede NM, Peltoniemi OAT. The effect of litter size, parity, farrowing duration on placenta expulsion and retention in sows. Theriogenology. 2017;92:36–44.

Canario L, Bidanel JP, Rydhmer L. Genetic trends in maternal and neonatal behaviors and their association with perinatal survival in French Large White swine. Front Genet. 2014;5:410.

Castrén H, Alger B, De Passille AM, Rushen J, Uvnäs-Moberg K. Preparturient variation in progesterone, prolactin, oxytocin and somatostatin in relation to nest building in sows. Appl Anim Behav Sci. 1993;38:91–102.

Castrén H, Algers B, De Passillé AM, Rushen J, Uvnäs-Moberg K. Nest-building in sows in relation to hormone release. Appl Anim Behav Sci. 1994;40:74–5.

Cronin GM, Simpson GJ, Hemsworth PH. The effects of the gestation and farrowing environments on sow and piglet behaviour and piglet survival and growth in early lactation. Appl Anim Behav Sci. 1996;46:175–92.

Damm BI, Vestergaard KS, Schrøder-Petersen DL, Ladewig J. The effects of branches on prepartum nest-building in gilts with access to straw. Appl Anim Behav Sci. 2000;69:113–24.

Damm BI, Lisborg L, Vestergaard KS, Vanicek J. Nest-building behavioural disturbances and heart rate in farrowing sows kept in crates and Schmid pens. Livest Prod Sci. 2003;80:175–87.

Edwards SA, Fraser D. Housing systems for farrowing and lactation. Pig J. 1997;39:77–89.

Gu Z, Gao Y, Lin B, Zhong Z, Liu Z, Wang C, Li B. Impacts of freedom farrowing pen design on sow behaviours and performance. Prev Vet Med. 2011;102:296–303.

Gustafsson M, Jensen P, de Jonge FH, Illman G, Špinka M. Maternal behaviour of domestic sows and crosses between domestic sows and wild boar. Appl Anim Behav Sci. 1999;65:29–42.

Hales J, Moustsen VA, Nielsen MB, Hansen CF. Higher preweaning mortality in free farrowing pens compared with farrowing crates in three commercial pig farms. Animal. 2014;8:113–20.

Hansen KE, Curtis SE. Prepartal activity of sows in stall or pen. J Anim Sci. 1980;51:456–60.

Harris MJ, Li YZ, Gonyou HW. Savaging behaviour in gilts and sows. Can J Anim Sci. 2003;83:819–21.

Hellbrügge B, Tölle KH, Bennewitz J, Henze C, Presuhn U, Krieter J. Genetic aspects regarding piglet losses and the maternal behaviour of sows. Part 2. Genetic relationship between maternal behaviour in sows and piglet mortality. Animal. 2008;9:1281–8.

Herpin P, Hulin JC, Le Dividich J, Fillaut M. Effect of oxygen inhalation at birth on the reduction of early postnatal mortality in pigs. J Anim Sci. 2001;79:5–10.

Islas-Fabila P, Mota-Rojas D, Martínez-Burnes J, Mora-Medina P, González-Lozano M, Roldan-Santiago P. Physiological and metabolic responses in newborn piglets associated with the birth order. Anim Reprod Sci. 2018;197:247–56.

Jarvis S, Lawrence AB, McLean KA, Deans LA, Chirnside J, Calvert SK. The effect of environment on behavioural activity, ACTH, beta-endorphin and cortisol in prefarrowing gilts. Anim Sci. 1997;65:465–72.

Jarvis S, Reed BT, Lawrence AB, Calvert SK, Stevenson J. Pituitary-adrenal activation in pre-parturient pigs (Sus scrofa) is associated with behavioural restriction due to lack of space rather than nesting substrate. Anim Welf. 2002;11:371–84.

Jarvis S, Reed BT, Lawrence AB, Calvert SK, Stevenson J. Peri-natal environmental effects on maternal behaviour, pituitary and adrenal activation, and the progress of parturition in the primiparous sow. Anim Welf. 2004;13:171–81.

Jensen P. Observations on the maternal behaviour of free-ranging domestic pigs. Appl Anim Behav Sci. 1986; https://doi.org/10.1016/0168-1591(86)90105-X.

Jensen P. Nest building in domestic sows: the role of external stimuli. Anim Behav. 1993;45:351–8.

Jensen P, Broom DM, Csermely D, Dijkhuizen AA, Hylkema S, Madec F, Stamataris C, Von Borell E. The welfare of intensively kept pigs. Report of the Scientific Veterinary Committee of the European Union, 1997, p. 31–4.

Johnson AK, McGlone JJ. Behaviour genetics of the domestic pig. In: The genetics of the pig, vol. 21. Wallingford: CABI; 2011. p. 200–17.

Kaiser M, Jacobsen S, Haubro-Andersen P, Bækbo P, Cerón J, Dahl J, Escribano D, Jacobson M. Inflammatory markers before and after farrowing in healthy sows and in sows affected with postpartum dysgalactia syndrome. BMC Vet Res. 2018;14:83. https://doi.org/10.1186/s12917-018-1382-7.

Lawrence AB, Petherick JC, McLean KA, Deans LA, Chirnside J, Vaughan A, Clutton E, Terlouw EMC. The effect of environment on behaviour, plasma cortisol and prolactin in parturient sows. Appl Anim Behav Sci. 1994;39:313–30.

Manjarin R, Montano YA, Kirkwood RN, Bennet DC, Petrovski KR. Effect of piglet separation from dam at birth on colostrum uptake. Can J Vet Res. 2018;82:239–42.

Mayer JJ, Martin FD, Brisbin IL Jr. Characteristics of wild pig farrowing nests and beds in the upper Coastal Plain of South Carolina. Appl Anim Behav Sci. 2002;78:1–7.

Oliviero C, Heinonen M, Valros A, Hälli O, Peltoniemi OA. Effect of the environment on the physiology of the sow during late pregnancy, farrowing and early lactation. Anim Reprod Sci. 2008a;105:365–77.

Oliviero C, Pastell M, Heinonen M, Heikkonen J, Valros A, Ahokas J, Vainio O, Peltoniemi OAT. Using movement sensors to detect the onset of farrowing. Biosyst Eng. 2008b;100:281–5.

Oliviero C, Heinonen M, Valros A, Peltoniemi OAT. Environmental and sow-related factors affecting the duration of farrowing. Anim Reprod Sci. 2010;119:85–91.

Peltoniemi O, Björkman S, Maes D. Reproduction of group-housed sows. Porcine Health Manag. 2016; https://doi.org/10.1186/s40813-016-0033-2.

Peltoniemi O, Björkman S, Oropeza-Moe M, Oliviero C. Developments of reproductive management and biotechnology in the pig. Anim Reprod. 2019;16:524–38.

Peltoniemi O, Han T, Yun J. Coping with large litters: management effects on welfare and nursing capacity of the sow. J Anim Sci Technol. 2021;63:199.

Price EO, Hutson GD, Price MI, Borgwardt R. Fostering in swine as affected by age of offspring. J Anim Sci. 1994;72:1697–701.

Quilter CR, Blott SC, Wilson AE, Bagga MR, Sargent CA, Oliver GL, Southwood OI, Gilbert CL, Mileham A, Affara NA. Porcine maternal infanticide as a model for puerperal psychosis. Am J Med Genet B Neuropsychiatr Genet. 2007;144B:862–8.

Rushen J, Fraser D. Nutritive and nonnutritive sucking and the temporal organization of the sucking behavior of domestic piglets. Dev Psychobiol J Int Soc Dev Psychobiol. 1989;8:789–801.

Taverne MAM, Noakes DE. Pregnancy and its diagnosis. In: Noakes DE, Parkinson TJ, England GCW, editors. Veterinary reproduction and obstetrics. Saunders Elsevier: London; 2009. p. 76–122.

Thodberg K, Jensen KH, Herskin MS, Jørgensen E. Influence of environmental stimuli on nest-building and farrowing behaviour in domestic sows. Appl Anim Behav Sci. 1999;63:131–44.

Van der Steen HAM, Schaeffer LR, de Jong H, de Groot PN. Aggressive behavior of sows at parturition. J Anim Sci. 1988;66:271–9.

Van Dijk AJ, Van Rens BT, Van der Lende T, Taverne MA. Factors affecting duration of the expulsive stage of parturition and piglet birth intervals in sows with uncomplicated, spontaneous farrowings. Theriogenology. 2005;64:1573–90.

Vestergaard K, Hansen LL. Tethered versus loose sows: ethological observations and measures of productivity. I. Ethological observations during pregnancy and farrowing. Ann Rech Vet. 1984;15:245–56.

Yun J, Valros A. Benefits of prepartum nest-building behaviour on parturition and lactation in sows – a review. Asian-Australas J Anim Sci. 2015;28:1519.

Yun J, Swan KM, Vienola K, Farmer C, Oliviero C, Peltoniemi O, Valros A. Nest-building in sows: effects of farrowing housing on hormonal modulation of maternal characteristics. Appl Anim Behav Sci. 2013;148:77–84.

Yun J, Swan KM, Farmer C, Oliviero C, Peltoniemi O, Valros A. Prepartum nest-building has an impact on postpartum nursing performance and maternal behaviour in early lactating sows. Appl Anim Behav Sci. 2014a;160:31–7.

Yun J, Swan KM, Vienola K, Kim YY, Oliviero C, Peltoniemi OAT, Valros A. Farrowing environment has an impact on sow metabolic status and piglet colostrum intake in early lactation. Livest Sci. 2014b;163:120–5.

Yun J, Swan KM, Oliviero C, Peltoniemi O, Valros A. Effects of prepartum housing environment on abnormal behaviour, the farrowing process, and interactions with circulating oxytocin in sows. Appl Anim Behav Sci. 2015;162:20–5.

Yun J, Han T, Björkman S, Nystén M, Hasan S, Valros A, Oliviero C, Kim Y, Peltoniemi O. Factors affecting piglet mortality during the first 24 h after the onset of parturition in large litters: effects of farrowing housing on behaviour of postpartum sows. Animal. 2019; https://doi.org/10.1017/S1751731118002549.

Rabbit Maternal Behavior: A Perspective from Behavioral Neuroendocrinology, Animal Production, and Psychobiology

Gabriela González-Mariscal, Steffen Hoy, and Kurt L. Hoffman

Abstract Rabbit maternal behavior (MB) impacts meat and fur production on the farm, survival of the species in the wild, and pet welfare. Specific characteristics of rabbit MB (i.e., three-step nest building process; single, brief, daily nursing bout) have been used as models for exploring particular themes in neuroscience, like obsessive-compulsive actions, circadian rhythms, and cognition. Particular hormonal combinations regulate nest building by acting on brain regions controlling MB in other mammals. Nonhormonal factors like type of lodging and the doe's social rank influence nursing and milk production. The concurrency of pregnancy and lactation, the display of nonselective nursing, and the rapid growth of altricial young – despite a minimal effort of maternal care – have prompted the study of mother-young affiliation, neurodevelopment, and weaning. Neurohormonal mechanisms, common to other mammals, plus additional strategies (perhaps unique to rabbits) allow the efficient, adaptive display of MB in multiple settings.

Keywords Animal production science · Animal welfare · Behavioral neuroendocrinology · Behavioral neuroscience · Breeding · Lactation · Neuropharmacology · Neuropsychiatric disorders · Psychobiology · Rabbit

G. González-Mariscal (✉) · K. L. Hoffman
Centro de Investigación en Reproducción Animal, CINVESTAV-Universidad Autónoma de Tlaxcala, Tlaxcala, Mexico

S. Hoy
Department of Animal Breeding and Genetics Justus Liebig University Giessen, Giessen, Germany

G. González-Mariscal (ed.), *Patterns of Parental Behavior*, Advances in Neurobiology 27, https://doi.org/10.1007/978-3-030-97762-7_5

1 Introduction

Doe rabbits have been the object of study in reproductive neuroendocrinology since the early studies that explored the ways by which copulation induces reflex ovulation (for review, see Ramírez and Beyer 1988). Rabbits have also been investigated in relation to lactation, due largely to their peculiar nursing pattern (*circa* one short-lasting bout per day; Findlay and Roth 1970; Findlay and Tallal 1971; Mena et al. 1990a, b). The behavioral aspects of maternal behavior – which include nest building, in addition to nursing – were initially studied by the group of Zarrow and Denenberg in a series of works exploring the endocrine basis of such activities under laboratory conditions (Zarrow et al. 1961, 1963, 1965). Other investigators have observed the behavior of wild doe rabbits in the field or kept within outdoor enclosures (Hoy and Selzer 2002). A different perspective comes from the field of animal science, where investigators have sought to optimize the production of rabbit meat, while considering issues of animal welfare, cost of supplies, environmental factors, etc. Global rabbit meat consumption peaked in 2017 (Anonymous 2021), and it is expected to retain its growth in the short term, largely because of its recognized nutritional value (Hernández and Gondret 2006). Rabbits have also been kept as pet animals worldwide, from the dwarf breeds to the giants. Rabbit owners can be very committed to their pets, but there is little knowledge about the behavior of the different rabbit varieties, particularly in the home environment where they are kept. Recently, the field of psychobiology has incorporated the behavior of mother rabbits as a model to study motivated, hormone-dependent behaviors that are innate and adaptive to a changing environment (Hoffman and Rueda Morales 2009).

This background shows that rabbit maternal behavior constitutes a rich topic of study that has been addressed from diverse perspectives. Sadly, such information is scattered across a scientific literature that encompasses the fields of neuroendocrinology, animal science, comparative ethology, psychobiology, and behavioral endocrinology, among others. Consequently, researchers working on rabbits are often unaware of the findings of studies that were published outside their usual "niche." In this chapter, we have attempted to overcome this problem by presenting a broad view of rabbit maternal behavior with the following objectives: (i) to acquaint neuroscientists with specific aspects of the doe's behavior that can be used to explore particular functions of the nervous system, such as circadian rhythms, cognition, and innate motor patterns; (ii) to show that results coming from the laboratory can be *translated* into useful information for raising rabbits on the farm; (iii) to engage biologists into the investigation of the underpinnings of the mechanisms governing the behavior of wild animals; and (iv) to prompt the interpretation of laboratory findings within an ecological-evolutionary perspective.

2 The Doe's Behavior Throughout Pregnancy and Parturition: Mother-Offspring Activities During Lactation

Nest Building in the Laboratory, on the Farm, and in the Wild

Hormones and Brain Targets Regulating Specific Aspects of Nest Building

Rabbit does build a nest across pregnancy, where they will deliver the litter and nurse the kits. Nest building has been described in several domestic rabbit breeds, under a variety of housing conditions, in the laboratory and on the farm (Denenberg et al. 1963; González-Mariscal et al. 1994; Ross et al. 1956, 1963). Wild rabbits, and domestic breeds kept within outdoor enclosures, also engage in nest building during pregnancy (Deutsch 1957; Lloyd and McCowan 1968). This process begins by digging a burrow that, in nature, is located away from the rabbit colony and – ideally – in soft soil grounds not prone to flooding. Little is known about the complex cognitive processes that determine the selection of the maternal burrow site. Yet, it has been observed that high-ranking does have the best nest sites, a factor that greatly increases their fecundity and lifetime fitness (von Holst et al. 2002).

Nest building has been described and quantified under laboratory conditions in several domestic breeds, kept in cages within vivariums. The earliest studies were performed by the group of Zarrow and Denenberg, who compared the nest building process among different rabbit strains, determined the effect of maternal experience, and laid the ground work for the detailed exploration of the endocrine regulation of this behavior (Denenberg et al. 1958; Ross et al. 1956; Zarrow et al. 1961, 1963, 1965). Later studies (from the group of González-Mariscal, Rosenblatt, and Beyer) developed the simple (but reliable) methods that have allowed the investigation of the hormones and brain sites where these agents act to regulate the sequential expression of digging, straw-carrying, and hair-plucking, which are the three activities that make up the nest building process.

Figure 1 compares nest building across gestation (González-Mariscal et al. 1994) with the one induced in ovariectomized (ovx) females by subcutaneous (s.c.) injections of estradiol benzoate (EB) and progesterone (P; González-Mariscal et al. 1996). In both models, the onset of digging occurs, while both hormones (either endogenous or given exogenously) are present. The decline in (or removal of) P provokes a reduction (and eventual cessation) of this activity *and* an onset of straw-carrying. Hair-plucking emerges as straw-carrying declines (González-Mariscal et al. 1994, 2000), associated with a sudden rise in endogenous prolactin (PRL; McNeilly and Friesen 1978). Further support for a participation of PRL in straw-carrying and hair-plucking comes from the observation that antagonizing pituitary PRL release (by injecting bromocriptine) abolishes these activities (without modifying digging; González-Mariscal et al. 2004b). Testosterone (T), present in blood across pregnancy, promotes a specific aspect of nest building: hair-plucking (which

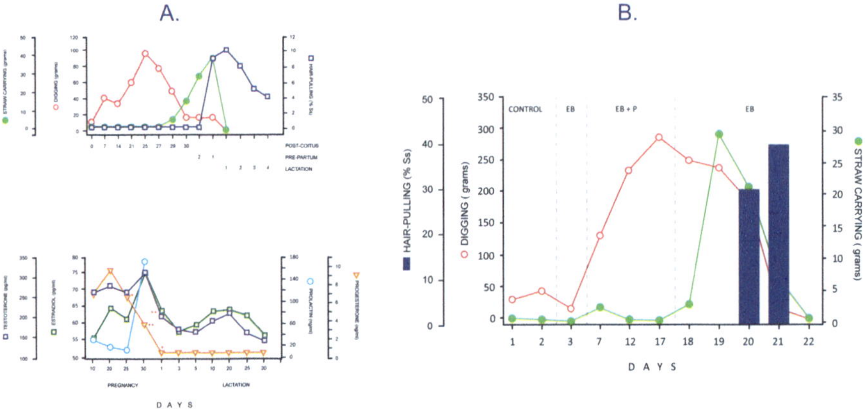

Fig. 1 Nest building observed in (**a**) pregnant does and (**b**) ovariectomized does given estradiol benzoate + progesterone. (Reproduced from (**a**) González-Mariscal et al. 1994. Physiol Behav. 55:1081–9; (**b**) González-Mariscal and Rosenblatt 1996)

occurs following hair-loosening). Its metabolite, 5-alpha-dihydrotestosterone, is even more potent in this regard (González-Mariscal et al. 2003).

The concurrence of pregnancy and lactation is common in the wild and on rabbit farms, as a consequence of mating at postpartum estrus or following the suppression of a nursing episode in early lactation (see following section). Under this condition, pregnant-lactating (PL) does have higher concentrations of estradiol in blood in early and midpregnancy than do pregnant-only rabbits. P levels, in contrast, are lower in pregnant-lactating does on pregnancy days 7 and 14 than in pregnant-only animals. These differences in the concentration of estradiol and P have a *quantitative and a qualitative* impact on the expression of nest building: while the *total* amount of digging is similar between PL and pregnant-only does, the former concentrate this activity at the end of pregnancy and dig little at the beginning. In contrast, PL does carry less *total* straw than do pregnant-only rabbits (González-Mariscal et al. 2009a).

Investigations into the brain sites where estradiol and P act to regulate nest building began by implanting these hormones in the anterior hypothalamus-preoptic area (González-Mariscal et al. 2005), a region that is rich in estradiol receptor alpha (ERα; Caba et al. 2003a) and P receptor (PR; Caba et al. 2003b). Figure 2 shows that digging is induced in ovx does by implanting EB (but not cholesterol) into the medial preoptic area (MPOA) *and* injecting P s.c. Digging declines when P injections are interrupted, but this, in turn, allows the onset of straw-carrying. These results indicate that the action of estradiol on the MPOA is necessary and sufficient to induce digging and straw-carrying *so long as P is present systemically and then removed*. In contrast, the location of the brain site(s) where P acts to regulate nest building has remained elusive: P implants into the MPOA of ovx does treated systemically with EB did not induce digging. Perhaps P has to act simultaneously at several brain sites to promote this activity.

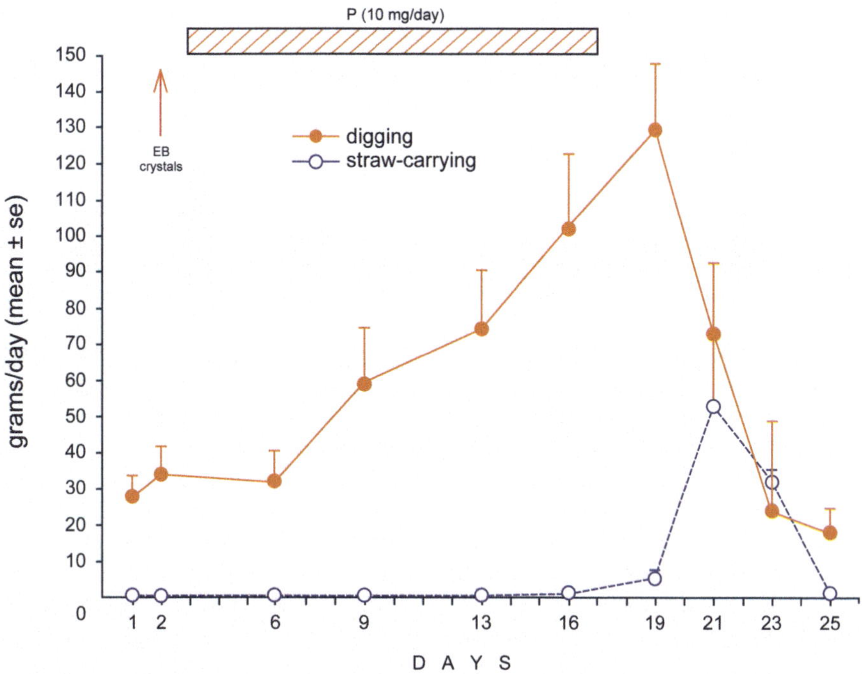

Fig. 2 Implants of estradiol benzoate (EB) into the medial preoptic area (MPOA) induce digging in ovx does injected s.c. with P. Removal of this hormone leads to a reduction in digging and onset of straw-carrying. (Modified from González-Mariscal et al. 2005. Horm Behav. 47:272–9)

The MPOA is pivotal for the expression of maternal behavior in several mammals (for review, see Numan et al. 2006). In rabbits, electrolytic lesions to the anterior hypothalamus-MPOA reduced digging, straw-carrying, and hair-plucking in *intact* nonpregnant does (exposed to their own ovarian estrogens) injected s.c. with P (Fig. 3a). In ovx rabbits given EB + P s.c., similar electrolytic lesions nearly abolished straw-carrying and hair-plucking, though digging was unaffected (Fig. 3b; Basurto et al. 2018). Taken together, these studies (González-Mariscal et al. 2005; Basurto et al. 2018) indicate that the straw-carrying and hair-plucking components of nest building require estradiol signaling within the MPOA and a rise and fall of circulating P, while P-responsive regions that promote digging behavior lie outside of the MPOA.

Effect of Environmental Factors (Housing, Available Materials, Seminatural Enclosures)

In 75% of all cases, nest building in wild rabbits starts one day or one night before kindling. In a third of all litters, nest building does not begin until 4–5 h before birth. Occasionally, collecting of nest material begins 5 days before parturition

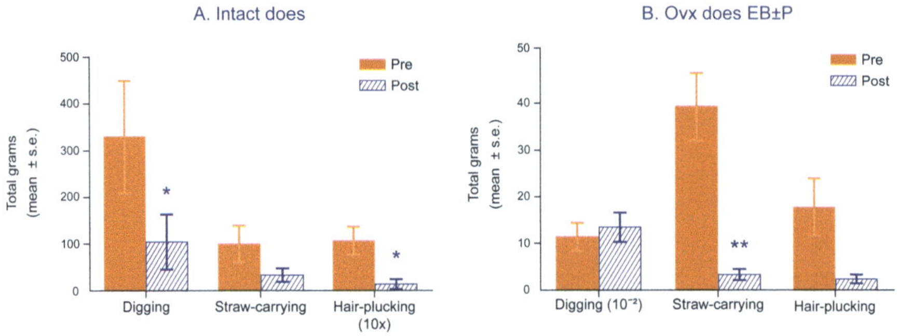

Fig. 3 (**a**) Electrolytic lesions to the anterior hypothalamus-POA of *intact* does injected s.c. with P significantly reduce digging, straw-carrying, and hair-plucking. (**b**) In ovx does given EB+P s.c. similar lesions nearly abolished straw-carrying and hair-plucking, but digging was unaffected. (Reproduced from Basurto et al. 2018. Horm Behav. 102:48–54)

(Selzer 2000). In preparation of kindling, wild rabbits use underground burrows with tunnels. These rabbit warrens can consist of a large number of tunnels (up to 150) and chambers (up to 111), and they are used over several generations (Parer et al. 1987). Rabbits carry hay, grass, straw, or other organic matter as nest material into the previously dug underground site – depending on what is available. In addition, the nest is padded with hair, which is plucked out in tufts from the ventrum, abdomen, and flanks. Wild rabbit does select among available nest materials, preferring dry grass over green, long grass rather than short, soft rather than hard (Hudson et al. 1996). Straw and hay resemble most the materials rabbits use in the wild (Hudson et al. 2000). In commercial rabbit farms, the most commonly used nest materials are wood shavings, hay, straw, wool or cotton waste, or even sawdust. In choice tests, most rabbit does preferred a commercial long thin wooden fiber nest material (Lignocel®), either alone or mixed with other nest materials (straw, hay). Straw and hay were less preferred than fiber nest material, respectively. The frequency of nest material carrying was highest on the day of parturition (Farkas et al. 2016a). The type of nest material used is important because the suckling kits can ingest it, in addition to the fecal pellets left by does at each nursing bout. The former can influence the kits' food choices after weaning, while the latter promotes the development of their gut flora (Hudson et al. 2000). Although the poorest quality nests were made using wood shavings, the type of nest material and the quality of the nest did not influence the productive performance of does (Farkas et al. 2016b). Yet, the quality of the nest can influence the survival of kits, and, indeed, the nest quality is improved with increasing experience of the doe (Ross et al. 1956; Verga and Luzi 2006). The season of the year, the breed, and the rabbit doe's diet do not seem to have any influence on nest building behavior and nest quality under farm conditions (literature in Schulte 1998).

Shortly before kindling, with the help of their snout and forelegs, the rabbit does create a hollow in the nest material. The nest building behavior of domestic rabbits is very similar to that of wild rabbits. While some wild rabbit does accept burrow tubes and nest boxes, offered in an outdoor enclosure, domestic rabbit does accept artificial nest boxes in 100% of cases. The time course of the nest building behavior of domestic rabbit does kept individually in concrete hutches or in get-away cages (i.e., enriched cages with an elevated platform) largely matches that of domestic rabbits in open-air enclosures (Selzer 2000). The lack of nest building behavior before parturition, or poor quality nest building, reflects a disordered maternal behavior. In such situations, the young are often born scattered outside the nest box and soon die of hypothermia (Denenberg et al. 1959).

Effect of Social Rank

Wild rabbits (and domestic breeds too) live in social groups that consist of at least one buck, one doe, and their offspring. In most cases, however, adult animals of both sexes live together in such associations (Gibb 1993). Within the group, there is a gender-separated hierarchy (von Holst 2001). To establish it, rabbits engage in agonistic interactions: they chase each other and the pursued animal may be bitten in the back. In most cases, the ranking fights in the wild go on without serious injury, as the inferior animals can retreat and evade from attacks and chases. In outdoor enclosures, on the other hand, due to the spatial restrictions and insufficient minimum distances, injuries and, in extreme cases, deaths as a result of aggressive confrontations can occur (Kraft 1976). In bucks, the rank order determines access to females ready to mate, and this is maintained over the entire reproductive period through fights. In the case of females, rank determines access to particularly good burrows. Dominant wild rabbit does choose the main burrow to build their nests; sometimes they allow the second highest doe to do so too (Selzer 2000). All other subdominant rabbit does are only allowed to place their nest burrows at some distance from the main burrow accommodating the colony (Gibb 1993). Fighting among females in wild rabbits is less common and is limited to the beginning of the breeding season (von Holst 2001). In wild and domestic rabbits kept in open-air enclosures, the relationships among adult females follow a linear hierarchy, and the buck is not involved in the interactions among the does (Hoy and Schuh 2004). The largest proportion of aggressive behaviors comes from the highest-ranking doe, while most defensive responses are shown by the lowest-ranking doe.

Investigations in a large outdoor enclosure showed that the social rank of rabbits had an impact on the success of reproduction: in years with a low population density, around 90% of all kits came from dominant males, while in years with a high population density (21–29 males, 24–34 females kept on 22,000 m^2), the progeny of dominant bucks represented only 60% of the total born. Dominant females had more litters – and thus also more kits – than subdominant individuals (von Holst 2001). This was due to two factors: an earlier start of annual reproduction and lower intrauterine losses compared to the inferior females. The social rank of individuals had a dramatic impact on life expectancy in wild rabbits: the higher the rank of males and females, the longer they lived.

Delivery, Placentophagia, Postpartum Estrus, and Their Management on the Farm

Duration of Kindling and Kit Survival in Relation to Maternal Social Rank

Kindling is announced by an increased restlessness of the doe – she often changes her whereabouts. During parturition, the doe sits in front of or in the nest and licks her genital region. Wild rabbits (Hoy 2009) and domestic breeds (González-Mariscal et al. 2013a) can give birth at any time during the day or night. The interval between the birth of the first and the last young of a litter (duration of kindling) is on average 12 min for wild rabbits (Hoy 2009) and up to 20 min for domestic ones (Hudson et al. 1999). At parturition, doe rabbits consume the fetal placentas and, as each kit emerges from the birth canal, they lick it (Melo and González-Mariscal 2003). Placentophagia represents a major – though transitory – change in the mother's eating habits as rabbits are strict herbivores (for review, see Lukefahr et al. 2021). When experimentally given the option to eat placenta or raw chicken liver, all does ate placentas *and* liver at the moment of parturition; this behavior continued until postpartum day 5 in approximately 50% of females. By contrast, at all other reproductive stages, the does would not eat either placenta or liver. After birth, the newborns dig into the nest material. Kits born outside the nest quickly try to join the huddle of their newly born siblings inside the nest. Immediately after giving birth, the doe leaves the litter. The does do not retrieve kits that stray away from the nest (Ross et al. 1959). In contrast to literature information (e.g., Seitz 1997; Schulte 1998), Selzer (2000) could not in any case determine that the nest or the tunnel was closed by the doe after giving birth, under natural conditions.

The mortality of nestlings from dominant does in wild rabbits is significantly lower than that of subdominant mothers. Dominant does have more and heavier kits at weaning, which can be explained by differences in lactation performance. Even after leaving the nest, the survival probability of the young of dominant females is better than that of subdominant mothers (von Holst 2001). The growth of the kits of dominant females is also significantly better than that of the offspring of inferior rabbits.

Mating Postpartum in the Lab, on the Farm, and in the Wild: Behavioral and Neuroendocrine Consequences for the Doe

If the buck has access to the nest or nest box, he is present at parturition, as has been observed in 24 h video recordings of wild rabbits kept in outdoor enclosures (Selzer 2000). Even immediately before kindling, an increasing interest of the buck in the doe can be observed. During kindling, he makes copulation attempts, which are hardly repelled by the doe. Immediately after the birth of the last kit, copulation is already successful. This behavior can be observed in both wild and domestic rabbits. In the former, the buck often waits for the doe, which has just finished kindling, at the entrance of the nest burrow. These observations indicate that, following delivery, doe rabbits enter a state of postpartum estrus, during which they can mate, ovulate, and become pregnant again. This has been documented in the wild, in the laboratory, and on the farm (Beyer and Rivaud 1969; Hammond 1925). Although the average duration of pregnancy is 31 days, the next litter (born from a lactating doe) lasts between 29 and 34 days (Selzer 2000). The concurrency of pregnancy and lactation accelerates and maximizes reproduction, strategies that favor the survival of a species that cannot fight predators but, rather, hides from them. Indeed, in wild rabbits, most adult females are concurrently pregnant-lactating during the breeding season (Brambell 1944). Rabbit breeders have tried to use postpartum estrus as a tool to accelerate their production of meat. However, the concurrency of pregnancy and lactation has a drawback: it provokes an early, abrupt decline in milk output from day 20 onward (González-Mariscal et al. 2009a; Lebas et al. 1972; Partridge et al. 1986). This, of course, provokes a lower body weight gain in the kits of the "first" litter, which is a serious inconvenience for rabbit breeders. To overcome this problem, does could, theoretically, be mated at around lactation day 10. Postponing postpartum mating by 10 days would "push" the decline in milk output to lactation day 30, when weaning (of the first litter) normally occurs. Yet, lactating females are in a state of lactational anestrus, during which they are not sexually receptive, they do not ovulate in response to the male mount, and they do not express chinning, a form of scent-marking displayed by estrous does (Beyer and Rivaud 1969; García-Dalmán and González-Mariscal 2012). To overcome this problem, rabbit farms have used the so-called "biostimulation" strategy to restore estrus and achieve ovulation in lactating rabbits. At some point between lactation days 9 and 12, does are separated from their litter for 48 h, after which suckling is allowed again. This is followed by natural mating or artificial insemination, which leads to a second pregnancy concurrent – but not aligned – with the first lactation (Castellini 2007; Fortun-Lamothe and Lebas 1996; Martínez-Gómez et al. 2004).

Despite the relative brevity of the mother-litter separation, "biostimulation" has specific neuroendocrine consequences on the lactating doe, which may underlie the restoration of estrus, specifically higher concentrations of prolactin and estradiol in blood compared with does in which lactation was not interrupted (Ubilla et al. 2000). Moreover, the reactivity of the doe to the stimuli received at mating is different among virgin, biostimulated, and lactating rabbits. This was determined by quantifying the number of cells immunoreactive (IR) to the c-FOS protein, commonly used as a proxy for visualizing regional brain activity, i.e., more c-FOS-IR cells indicate more neuronal activity. In the POA, copulation markedly increased (relative to unmated females) the number of c-FOS-IR cells in virgins but provoked no changes in biostimulated or lactating animals (Fig. 4a). By contrast, in the paraventricular nucleus (PVN), an increased number of c-FOS-IR cells was seen in virgins following mating, while a clear reduction occurred in biostimulated and lactating does (Fig. 4b; González-Mariscal et al. 2015).

The relevance of these findings for the expression of sexual receptivity and the doe's fertility warrants future investigation, as highlighted by Theau-Clèment et al. (2006).

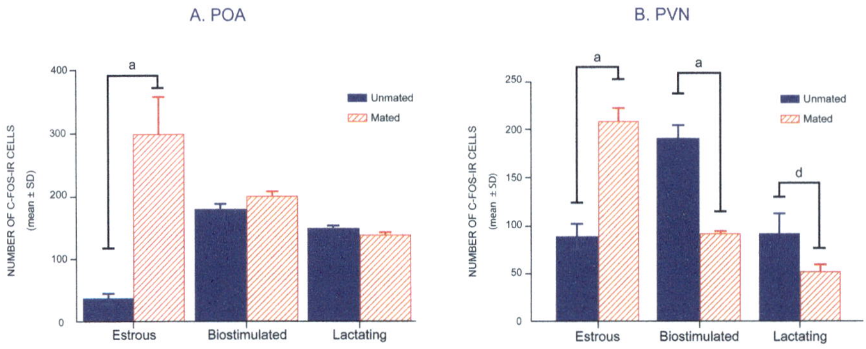

Fig. 4 Changes in the number of c-FOS immunoreactive (IR) cells observed after mating, relative to the doe's physiological state, in (**a**) preoptic area and (**b**) paraventricular nucleus. (Reproduced from González-Mariscal et al. 2015. Brain Res. 1608:66–74)

Fostering Kits

The brief contact between mother and kits, combined with a short nursing time (Hudson et al. 1996; Lincoln 1974; González-Mariscal et al. 2013b), provides potentially good conditions for fostering the kits in the case of domestic animals. Young rabbits can distinguish between their own mother and an alien one, largely via olfactory perception (Schaal et al. 2006; Val-Laillet and Nowak 2008). Yet, they can easily be raised by almost any lactating doe. The necessity of fostering and rearing by an unfamiliar doe, in the laboratory or on the farm, can ensue from (a) the death of the mother or (b) a litter with too many kits for the mother to raise. It is also ethically desirable to allow the newborn to be raised by another lactating doe, rather than letting them starve. Mothers readily adopt alien kits and nurse them in the same way as their own, in terms of latency to enter the nest box, milk output, and duration of nursing bouts. Moreover, this nonselective nursing is still observed regardless of the type of nest the litter is put into, namely, the mother's own nest, another doe's nest, or a "human-made" one, constructed with straw and artificial hair (González-Mariscal and Gallegos 2007).

Lactation: Behavioral Aspects

Periodicity of Nursing, Suckling Bout Duration, and Impact of Litter Size

The initiation and maintenance of lactation depends on the display of three behaviors: (i) the mother's willingness to enter the maternal nest holding the kits; (ii) her adoption of a crouching posture over them, which facilitates the kits' access to the maternal nipples; and (iii) the suckling stimulation provided by the young. Each of these behaviors is regulated by a complex array of sensory stimuli, hormones, and brain regions in both mother and kits. Specific evidence has shown that the hormones of pregnancy prime the maternal brain to respond "appropriately" to the new stimulus "encountered" at delivery, i.e., the kits. As described earlier, the changing levels of estradiol, P, and testosterone across pregnancy regulate the construction of the maternal nest. Close to parturition, as P levels decline, estradiol rises, and this – in turn – provokes a rise of PRL in blood (McNeilly and Friesen 1978), *before* the doe has received any suckling stimulation (Fig. 1a). This prepartum secretion of PRL is essential for initiating maternal responsiveness: injections of bromocriptine (a dopaminergic agonist that blocks PRL secretion) from *prepartum* day 26 to delivery prevent the onset of nursing behavior in 70% of primiparous does. By contrast, bromocriptine injections only across postpartum days 1–5 antagonize maternal behavior only in 20% of primiparous rabbits, although they do prevent milk production (Fig. 5; González-Mariscal et al. 2000).

Despite a major role of the hormones of pregnancy for *initiating* maternal behavior, are they sufficient to *consolidate* and *maintain* it from delivery onward? The intense contact of the doe with her kits occurring in the immediate postpartum period involves licking them, ingesting amniotic fluid, consuming placentas, and – consequently – receiving a plethora of tactile, olfactory, and gustatory stimuli that play a major role in consolidating her maternal responsiveness. If kits are removed as they emerge from the birth canal and given back to the mother on the following day, 70% of does will not behave maternally toward them (Fig. 6a). This is true of primiparous mothers, but experienced ones are not affected by such experimental procedure, a finding indicating that maternal experience somehow renders does less susceptible to disruptive factors. Contact with the litter in early lactation is also essential to maintain maternal responsiveness: anesthetizing mothers only during

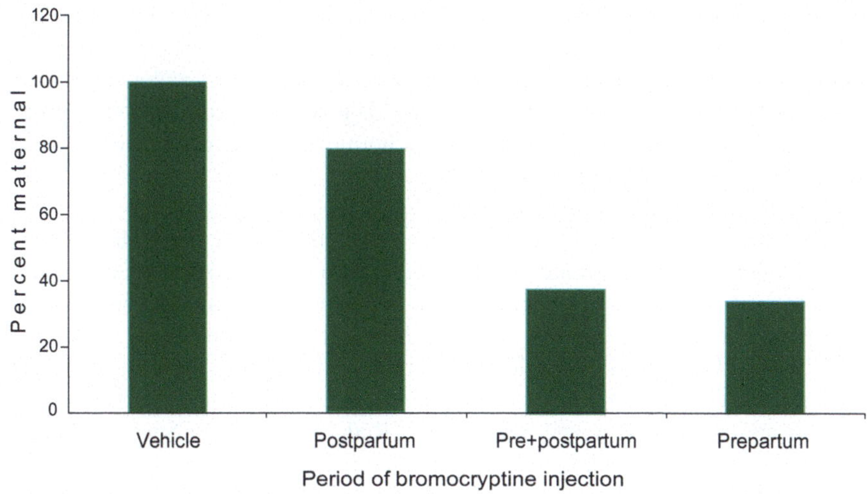

Fig. 5 Effect of injecting bromocriptine from prepartum day 26 to delivery, across postpartum days 1–5, or during both periods on the expression of maternal behavior in primiparous does, relative to females receiving vehicle. (Modified from González-Mariscal et al. 2000. J Neuroendocrinol 12:983–92)

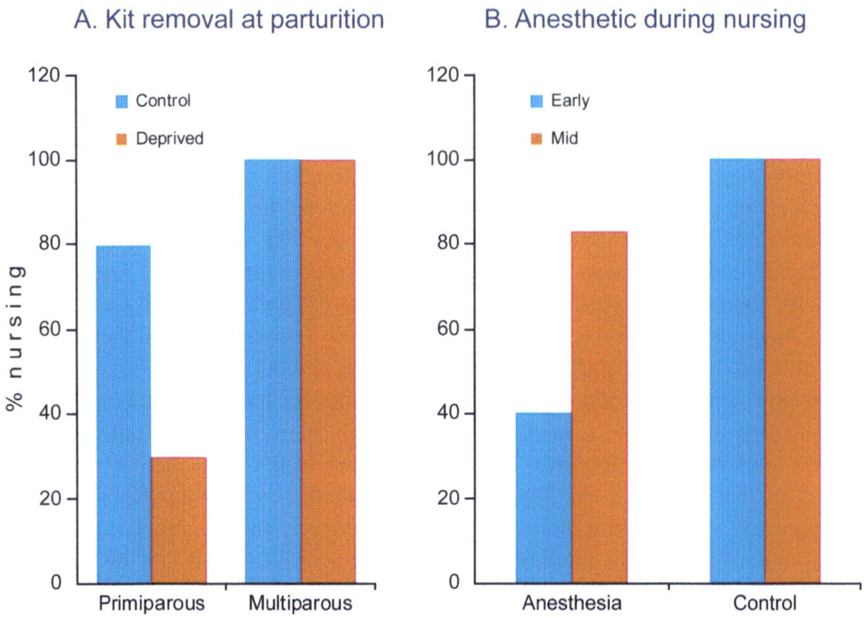

Fig. 6 (**a**) Effect of removing kits at parturition on the display of maternal responsiveness toward them on the following day: note differences between primiparous and multiparous rabbits. (**b**) Effect of injecting an anesthetic, to prevent the conscious perception of stimuli from the kits, across early or mid-lactation on maternal responsiveness 1–3 days following cessation of anesthetic. (Reproduced from González-Mariscal et al. 1998. Dev Psychobiol. 32:101–11)

the suckling bout across lactation days 2–7 reduces the proportion of does behaving maternally on the following days. This procedure is ineffective when performed in mid-lactation (Fig. 6b; González-Mariscal et al. 1998). Together, the above evidence supports the idea that a "maternal brain" is gradually "built" from late pregnancy into early lactation by a combination of hormones and somatosensory stimuli that allow does to recognize kits as a "meaningful" stimulus and to display the "adequate" set of behaviors toward them.

What does an "adequate" maternal behavior entail? Several studies have documented that in various domestic breeds, kept under a controlled photoperiod and a constant food supply, nursing occurs with a frequency approximating once/day in most does (Drewett et al. 1982; González-Mariscal 2007; Lincoln 1974; Matics et al. 2004). If kept in outdoor enclosures, domestic strains and wild rabbit does also nurse at a frequency of approximately once/24 h, with most nursing episodes occurring during darkness (Hoy and Selzer 2002; Rödel et al. 2012). Such a reliable, predictable nursing periodicity relies on numerous factors of control that are only beginning to be unveiled. However, it is already clear that a major contributing factor is the stimulation the doe receives during suckling. In New Zealand White does, nursing frequency is close to once every ca. 24 h if the suckling litter includes at least six kits. Nursing typically occurs during the dark phase, approximately 5 h

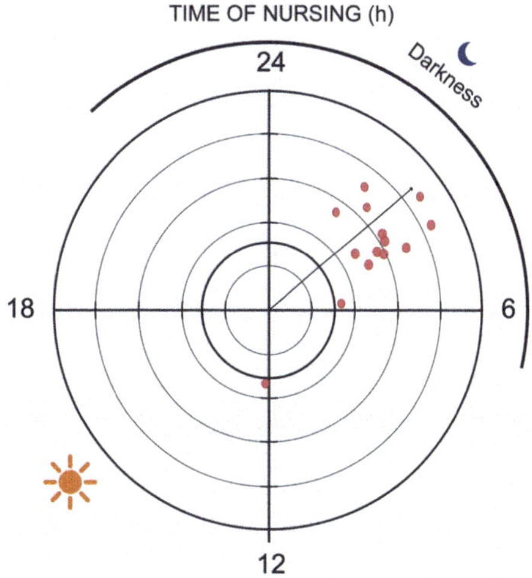

Fig. 7 A Rayleigh analysis performed on the nursing behavior of 15 does across lactation days 1–15 revealed a vector indicating the most likely time of nursing. The characteristics of this vector were phase angle =58°, rho = 0.78, $p < 0.001$. (Reproduced from González-Mariscal et al. 2013a. Chronobiol Int. 30:711–8)

before lights-on (Fig. 7). Smaller litters (i.e., 4 kits or less) lead to several entrances of the doe into the nest box across the day and, eventually, to a loss of maternal behavior (González-Mariscal et al. 2013a). Interestingly, wild rabbit does kept within an enclosure also show several entrances into the nursing burrow if they nurse small litters (Rödel et al. 2012).

Milk output is not essential for the display of *nursing behavior*: non-lactating *virgin* rabbits induced to behave maternally toward a foster litter by lesions to the accessory (González-Mariscal et al. 2004a) or the main (Chirino et al. 2007) olfactory systems enter the nest box, adopt a crouching posture over the kits, and allow suckling (Fig. 8a, b). These findings agree with the early report (Findlay and Roth 1970) that removing milk *under anesthesia* before does are exposed to their litter *does not* prevent them from entering the nest box, crouching over the kits, and allowing suckling. Thus, although normally concurrent, nursing behavior and milk output can be experimentally dissociated to reveal different mechanisms of control (see next section).

The above findings indicate that removing a tonic inhibitory influence (exerted by the olfactory systems) is sufficient to promote nursing behavior in a variable proportion of virgin rabbits that are – obviously – unable to produce milk. Yet, the conditions used in those studies involved introducing the kits into the nest box *only* at the time of testing. This precluded the expression of "spontaneous" nursing. However, a later study in which anosmic virgins were exposed to (foster) kits throughout the day and observed uninterruptedly for 2 weeks determined that nursing behavior occurred with circadian periodicity in 55% of cases (González-Mariscal et al. 2015). Together, the above findings indicate that the "fine tuning" of nursing behavior involves the hormones of pregnancy and the suckling stimulation from the kits.

The single daily nursing bouts displayed throughout lactation are short, lasting around 3–5 min in all studied breeds (Drewett et al. 1982; González-Mariscal et al.

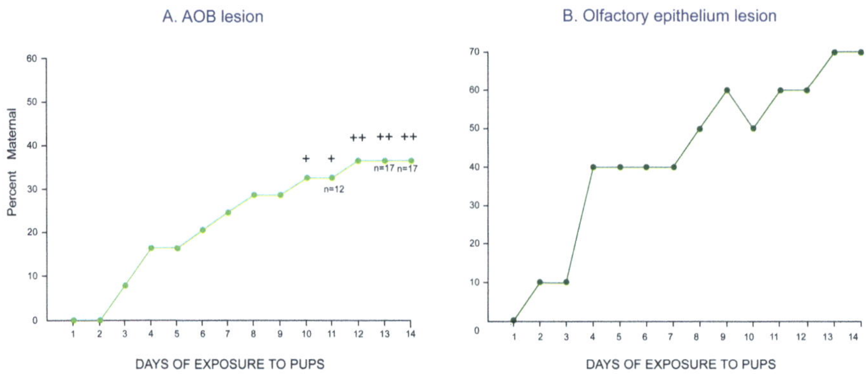

Fig. 8 Cumulative percentage of virgin rabbits that show nursing behavior toward a foster litter following lesions to the accessory olfactory bulbs (**a**) or the olfactory epithelium (**b**). (Reproduced from (**a**) González-Mariscal et al. 2004a. Behav Brain Res. 152:89–95; (**b**) Chirino et al. 2007. Behav Brain Res. 180:127–32)

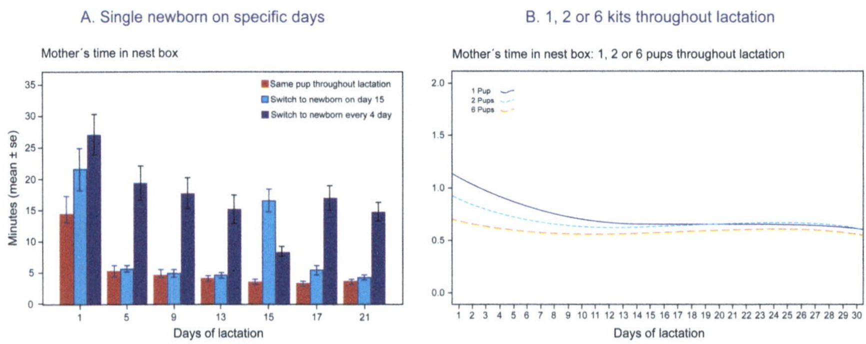

Fig. 9 Effect of providing (**a**) a single newborn on specific days of lactation; (**b**) 1, 2, or 6 continuously growing kits, on the time the mother spends inside the nest box. (Reproduced from González-Mariscal et al. 2013b. Dev Psychobiol. 55:809–17)

1994, 2013b; Lincoln 1974), so long as a threshold amount of suckling stimulation is received. The amount of time the mother spends inside the nest box on a given day critically depends on the number and age of the kits suckled *on that day*. Thus, does given newborn kits for suckling spend a longer time inside the nest box, regardless of the day of lactation when this manipulation is made (Fig. 9a). Moreover, does spend a longer time inside the nest box if they suckle one or two kits (Findlay and Tallal 1971) than if they suckle more, even if the kits are growing as lactation progresses (Fig. 9b; González-Mariscal et al. 2013b).

Despite its short duration, each suckling bout induces a massive secretion of oxytocin (OT); the magnitude of this secretion is directly related to the number of suckling kits (Fuchs and Wagner 1963). Accordingly, the number of OT-immunoreactive (IR) neurons, located in the PVN and supraoptic nucleus (SON), increases across pregnancy and is greatest in lactating does (Caba et al.,

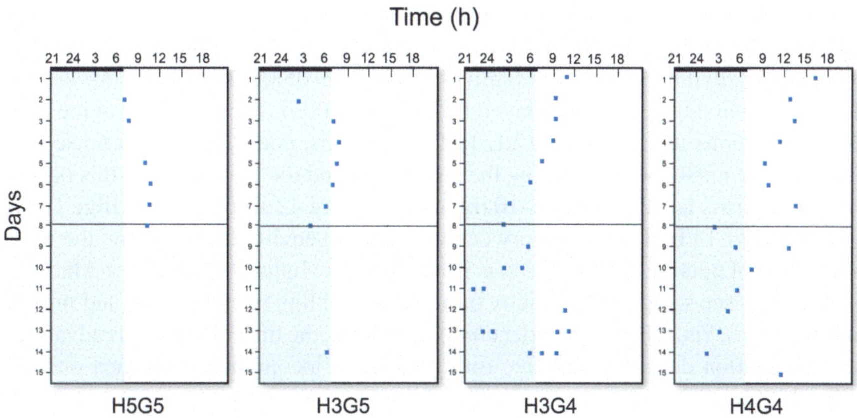

Fig. 10 Actograms of four lactating rabbits that showed either a loss of nursing behavior (H5G5 and H3G5) or a disruption of nursing periodicity (H3G4 and H4G4) following bilateral lesions to the PVN. The dots represent entrances into the nest box; the black bars on top indicate darkness. Note that, before the lesions, mothers entered the nest box only once/day. After lesions (indicated by the horizontal lines), the pattern of nest box entrances/day was disrupted or lost. (Reproduced from Domínguez et al. 2017. Eur J Neurosci 46:2133–40)

1996). Yet, the relevance of OT for the expression of rabbit maternal behavior is still uncertain, even though OT binding sites have been detected in the prefrontal cortex, preoptic area (POA), and lateral septum of does across several stages of the reproductive cycle (Jiménez et al. 2015). Moreover, bilateral lesions of the PVN disrupt nursing or alter its circadian periodicity (Fig. 10; Domínguez et al. 2017).

However, since the PVN contains a plethora of peptides and neurotransmitters in addition to OT (Sofroniew 1985) and the chemical identity of the cells lesioned in the above study was not established, it remains to be determined whether the effects observed can be attributed to damage to OT neurons.

Photoperiod and "Clock Genes"

Rabbits are crepuscular mammals, as supported by the daily fluctuations reported for a variety of behavioral, neuroendocrine, and physiological parameters measured in this species (for review, see Aguilar-Roblero and González-Mariscal 2020). The nursing behavior of both wild and domestic rabbits has been investigated in does kept in outdoor enclosures. The observations of Seitz (1997) and Selzer (2000) – reported in two PhD theses – indicate that most of the nursing events (84.0% in wild rabbits and 85.8% in domestic rabbits) take place after the onset of dusk: 3–6 h in wild rabbits and 2 h in domestic rabbits. Only a small rise in nursing activity was seen in the early morning. Wild rabbits spend the time between dawn and dusk mainly in the warren, without food and water and probably without the possibility of urination or defecation. They leave the burrows at the onset of dusk, when they start eating and elimination. After these activities, they nurse their kits (Selzer 2000).

When kept under artificial lighting conditions, more than 25% of the nursing events displayed by domestic does occur within the first 2 h of darkness. If the light (L)-dark (D) rhythm (12 L:12 D) is shifted by 1 h (from 05:00–17:00 to 06:00–18:00), the peak in nursing activity is delayed by 1 h (Hoy 2006). These findings agree with the observations that (i) under 12 L:12 D conditions, does housed in a noise-free environment nurse the litter during the dark phase and the time at which this occurs advances across lactation days 1–10 (by an average of 42 ± 16 min/day; Jilge 1993) and (ii) under 14 L:10 D laboratory conditions, with environmental noise, the most likely time of nursing occurs between 5 and 6 h before lights-on (González-Mariscal et al. 2013a; see section "Periodicity of nursing, suckling bout duration, and impact of litter size"). Yet, when kept under continuous light, the time of nursing is advanced across lactation days 7–8 and free-runs parallel to locomotion from then onward (Jilge 1995).

In addition to photoperiod *duration*, the intensity and color of light can also impact the proportion of sexually receptive does, the kindling rate, and the growth of the litter (Szendrö et al. 2016; Wu et al. 2021). Future studies are needed to explore the ways by which the characteristics of light impact these aspects of the doe's reproduction.

As suckling induces PRL secretion (Fuchs et al. 1984; see section "Kits' behavior") largely through the inhibition of tuberoinfundibular dopamine (TIDA) neurons (Gregerson 2006), the possibility that circadian nursing modifies the expression of the *per1* "clock gene" in those cells was investigated by Meza et al. (2011). Maximal levels of the PER1 protein were, indeed, found in TIDA neurons at 4 and 8 h post-nursing. Later studies (Aguirre et al. 2017) investigated whether dopaminergic cells of the A10 group and neurons of the mesolimbic system become activated in *anticipation* of "the next" nursing episode, concomitant with the mother's increased locomotor activity at such time (Jilge 1995). c-FOS protein expression on lactation day 7 was unmodified 12 h before nursing, but it increased immediately before the "usual" nursing time and was still elevated 90 min later in the ventral tegmental area (VTA), nucleus accumbens (core and shell), prelimbic and cingulate cortex, lateral septum (pars ventralis), and the A10 dopaminergic cells. These findings agree with the idea that efferents from the MPOA to the VTA play a major role in regulating maternal motivation in mammals (Numan and Stolzenberg 2009; Olazábal et al. 2013). Indeed, PER1 protein expression shifts across the day in the MPOA in relation to the time of scheduled nursing (Meza et al. 2015). Moreover, changing the time of nursing provokes a concomitant change in the time of maximal PER1 expression in the PVN, though not in the suprachiasmatic nucleus (SCN; Meza et al. 2008). These findings coincide with the observations that suckling (i) activates the PVN (but *not* the SCN), as per c-FOS protein expression (González-Mariscal et al. 2009b), and (ii) induces PER1 protein in OT cells from the PVN (Meza et al. 2008). Furthermore, does given kits at 18 or 24 h after the previous nursing showed more c-FOS-IR cells in the MPOA and lateral septum than rabbits not provided with a litter on the same day; these effects were not observed in the SCN (Jiménez and

González-Mariscal 2019). Interestingly, does given kits at 6 h after the previous suckling (but not mothers given a litter at 3 h) displayed nursing behavior, but the number of c-FOS-IR cells remained unchanged in all brain regions explored. Together, these pieces of evidence open the possibility that a suckling-entrained oscillator – not involving the SCN – may operate to regulate the periodicity of nursing. They also indicate that the responsiveness of the neural circuits regulating maternal motivation increases relative to the time elapsed since the previous suckling stimulation, being maximal 24 h later.

Domestic Breeds vs. Wild Animals

By using infrared video and time lapse recordings of wild (n = 6) and domestic (n = 8) does (kept in outdoor enclosures or in cages), a mean frequency of 1.2 and 1.1 nursing events/24 h, respectively, was found (summarized in Hoy 2006 – Table 1). The mean litter size in wild rabbits was 6.8 and in domestic rabbits 8.3 kits.

The highest nursing frequency combined with the lowest mean duration of a nursing event took place in the second week of lactation (Seitz 1997; Schulte and Hoy 1996; Selzer 2000; Table 2).

Pet breeders have reported differences in the frequency and duration of nursing events across rabbit varieties. The heaviest ones, e.g., German Widder Blue (mean weight of does: 5.5 kg, 1.67 nursings/24 h) and Light Large Silver (5.0 kg, 1.76 nursings/24 h), showed a higher frequency of nursing than New Zealand White or Red (4.5 kg, 1.26, 1.30 nursings/24 h, respectively), Rhoen rabbits (3.0 kg, 1.0 nursing/24 h), and Widder Dwarf (1.7 kg, 1.59 nursings/24 h). The mean duration of nursing bouts was longest in does of the German Widder Blue and Light Large Silver breeds (226 and 223 s, respectively) and shortest (193 s) in Widder Dwarfs (Selzer 2000). New Zealand White does showed a longer mean nursing bout duration (205 s) compared with ZIKA hybrids (193 s).

Table 1 Frequency of nursing events recorded across intervals of 24 h in wild and domestic rabbits kept in two free-range areas (modified from Selzer 2000)

	Number of 24 h intervals[a]	Frequency of nursing events in 24 h[a]	Number of does[b]	Frequency of nursing events in 24 h[b]
		mean ± s.d.		mean ± s.d.
Wild rabbits[c]	104	1.28 ± 0.54	6	1.26 ± 0.20
Domestic rabbits[c]	257	1.12 ± 0.49	8	1.16 ± 0.12
		$P < 0.05$		$P > 0.05$

[a]Derived from all 24 h intervals recorded: n = 104 in wild rabbits; n = 257 in domestic rabbits
[b]Derived from the average nursing frequency of all does studied in each group: n = 6 in wild rabbits; n = 8 in domestic rabbits
[c]Average litter size was 6.8 in wild rabbits and 8.3 in domestic rabbits

Table 2 Duration and frequency of nursing events in wild and domestic rabbits depending on week of lactation (Selzer 2000; Selzer et al. 2001)

	Wild rabbits (n = 6)			Domestic rabbits (n = 8)		
Week of lactation	Duration of nursing event (s)	Percentage of days with ≥2 nursing events (%)	Frequency of nursing events/24 h	Duration of nursing event (s)	Percentage of days with ≥2 nursing events (%)	Frequency of nursing events/24 h
1	184.4 ± 30.3[d]	21.2	1.24	229.9 ± 56.9[c]	9.2	1.09
2	169.2 ± 35.2[d]	44.8	1.48	200.5 ± 32.0[c]	22.2	1.27
3	185.0 ± 42.0	34.8	1.35	205.8 ± 36.3	15.1	1.15
4	186.3 ± 21.2	10.5	0.95	211.9 ± 30.4	2.8	0.99
Means ± s.d.	178.5 ± 34.4[b]		1.28[a] ± 0.54	211.8 ± 41.6[b]		1.12[a] ± 0.49

Means with the same letters (a, b, c, d) are significantly different from each other (P < 0.05)

Table 3 Mean frequency of nursing events/24 h depending on size and structure of cages, considering all 24 h observation periods (n = 400) (Selzer 2000)

	Unstructured cages[a] (20 does)		Structured cages[b](19 does)	
Relative cage size	Number of 24 h periods studied	Frequency of nursing events in 24 h means ± s.d.	Number of 24 h periods studied	Frequency of nursing events in 24 h means ± s.d.
Onefold	43	1.37 ± 0.65	61	1.32 ± 0.53[d]
Twofold	92	1.26 ± 0.48	91	1.25 ± 0.46
Threefold	85	1.25 ± 0.52	97	1.11 ± 0.31[d]
Total	220	1.29 ± 0.53[c]	249	1.21 ± 0.44[c]

[a]Onefold cage dimensions: width = 50 cm, depth = 70 cm, height = 70 cm
[b]Same dimensions, elevated platform made of plastic slats, tube at the entrance to the nest box
[c]No significant differences between the means of unstructured and structured cages
[d]Significantly different from each other (P < 0.01)

Effect of Different Housing Systems

Table 3 shows the effect of cage size and type (i.e., with or without an elevated platform within the cage) on the number of nursing events observed in 24 h. Significant differences were noted in relation to cage size, i.e., onefold vs. threefold, though only in structured cages, i.e., those provided with an elevated platform: the does climb onto it more frequently after the kits have left the nest boxes. As the kits become older (21 days of age) and start hopping onto the platform, the does' preference for it decreases (Selzer et al. 2001).

Milk Production Across Early, Middle, and Late Lactation (with and Without Concurrent Pregnancy)

Impact of Breed, Litter Size, and Environmental Factors

The daily milk yield is usually determined after the mother has been separated from the young overnight and, on the following morning, the kits or the doe are weighed individually *before suckling*. Immediately after the end of the nursing event, the kits (or the doe) are weighed again individually. The difference in weight (i.e., an increase in litter weight or a decrease in doe's) indicates the total milk yield in a day. The first nursing bout takes place, on average, 12–24 h postpartum. The milk yield of rabbit does depends on several factors. The most important ones are rabbit breed and week of lactation (Iraqi et al. 2007; Savietto et al. 2013, 2014; El Nagar et al. 2014; Pałka et al. 2017), age and size of the litter (Schlolaut 1998, Ludwiczak et al. 2020), doe's body weight (Schlolaut 1998), number of teats (Mohamed and Szendrö 1992), nutrition of the doe (Pascual et al. 1996, 2003; Xiccato et al. 1995), climatic conditions (Rafai and Papp 1984; Askar and Ismail 2012), and the concurrency of pregnancy and lactation (González-Mariscal et al. 2009a; Hudson et al. 1996; Lebas et al. 1972; Partridge et al. 1986). Table 4 reviews the impact of rabbit breed, week of lactation, age/parity/litter number, litter size, number of teats, and ambient temperature on milk production.

Main Hormones Involved

As in all mammals, OT is the main hormone that allows milk output, by promoting the contraction of the myoepithelium surrounding the milk ducts (Wakerley 2006). Table 5 shows that the concentration of OT in plasma increases by around threefold (relative to baseline levels) already at 1 min after the beginning of suckling, reaches maximal levels at 30 min, and declines to baseline by 60 min. PRL (the main hormone that promotes milk synthesis; Neville 2006) also increases above baseline already at 1 min after suckling initiation, reaches maximal concentrations at 10 min, and remains at practically the same levels at 60 min. The magnitude of the secretion of both OT and PRL also depends on the size of the litter and the stage of lactation (Fuchs et al. 1984).

Weaning

The ways by which maternal behavior ends have been far less studied than the mechanisms that mediate its onset. Most mammals display an intense interaction with their young and several nursing episodes/day during early lactation. Later on,

Table 4 Factors influencing milk yield in rabbit does

<table>
<tr><th>Factor, author(s)</th><th colspan="5">Effect</th></tr>
<tr><td colspan="6">Genotype</td></tr>
<tr><td>McNitt and Lukefahr (1990)</td><td colspan="5">Genotype of doe tended (P > 0.07) to be important for milk yield; Californian does had higher production than does of other breeds</td></tr>
<tr><td>Schlolaut (1998)</td><td colspan="5">In Angora rabbits, the longer the hair grows, the less feed they ingest and the milk yield is lowered</td></tr>
<tr><td>El Nagar et al. (2014)</td><td colspan="5">Differences among three Spanish maternal lines (A, V, LP)</td></tr>
<tr><td>Ludwiczak et al. (2020)</td><td colspan="5">Does of large and heavy breeds produce more milk than those of small, light breeds</td></tr>
<tr><td colspan="6">Week of lactation</td></tr>
<tr><td>Schley (1985)</td><td colspan="5">Milk yield of the doe increases to around 100–150 g per day in the first 2 weeks of life of the kits. It doubles by the third week, reaching its peak (up to 300 g) between the 18th and 23rd day after the birth of the kits</td></tr>
<tr><td>Maertens et al. (2006)</td><td colspan="5">In multiparous does subjected to intensive reproduction rhythms (mating within 3 days after parturition), the milk output peaks 2–3 days earlier</td></tr>
<tr><td>Lebas et al. (1972)</td><td colspan="5">From the 22nd day postpartum, and especially after the 28th day of lactation, the milk yield of pregnant does decreased very rapidly</td></tr>
<tr><td>Lebas et al. (1972), Xiccato et al. (1995)</td><td colspan="5">The intensive reproduction rhythm (mating the does within 3 days postpartum) decreases the production of milk even after 17–19 days of lactation. A semi-intensive reproduction rhythm (mating on postpartum day 11) allows them to maintain high production of milk even after 25 days</td></tr>
<tr><td>Seitz (1997)</td><td colspan="5">Individual milk intake of kits per suckling event increases up to the 2nd week of lactation and decreases until the 4th week</td></tr>
<tr><td rowspan="5">El Nagar et al. (2014)</td><td colspan="5">Milk output: grams/week</td></tr>
<tr><td>Line</td><td>Week 1</td><td>Week 2</td><td>Week 3</td><td></td></tr>
<tr><td>A</td><td>872</td><td>1.503</td><td>1.865</td><td>Three Spanish</td></tr>
<tr><td>V</td><td>919</td><td>1.633</td><td>2.004</td><td>Maternal</td></tr>
<tr><td>LP</td><td>1.043</td><td>1.819</td><td>2.254</td><td>Lines</td></tr>
<tr><td colspan="6">Age/parity/litter number</td></tr>
<tr><td>Schlolaut (1998)</td><td colspan="5">In does inseminated 1–2 days after kindling and with free intake of concentrated feed, the milk yield increased up to an age of 16 months (i.e., up to the 8th litter) and remained at about the same level up to an age of 25 months</td></tr>
<tr><td>Zimmermann (1990)</td><td colspan="5">Does in their first litter have a lower milk production than in later litters</td></tr>
<tr><td>Schulte (1998)</td><td colspan="5">Significant relationship between litter number and mean milk intake per young and suckling bout. Milk intake increases by 30–50% from 1st to 3rd litter; from 4th litter onward, milk intake of the kits decreases</td></tr>
<tr><td>El Nagar et al. (2014)</td><td colspan="5">No effect of parity on milk production traits</td></tr>
<tr><td colspan="6">Litter size</td></tr>
<tr><td>Schlolaut (1998)</td><td colspan="5">With increasing litter size (until 8–9 kits), the milk yield increases</td></tr>
</table>

(continued)

Table 4 (continued)

Factor, author(s)	Effect
Seitz (1997)	With increasing litter size, the suckling frequency increases and the individual milk intake per suckling event decreases
Ludwiczak et al. (2020)	Daily milk production is higher for does nursing 10 kits per litter compared to those nursing 8 kits per litter
Number of teats	
Mohamed and Szendrö (1992)	Milk yield seems to increase with a larger number of teats
High temperature	
Rafai and Papp (1984), Askar and Ismail (2012)	High temperature impairs milk yield in rabbit does
Szendrö et al. (1998), Fernández-Carmona et al. (2003)	The influence of high temperature on milk production was studied under constant (high) ambient temperature in climatic chambers
Szendrö et al. (1998), Fernández-Carmona et al. (2003)	Under natural conditions, the impact of heat stress on doe milk yield is reduced because there is varying temperature between day and night, with relatively low temperatures early in the morning and at night, when rabbits consume most of the daily feed intake
Maertens and De Groote (1990), Pascual et al. (1996)	Effect of heat stress on milk yield under natural conditions: the higher the outside temperature, the more intense are the decline in feed intake and, thus, milk yield
Bakr et al. (2015)	Rabbit does kept either in a comfort room with the temperature kept constantly between 18 °C and 22 °C or in a heat room, where the environmental temperature pattern mimicked the daily temperature cycles characteristic of summer in Mediterranean countries: no significant difference for daily milk yield between rooms
Mahmoud (2013)	Under Egyptian conditions: milk yield was significantly increased during winter compared with summer season and was significantly higher with increasing doe's body weight
Szendrö et al. (2018)	Four climatic conditions were compared: temperatures of 5, 15, 23, or 30 °C. Heat stress, provoked by a constant temperature of 30 °C, reduced milk yield (148, 152, 150, and 106 g/day, resp.)

Table 5 Plasma concentrations of oxytocin (OT; pg/ml) and prolactin (PRL; ng/ml) at different times post-suckling[a, b, c]

		Time after suckling began (min)									
Baseline		1		3		10		30		60	
OT	PRL	OT	PRL	OT	PRL	OT	PRL	OT	PRL	OT	PRL
19 ± 11	4 ± 2	57 ± 20	37 ± 19	112 ± 38	49 ± 15	44 ± 19	73± 15	59 ± 26	61 ± 20	26 ± 11	70 ± 22

[a]Lactation day 9
[b]Litter size = 5–7 kits
[c]Data show means ± SE; derived from Fuchs et al. (1984)

the mother-young interactions become more sparse, and the number of daily suckling bouts gradually decreases (for reviews, see González-Mariscal and Poindron 2002; Numan et al. 2006). Are milk output and maternal behavior linked? As described earlier, milk production in rabbits increases steadily across lactation days 1–20 in does that are lactating but not concurrently pregnant (i.e., *not* mated at postpartum estrus) and decreases thereafter, accompanied by concomitant reductions in PRL secretion (Fuchs et al. 1984). Yet, the frequency of *circa* one nursing bout/day remains unchanged. This observation would indicate that, in rabbits, the physiological component (i.e., milk production) and the behavioral component (nursing behavior) of maternal behavior are *not* linked. Indeed, young kits promote milk production, while "old" ones inhibit it, through the stimulation of a beta-adrenergic tone (Mena et al. 1990a, b). Yet, maternal behavior continues despite a clear decline in milk output: under laboratory conditions, lactating-only does continue to enter the nest box for approximately 40 days. An important contribution of suckling to the *maintenance* of maternal behavior is supported by the finding that thelectomized (i.e., nipple removed; thelx) does enter the nest box and adopt a crouching posture over the (foster) litter, despite their incapacity to provide milk. In contrast, does mated at postpartum estrus (thereby becoming concurrently pregnant-lactating) do stop entering the nest box a few days before the *second* litter is born (Fig. 11; González-Mariscal and Gallegos 2014; Hudson et al. 1996; Martínez-Gómez et al. 2004).

Together, the above findings indicate that the two components of weaning in rabbits, i.e., a decline in milk production and a cessation of nursing behavior, are regulated by different factors. The former process seems to involve suckling-induced changes in specific hormones (e.g., PRL) and neurotransmitter control at the spinal cord level (inhibitory noradrenergic tone). Conversely, little is known about the factors that determine the abrupt cessation of *nursing behavior*. We know that this occurs only in does that are concurrently pregnant-lactating, but the factors participating in this process remain unknown. Progesterone seems not to play a role because injections of this steroid to lactating-only rabbits do not modify the expression of nursing behavior, although they do reduce milk output (González-Mariscal and Gallegos 2014).

Kits' Behavior

In Anticipation of Nursing, During Suckling (Change in Teat Position), and After Milk Intake

Newborn kits can neither see nor hear and have no fur. Thus, they are tied to the nest (as nestlings). By lying close together, they can maintain a constant nest temperature and obtain somatosensory stimulation necessary for a normal neural, motor, and social development (see overview in González-Mariscal et al. 2016). Wild rabbit kits are nursed by their mother inside the nest until they are 9–10 days old. Does

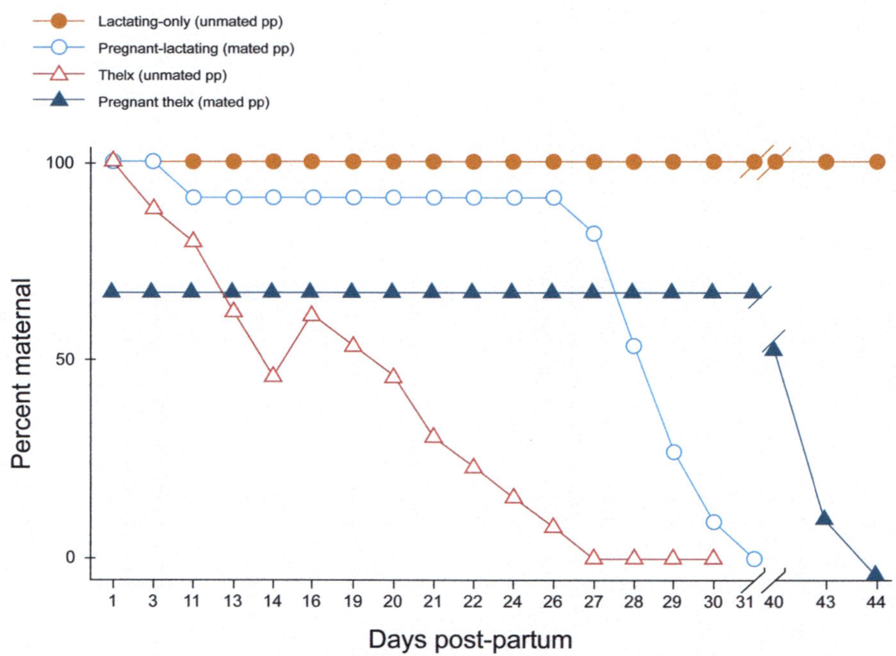

Fig. 11 Duration of maternal responsiveness postpartum is affected by concurrent pregnancy and suckling stimulation. (Reproduced from González-Mariscal and Gallegos 2014. Physiol Behav. 124:72–6)

kept within seminatural enclosures open the tunnel to the nest immediately before nursing and crawl into it. The doe stays inside the nest for 2–5 min for nursing and closes the entrance to the nest burrow after leaving. Like wild rabbits, domestic breeds occasionally (but not always) try to close the nest after nursing so that the young are protected from predators. The doe often defecates and urinates at the entrance to the warren to mask the smell of the nest (Myers and Poole 1961; Kraft 1976). Such behavior can occur up to 100 times in 24 h. Frequent attempts to close the nest are sometimes interpreted as a behavioral disorder (literature in Seitz 1997).

At the age of 7 days, the sense of hearing is largely developed, and the kits can perceive a light-dark alternation. Between days 7 and 11, they open their eyes and can orient themselves in their surroundings (Verga and Luzi 2006). After the kits have grown fur and have developed sight and hearing (10–11 days old), they begin to crawl around inside the nest and are nursed at its entrance. They begin to leave the nest at 3 weeks of age (in extreme cases with 12–16 days – Schulte 1998). One-week-old wild rabbit kits react with intensive vocalizations approximately 15 s before the mother enters the nest, i.e., when they feel the vibration inside the tunnel as the doe approaches. The mean number of vocalizations detected in litters having an average of 6.6 kits was 165 in 24 h, within a frequency range of 374–667 Hz, and a mean duration of 67 ms. The distribution of vocalizations across the day showed maximum values at night, especially at the time when nursing occurred or within 1 h prior to it (Hofmann and Hoy 2007).

In domestic rabbits, kept under laboratory conditions, an increased restlessness can be observed in the kits shortly before nursing (Caba et al. 2008; Hudson and Distel 1982; Jilge 1993). They start pushing the nest material aside and gradually uncover themselves from it. At this time, they are particularly sensitive to tactile stimuli and vibrations, and they react to even slight disturbances with increased activity, rearing movements, and vocalization. They respond to the doe entering the nest by rearing their heads up and pushing their muzzles deep into her belly fur to start the search for nipples (Hudson and Distel 1983). Competition for access to a teat is particularly strong, given that kits only obtain significant amounts of milk during the second minute of nursing. By viewing the behavior of the young during nursing in a glass-bottom nest box, Bautista et al. (2005) found that they compete for nipples in a vigorous scramble and without obvious signs of overt aggression. Despite the heat loss presumably ensuing from exposure of their skin following removal of the nest material, the anticipatory uncovering appears important as it enables kits to reach the doe's belly unhindered (Hudson et al. 1996). The timing of nursing also impacts the kits' circadian rhythm of corticosterone secretion as well as several metabolic parameters (reviewed in Caba and González-Mariscal 2009).

Kits show nipple-search behavior toward any lactating doe: immediately after she enters the nest box and adopts a crouching posture over the litter, the kits immediately start searching for her nipples. This behavior is guided by an olfactory cue, the so-called "mammary pheromone" (Hudson and Distel 1983), recently identified as 2-methyl-but-2-enal (Schaal et al. 2003). Kits do not have a fixed position for suckling but, rather, change teats quickly (about three times/min) during a nursing bout; this frequency changes with the age of kits (Hudson and Distel 1985; Karpati et al. 2000). Despite the short duration of nursing bouts, kits can consume between 25% and 35% of their body weight in milk through suckling (Hudson and Distel 1989; González-Mariscal et al. 1994; Lebas et al. 1972; Schulte 1998). The high energy and protein content of rabbit milk (Schlolaut 1998) is what enables kits to gain weight quickly.

When the doe jumps out of the nest at the end of nursing, the kits drop immediately from the nipples, which prevents being dragged away from the nest. They then urinate simultaneously, become very wet, and vigorously burrow back under the nest material and disperse throughout the nest. This activity lasts about 10–15 min, during which time the kits become dry and the nest material fluffed up again. The kits then gradually reassemble in the warmest part of the nest where they remain covered until the next nursing visit. This behavior clearly serves the thermoregulatory needs of these altricial young, which are neither groomed nor brooded by their mother (Hudson et al. 1996). Between suckling episodes, the young also compete for well-insulated, central positions within the litter huddle and expend considerable energy climbing over and burrowing under each other in a continuous effort to achieve and maintain such positions (Bautista et al. 2008).

Critical Assessment of Biostimulation on Kit Development

As described earlier, the so-called "biostimulation" strategy has been adopted worldwide to accelerate reproduction (and, thus, reduce production costs) in rabbit farms. A 48 h mother-litter separation, performed around lactation days 9–11, allows the survival of the young and does not significantly modify their weight at weaning (Alvariño et al. 1998; Castellini et al. 1998; Theau-Clément and Mercier 1999). Yet, the possibility that such procedure may impact the kits' developing brain has been investigated based on findings – obtained mainly in rodents – showing that maternal stimulation during the early postnatal period is critical for a normal neurodevelopment. Thus, processes like reproductive behavior, neuroendocrine reactivity to stress, learning, and memory are negatively affected by various forms of mother-young separation in early lactation (for reviews, see Curley and Champagne 2016; González-Mariscal and Melo 2017). The possibility that something similar might occur in rabbits was initially explored by Brecchia et al. (2009). These authors compared corticosterone (CORT) secretion following mildly aversive stimulation in rabbits that had been separated from their mothers for 48 h (i.e., subjected to "biostimulation") vs. control animals (i.e., separated from the mother for the usual 24 h between nursing bouts). CORT secretion at 48 h following maternal separation more than doubled in the "deprived" litters, relative to controls. Moreover, this increased CORT secretion in the early neonatal period had a strong impact on the rabbits' response to mildly aversive stimulation in *adulthood*: "deprived" animals showed a blunted CORT response, relative to the control group, but both groups responded equally well to the injection of ACTH. These results suggest that, even though doe rabbits have reduced their maternal care to a minimum, such brief contact with the young is, nonetheless, essential to allow a normal development of some components of the stress-reactivity system in the kits. This possibility was further

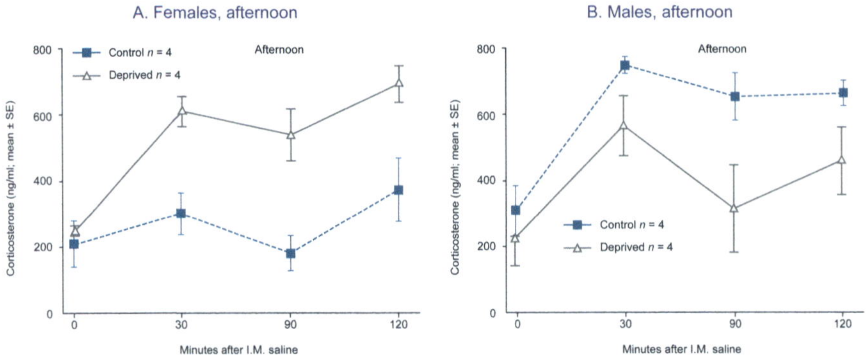

Fig. 12 Corticosterone secretion determined in the serum of control and deprived adult females (**a**) and males (**b**), sampled in the afternoon, before and at several time points after an i.m. injection of saline. (Reproduced from García-Fernández et al. 2019. Dev Psychobiol. 61:988–98)

Table 6 Behavioral tests measuring stress/anxiety in rabbits revealed no significant differences in the responses between control and mother-deprived adult animals nor between the sexes[a]

	Females		Males	
Test	Control (n = 10)	Deprived (n = 9)	Control (n = 9)	Deprived (n = 9)
Latency to righting reflex (sec)	42 ± (55, 22)	30 ± (57, 17)	35 ± (49, 23)	52 ± (93, 38)
Latency to approach human (sec)	37 ± (55, 29)	35 ± (59, 25)	38 ± (73, 28)	38 ± (52, 23)
# Contacts with human	3 ± (4, 3)	2 ± (3, 2)	3 ± (5, 2)	3 ± (3, 2)
Latency to exit dark box (sec)	34 ± (57, 28)	32 ± (136, 17)	36 ± (180, 23)	31 ± (81, 18)
Ambulation in open field[b]	8 ± (8, 6)	6 ± (9, 5)	8 ± (8, 5)	8 ± (9, 6)
Scent-marking[c]	1 ± (2, 0.4)	1.5 ± (3, 0.4)	7 ± (8, 4)	9 ± (9, 1)

[a]Modified from García-Fernández et al. (2019); data show medians ± interquartile ranges
[b]# Lines crossed
[c]# Chin marks made

investigated by García-Fernández et al. (2019) who compared the basal and reactive secretion of CORT and cortisol, plus a battery of behavioral tests, between control and "deprived" male and female adult rabbits. Figure 12 shows that reactive CORT secretion, following an i.m. injection of saline in the *afternoon*, was higher in "deprived" than in control females, while in males the opposite was true.

Similar results were found for cortisol (not shown). Interestingly, when saline was injected in the morning, no differences were found between the two experimental groups nor between the sexes (not shown). Likewise, behavioral tests commonly used to determine stress/anxiety in rabbits were unmodified by maternal deprivation or time of testing in males or females (see Table 6).

In contrast, male sexual behavior was altered in deprived individuals, who showed a large number of mounts not culminating in ejaculation. This was reflected in a large "miss rate" in deprived bucks (García-Fernández et al. 2019; Fig. 13).

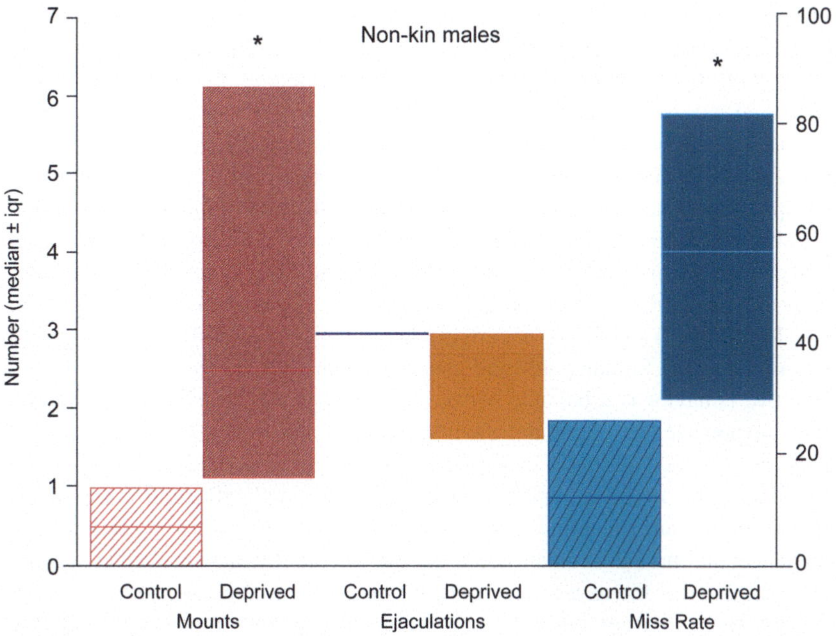

Fig. 13 Alterations in the sexual behavior of "deprived" bucks consisted of an increased number of mounts not culminating in ejaculation, which led to a large "miss rate." (Reproduced from García-Fernández et al. 2019. Dev Psychobiol. 61:988–98)

Sexual receptivity in does was unaffected by maternal deprivation or time of day when nonkin females were compared (not shown). Together, the above results indicate that a mother-litter separation that exceeds by 24 h the normal time between nursing bouts has long-lasting, complex consequences on the development of the particular neuronal networks that regulate the neuroendocrine reactivity to stress in both sexes, as well as sexual behavior in males.

3 Maternal Nest Building in the Rabbit as a Possible Model for Elucidating the Neurobiological Control of Compulsive Behavior

Obsessive-Compulsive Disorder (OCD): Dysregulation of Normal, Adaptive Behavior?

Many neuropsychiatric disorders can be conceptualized as extreme or dysregulated versions of normal, adaptive behavior. Obvious examples are the anxiety disorders, including generalized anxiety, social anxiety, panic disorder, and phobias. This diagnostic class of mental illness essentially involves maladaptive, exaggerated, and inappropriate expression of the normal fear responses: prolonged concern/attention to threat, maladaptive feelings and behaviors of social submission, inappropriate triggering of fight or flight response, and exaggerated fear of stimuli that once represented real day-to-day dangers (heights, snakes, spiders, open spaces), respectively. The relationship between normal, adaptive behavior and maladaptive neuropsychiatric symptoms provides a very useful theoretical framework on which to base preclinical studies aimed at understanding the neurobiology of the latter. The study of anxiety disorders has in effect pioneered this approach, with early studies of fear learning in rodents (e.g., Mowrer and Lamoreaux 1946) and later studies of Blanchard and Blanchard (1989) on the responsiveness of rodents to proximal (imminent) and distal (uncertain) threat, which as a result were recognized to be homologous to human panic and generalized anxiety, respectively.

Adaptive behavior is a balance between planned goal-directed actions that are flexibly responsive to changing environmental contingencies and automatic, innately programmed actions that are inflexible and more difficult to interrupt and modify. Examples of behaviors having both flexible and inflexible components include grooming and cleanliness behaviors, food caching, territorial marking and maintenance, and parental (maternal) behavior. Maternal behavior is an especially apt example because the new mother must be successful in the absence of any prior experience (necessitating innate, preprogrammed maternal responses); yet, the expression of preprogrammed behavioral routines must also maintain flexibility in order to adapt to variable and changing environmental and social contingencies.

Obsessive-compulsive disorder (OCD) is a mental illness that includes intrusive obsessive thoughts of harm to self, family, and home and compulsive behaviors that are persistent, repetitive, ritualistic, and ostensibly aimed at preventing harm. Obsessions and compulsions are distressing, time consuming, and highly debilitating. Interestingly, the content of obsessions and compulsions of OCD most often involves the aforementioned behaviors: grooming and cleanliness, hoarding, protection of self and home, and protection of family and loved ones. Szechtman et al. (2020) and Woody et al. (2019) have elaborated a model of OCD pathophysiology centered on a special innate "security motivation" that is activated when an animal is confronted with a situation where there is a potential for threat, but no threat is physically present. Security motivation promotes the performance of prophylactic security behaviors, such as washing, checking, and verifying. An important characteristic of security motivation is that there is no clear objective cue that can signal the *absence* of threat; therefore, the system must rely on an internally generated signal that quenches security motivation and stops security behaviors. It is this quenching mechanism that is proposed to be faulty in OCD: security motivation and prophylactic behaviors are "turned on" by potential threat, but cannot be "turned off" normally, resulting in prolonged security motivation (which manifests as obsessions) and excessive repetition of security behavior (compulsions).

Considering OCD within the theoretical framework of the security motivation, we have proposed that maternal nest building in the pregnant rabbit can serve as a neurobehavioral model for elucidating neurobiological mechanisms relevant to compulsive-like behavior (Hoffman and Rueda Morales, 2009, 2012).

Straw-carrying (described in Section "Nest building in the laboratory, on the farm, and in the wild") is a repetitive component of maternal nest building that is initiated in late pregnant rabbits by a clear, discrete objective stimulus (straw), and terminated in association with a subjective one (size and/or quality of nest). Given that the Szechtman and Woody model proposes that pathological compulsions arise from a dysfunctional stopping mechanism, defining neurobiological mechanisms that underlie the termination of straw-carrying behavior might yield important clues on neurobiological substrates that might underlie the quenching of security motivation.

Neurobiology of OCD and Straw-Carrying

Neurobiologically, how similar are straw-carrying behavior and human obsessive-compulsive symptoms? During the past 25 years, patients with OCD have been studied using ever-more sophisticated brain imaging techniques and analyses. Most studies have applied functional magnetic resonance imaging (fMRI), which is a means to visualize localized changes in cerebral blood flow that are associated with regional brain activation. This technique, along with positron emission tomography (PET) deoxyglucose imaging, identified brain regions that showed increased baseline (resting) activity in subjects with OCD compared to healthy controls. These regions include the orbitofrontal cortex (OFC), anterior cingulate cortex (ACC), striatum (caudate and putamen), and medial thalamus. fMRI has also been applied in experimental protocols in which the subject is exposed to obsession- and compulsive-provoking stimuli (e.g., photo of contaminated hands), revealing increased activation in these same regions in OCD compared to control (reviewed in Soriano-Mas 2021).

If straw-carrying accurately models compulsion, then exposing a pregnant rabbit to straw (the "compulsion-provoking stimulus") should result in the activation of the rabbit homologues of human brain regions that are activated in OCD symptom provocation paradigms. Cano-Ramírez and Hoffman (2017, 2018) carried out a series of studies in which nonpregnant and pregnant rabbits were either given straw or not and then sacrificed 1 h later and processed for c-FOS protein immunohistochemistry. Pregnant rabbits repetitively collected straw and carried it into the nest box, while nonpregnant ones simply nibbled on the straw or did not interact with it. Compared to nonpregnant groups (with or without straw), and to pregnant rabbits that were not given straw, pregnant straw-carrying rabbits showed increased c-FOS immunolabeling in the OFC, the ACC, the piriform cortex, the striatum (dorsal putamen and ventral caudate), and the somatosensory and primary motor cortex. Both pregnant and nonpregnant rabbits that were given straw showed increased activity in the medial and ventral putamen, indicating that activity in these areas was related to simply interacting with straw (Fig. 14).

In a subset of straw-carrying rabbits, we carried out an exploratory factor analysis (EFA) with cluster rotation on a dataset that encompassed c-FOS

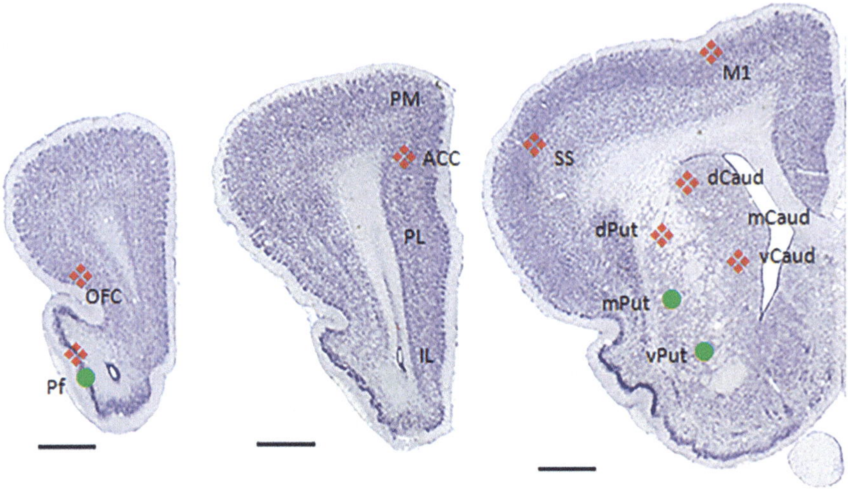

Fig. 14 Coronal hemisections of the rabbit brain from rostral (left) to caudal (right). Text labels denote brain regions that were studied in relation to straw-carrying behavior in pregnant female rabbits (Cano-Ramírez and Hoffman 2017, 2018). The red diamonds mark those regions that showed significantly increased activity (c-Fos label density) in association with straw-carrying, and the green circles mark those regions that showed increased activity associated with nonspecific interactions with straw. Abbreviations: *OFC* orbitofrontal cortex, *Pf* piriform cortex, *PM* premotor area, *ACC* anterior cingulate cortex, *PL* prelimbic area, *IL* infralimbic area, *SS* somatosensory cortex, *M1* primary motor cortex, *dPut, mPut, and vPut* dorsal, medial, and ventral putamen, respectively, *dCaud, mCaud, and vCaud* dorsal, medial, and ventral caudate, respectively. Scale bar = 0.2 cm. Images of brain sections were adapted from those available online from the Comparative Mammalian Brain Collections. (http://neurosciencelibrary.org/index.html)

immunohistochemistry data and several behavioral measures relevant to repetitive straw-carrying (Cano-Ramírez et al. 2020). Essentially, EFA is a means to identify groups of variables that tend to covary with each other and, when applied in the present context, would identify ensembles of brain regions that have correlated densities of c-FOS label. Importantly, distinct brain regions that have correlated activity can tentatively be assumed to comprise a functionally interconnected network: brain regions that "fire together, wire together." Likewise, behaviors that covary with activity in such ensembles of brain regions are likely to be modulated by these networks. This exploratory analysis revealed three ensembles of brain regions that displayed correlated activity during straw-carrying. Remarkably, these three ensembles mapped onto three functional corticostriatal-thalamocortical (CSTC) networks that have been described in rodents, nonhuman primates, and humans (Fig. 15).

These analyses place straw-carrying behavior within the framework of an emerging neurobiological model for the expression of flexible, goal-directed behavior and inflexible habit behavior. This model comprises the three CSTC networks described above (Fig. 16): (1) the limbic network is involved in motivation and long-term planning of a goal-directed behavior; (2) the associative network underlies the

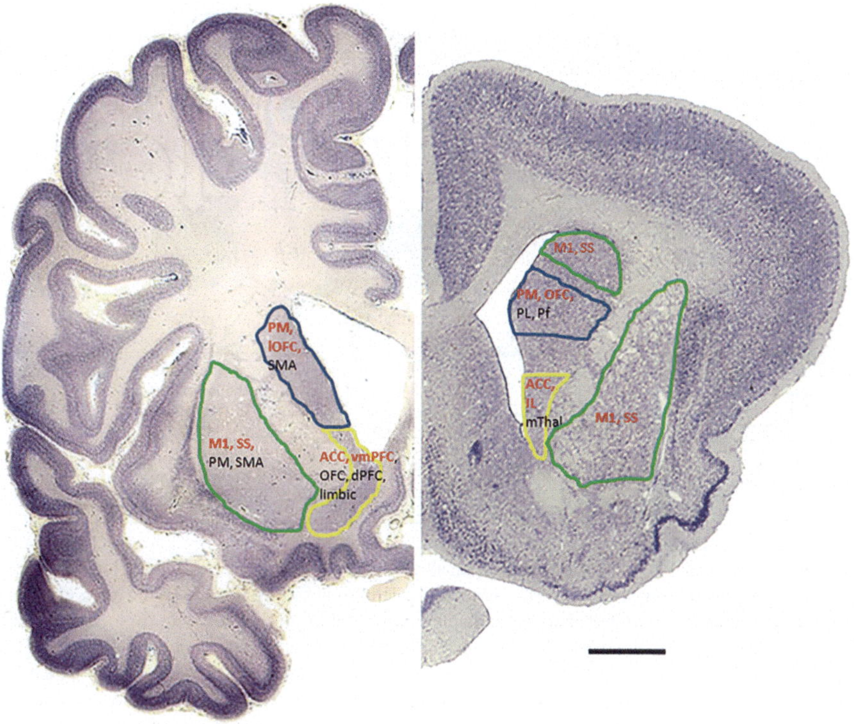

Fig. 15 Coronal hemisections of human (left) and rabbit (right) brain, showing striatal regions and cortical areas functionally associated with them. Image of human brain was reduced in size in order to facilitate comparisons with the rabbit brain. *Left (human) hemisection:* Cortical afferents to the putamen (outlined in green), the caudate (blue), and the ventral caudate/nucleus accumbens (yellow) are denoted by text labels, forming the corticostriatal components of the sensorimotor (green), associative (blue), and limbic (yellow) CSTC networks (Jaspers et al. 2017). *Right (rabbit) hemisection:* Three ensembles of rabbit brain regions that showed coactivation during straw-carrying: (i) "sensorimotor," dorsal caudate and putamen (green outline) coactivated with M1 and SS; (ii) "associative," medial caudate (blue outline) coactivated with PM, OFC, PL, and Pf; and (iii) "limbic," ventral caudate (yellow outline) coactivated with ACC, IL, and mThal. Corticostriatal afferents that show homology between human and rabbit are highlighted in red. Note that the rabbit IL is functionally and anatomically homologous to vmPFC. Activity of the hippocampus and amygdala during straw-carrying has not yet been investigated (Cano-Ramírez et al. 2020). Abbreviations: *M1* primary motor cortex, *SS* somatosensory cortex, *PM* premotor cortex, *SMA* supplementary motor area, *lOFC* lateral orbitofrontal cortex, *ACC* anterior cingulate cortex, *dPFC* dorsal prefrontal cortex, *Pf* piriform cortex, *mThal* medial thalamus. Scale bar = 0.2 cm. Images of brain sections were adapted from those available online from the Comparative Mammalian Brain Collections. (http://neurosciencelibrary.org/index.html)

flexible performance of short-term actions necessary to carry out the behavioral plan; and (3) the sensorimotor network is engaged when habitual, repetitive, or innately programmed subroutines are employed in order to achieve the goal. In response to triggering stimuli, these CSTC networks are proposed to be sequentially activated by feed-forward excitation (Haber et al., 2000).

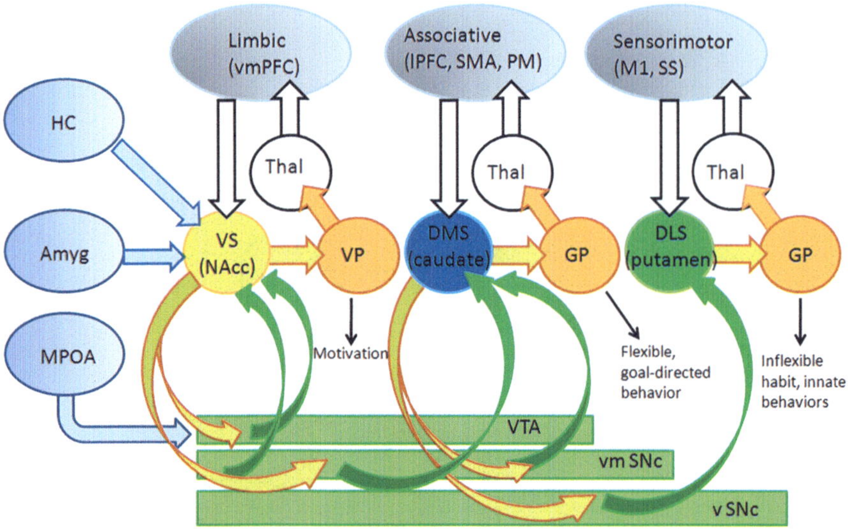

Fig. 16 CSTC networks involved in the expression of goal-directed and habit behavior. The "limbic" network (left) receives sensory-emotional and pneumonic information from the amygdala and hippocampus and is associated with initiating and maintaining motivation. The "associative" network (middle) mediates the expression of flexible behavior directed toward achieving a goal. The "sensorimotor" network (right) underlies the expression of innate behavioral patterns (such as grooming) and acquired habits. VS projections from the striatum to the midbrain VTA and vmSNc (yellow arrows) activate dopaminergic projections (green arrows) from these areas to the VS and DMS; the DMS, in turn, activates dopaminergic afferents from the vmSNc and vSNc to the DMS and DLS, respectively. Maternal motivation is facilitated by projections from the MPOA to the VTA, which in turn activates dopaminergic afferents to the VS. Abbreviations: *HC* hippocampus, *Amyg* amygdala, *MPOA* medial preoptic area, *VS* ventral striatum, *VP* ventral pallidum, *Thal* thalamus, *vmPFC* ventromedial prefrontal cortex, *lPFC* lateral prefrontal cortex, *SMA* supplementary motor area, *PM* premotor area, *DMS* dorsomedial striatum, *GP* globus pallidus, *M1* primary motor cortex, *SS* somatosensory cortex, *DLS* dorsolateral striatum, *VTA* ventral tegmental area, *vmSNc* ventromedial substantia nigra compacta, *vSNc* ventral substantia nigra compacta

Two influential hypotheses for OCD posit that there is a dysfunction in the control of these networks. One hypothesis proposes that there is a generalized imbalance between the activities of the associative and sensorimotor networks that favors the activation of the latter network and consequently the repetitive expression of inflexible behavioral routines ("compulsions") (Gillan and Robbins 2014). The second hypothesis is that of Szechtman and Woody, described above (Szechtman et al. 2020; Woody et al. 2019). In the Szechtman and Woody model, the limbic network is activated when the individual is confronted with situations of uncertainty and potential threat, activating "security motivation," which promotes activation of associative and sensorimotor networks. In healthy individuals, the expression of security behavior such as checking and hand-washing activates an unidentified mechanism that provides negative feedback to the limbic and associative networks, thereby quenching security motivation and terminating security behaviors.

Dopamine release within the striatal component of each of these networks by afferents from the ventral tegmental area (VTA) and substantia nigra compacta (SNc) is a key mechanism for promoting network activation, and dopamine release within each network is hierarchically modulated: the limbic network can modulate dopamine release in the medial striatum (thus controlling activation of the associative network), and the associative network can modulate dopamine release in the dorsolateral striatum (thus controlling activation of the sensorimotor network) (Fig. 16). Results of pharmacological studies of straw-carrying are consistent with dopamine-mediated sequential activation of these networks (Hoffman and Rueda Morales 2012). In that study, dopamine D1 and D2 receptor antagonists had no effect on the initiation of straw-carrying (presumably mediated by the initial activation of the limbic and associative networks). In the presence of these antagonists, straw-carrying waned within 5–10 min, while saline-treated rabbits persistently displayed this behavior for 30 min or more. Interestingly, straw-carrying bouts in saline-treated controls were expressed in a highly regular manner (average of 2 cycles every 5 min, with uniform variability), while those rabbits treated with dopamine antagonists displayed truncated and more variable bouts of straw-carrying. This finding is consistent with the proposal that the dopamine-mediated activation of sensorimotor CSTC circuits (see Fig. 17) mediates the transition from flexible expression of straw-carrying (presumably mediated by the associative network) to persistent and highly regular repetitive behavior.

Relationship Between OCD, Straw-Carrying, and Normal Maternal Behavior

Models of neurobiological control of maternal motivation and behavior (most of which are based on experiments where pup retrieval in rodents was used as an outcome measure) propose that pup-associated stimuli activate the medial preoptic area (MPOA), which sends projections to the ventral tegmental area (VTA) that in turn activate dopaminergic projections to the ventral striatum. Dopamine release in the ventral striatum then activates limbic networks that promote maternal motivation and the activation of flexible goal-directed behavior (associative networks) as well as innately programmed inflexible motor routines such as pup retrieval (likely mediated by sensorimotor networks). As described in section “Nest building in the laboratory, on the farm, and in the wild”, stereotaxic lesions of the MPOA, and encroaching on the ventral striatum (nucleus accumbens), significantly inhibited the expression of straw-carrying behavior during normal pregnancy as well as in ovariectomized rabbits treated with estradiol and progesterone (Basurto et al. 2018). Interestingly, stereotaxic lesioning and high-frequency electrical stimulation (deep brain stimulation (DBS)) of a very similar region in humans (the ventral striatum and ventral internal capsule) are often effective neurosurgical treatments for otherwise intractable OCD (Kahn et al. 2021; Miguel et al. 2019; Fig. 18).

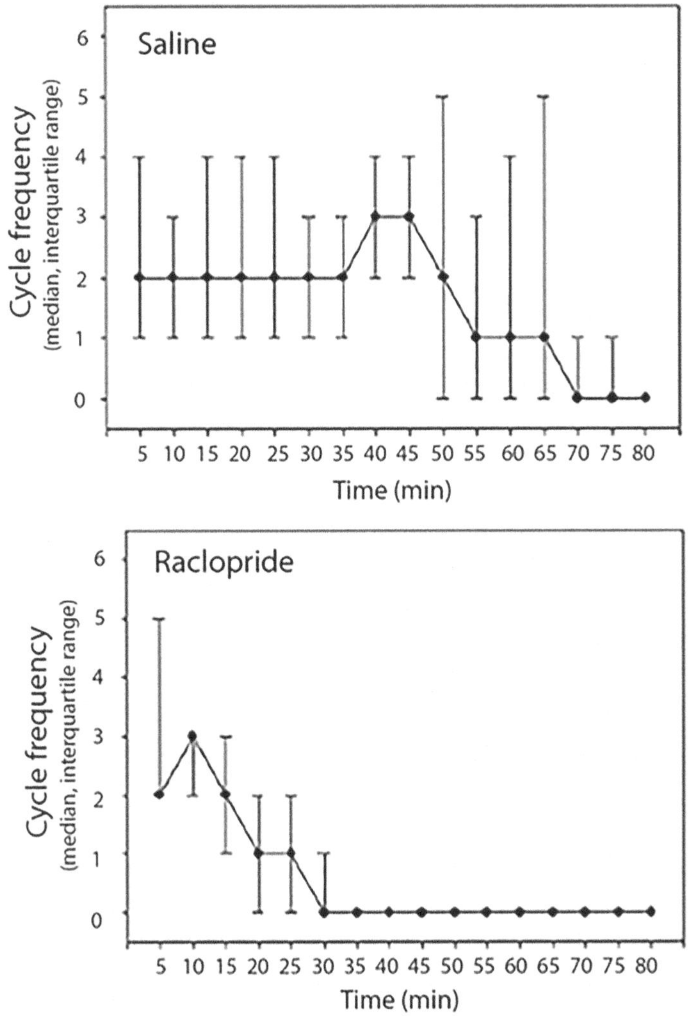

Fig. 17 Pretreatment with a dopamine receptor antagonist significantly shortened the duration of straw-carrying. Compared to saline (control; top panel), raclopride (D2/D3 receptor antagonist, bottom panel) significantly shortened the duration of bouts of repetitive straw-carrying. "Cycle frequency" (vertical axis) refers to the number of straw-carrying cycles displayed for each 5 min time bin (a complete cycle defined as collect straw, carry straw to nest box, and deposit it there and return to collect more straw). (Hoffman and Rueda Morales 2012)

As alluded to above, obsessions of OCD most often involve thoughts related to self-protection, protection of loved ones, and protection of one's home, while compulsions are often inflexible, ritualistic, and repetitive behaviors related to evolutionarily conserved behavioral programs. OCD associated with pregnancy and the postpartum period is an especially illustrative example in this regard. Becoming a

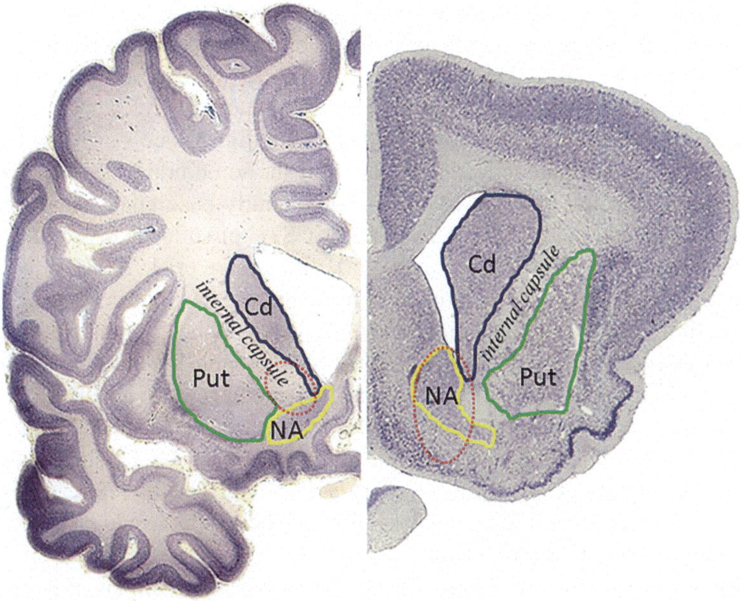

Fig. 18 Coronal sections of the human (left panel) and rabbit (right panel) brain. The human brain hemisection was reduced in size in order to facilitate comparisons with the rabbit brain. Caudate (Cd), putamen (Put), and nucleus accumbens (NA) are outlined in blue, green, and yellow, respectively. *Left (human) hemisection*: The red dashed oval marks the target area for neurosurgical lesions of the ventral striatum/ventral capsule (VS/VC) in treatment-resistant OCD. High-frequency electrical stimulation of this same area (deep brain stimulation (DBS)), which is believed to produce a reversible functional lesion, can also be an effective treatment for OCD. *Right (rabbit) hemisection*: The red dashed oval marks the approximate area within which experimental electrolytic lesions resulted in the selective loss of straw-carrying behavior (Basurto et al. 2018). Images of brain sections were adapted from those available online from the Comparative Mammalian Brain Collections. (http://neurosciencelibrary.org/index.html)

new parent is a life event often associated with experiences of intrusive thoughts and compulsive-like behaviors. Leckman et al. (1999) and Abramowitz et al. (2003) surveyed new mothers and fathers concerning their experiences of intrusive, obsession-like thoughts. Approximately 90% of new parents experienced intrusive thoughts about some horrible accident occurring to the infant, while 40% reported intrusive thoughts of doing intentional harm to the baby. Other obsessive thoughts were of the possibility of "misplacing" or "losing" the baby and thoughts of germs and illness. These thoughts were sometimes accompanied by compulsive-like behaviors, including checking on the baby, even though they "knew everything was fine." The content of the typical intrusive thoughts and compulsive-like behaviors in nonclinical populations of new parents is markedly similar to the obsessions and compulsions experienced by new parents with clinically diagnosed OCD. Indeed, a number of studies indicate that pregnancy and childbirth were associated with an exacerbation of existing clinical OCD symptoms, as well as with an increased risk for first onset of OCD (Guglielmi et al. 2014; Forray et al. 2010). However, other studies have questioned the significance of this apparent relationship (House et al. 2016; McGuinness et al. 2011).

Nevertheless, it is clear that in healthy mothers, maternally relevant stimuli such as a photographic image of one's own child or a recording of one's own child crying elicit fMRI regional brain activation in many of the same brain regions that have been mentioned in the previous sections and generally encompass many components of limbic, associative, and sensorimotor CSTC circuitry (Lorberbaum et al. 2002; Nitschke et al. 2004; Ranote et al. 2004; Strathearn et al. 2008). Infant cries activate a "security motivation-like" system in the mother: an alerting system – involving limbic CSTC circuits – that responds to cues indicating the possibility of threat, in this case, possible harm to the infant (Lorberbaum et al. 2002). According to the Szechtman and Woody model, in healthy mothers, activity in the security motivation system would be quenched by checking the infant and performing innate maternal care behaviors. By contrast, in individuals with OCD, such behaviors would be ineffective at quenching security motivation, and therefore motivation would remain (taking the form of an obsession) and behaviors would be unnecessarily repeated (becoming a compulsion).

When modeling neuropsychiatric disorders in laboratory animals, it is essential to work within a theoretical framework that allows for evidence-based, valid neurobiological and behavioral cross-species comparisons. In the present context, both hypotheses for OCD pathophysiology mentioned above (imbalance in goal-directed/habit behavior and the security motivation hypothesis) are based on evolutionarily conserved neurobiological processes that may be disrupted in OCD: imbalance in relative activity of sensorimotor and associative CSTC circuits that favors inflexible habit behavior and deficient negative feedback onto limbic and associative CSTC circuits. These circuits encompass brain regions that are known to exhibit altered activity in OCD, such as the OFC, ACC, ventromedial PFC, striatum, and thalamus, as well as being fundamental neurobiological substrates for the expression of maternal motivation and behavior in humans and in nonhuman animals alike. Our studies indicate that straw-carrying in the pregnant rabbit also engages these same circuits. Within this theoretical framework, straw-carrying by the pregnant rabbit might be considered a compulsive behavior that is adaptively expressed within a specific physiological and reproductive context. Using this neurobehavioral model in order to address targeted questions such as *How is repetitive straw-carrying turned off?* should generate hypotheses on neurobiological mechanisms that may be dysfunctional in OCD.

4 Final Reflections

Throughout this chapter, we have provided a landscape depicting how the study of rabbit maternal behavior can illuminate a more complete understanding of the "parental brain." When viewed as a whole, findings arising from disciplines as varied as neuroendocrinology, animal science, behavioral ecology, and psychobiology reveal that rabbits share commonalities with other mammals regarding the neuroendocrine regulation of maternal behavior (for reviews, see González-Mariscal and Poindron 2002; Numan and Insel 2003; Numan et al. 2006). On the other hand, it is also evident that mother rabbits show some "peculiar" behaviors that are specific to lagomorphs, such as an elaborate nest building process and brief nursing bouts that are displayed with predictable once-a-day periodicity. These innate behavioral patterns are both fixed and flexible, a property that allows rabbits to adapt to a changing environment. Both in the laboratory and on the farm, the behavior of maternal rabbits has been observed in response to a number of experimental challenges (e.g., type of nest material, exposure to liver vs. placenta, number of kits nursed, type of cage used, own vs. alien kits, lactating-only vs. pregnant-lactating states). Their responses to these challenges both emphasize that mother rabbits have a remarkable capacity to modify their behavior in ways that optimize the survival of their progeny as well as provide important clues on the neurobiological and neuroendocrinological underpinnings of maternal behavior. Moreover, studies of wild and domestic rabbits, kept under seminatural conditions, have revealed an important contribution of social hierarchy to the success of a doe in producing overlapping litters across a breeding season.

Nevertheless, several pertinent questions on rabbit maternal behavior remain underexplored including: How is the nipple stimulation, received during suckling, "transformed" into a neural signal that determines the duration and frequency of a nursing bout? Which factors determine weaning? Is an evolutionary perspective possible from studying maternal behavior in other lagomorphs (i.e., hares, pikas; Broekhuizen et al. 1986; Broekhuizen and Mulder 1983)? These questions touch on central themes within the fields of the neurosciences, behavioral neuroendocrinology, and evolutionary biology, and rabbits are a species particularly well suited for exploring them. It is encouraging that, within the field of biological psychiatry, rabbit nest building is being used as a model for exploring the neurobiological substrates of obsessions and compulsions, which are neuropsychiatric symptoms unique to humans. As nest building is objectively and easily quantified in doe rabbits, it can be used to test drugs that target particular neurotransmitter systems, presumably involved in the expression of OCD.

A fruitful interaction is emerging between cuniculture (i.e., rabbit breeding) and behavioral neuroendocrinology/psychobiology. Although the former field is mainly interested in productivity (rabbits are considered livestock), this aim depends largely on optimizing reproduction, in agreement with animal welfare. Consequently, reliable information on the ways through which farm conditions and animal management impact the adult rabbit reproductive axis, the responses to stress, and the development of kits can contribute positively to this important food industry. Conversely, "laboratory researchers" are finding conditions that are frequent on the farm but unusual in behavioral neuroendocrinology (e.g., concurrently pregnant-lactating does) which allow the study of main topics in the field, such as reflex ovulation and weaning.

We hope this review chapter encourages our readers to broaden their outlook on rabbit maternal behavior beyond the usual boundaries of the field. This topic has been enriched largely through the recognition that findings from "peripheral" fields are connected around a common, complex function of the nervous system: the regulation of mammalian behavior.

References

Abramowitz JS, Schwartz SA, Moore KM, Luenzmann KR. Obsessive-compulsive symptoms in pregnancy and the puerperium: a review of the literature. J Anxiety Disord. 2003;17:461–78. https://doi.org/10.1016/s0887-6185(02)00206-2.

Aguilar-Roblero R, González-Mariscal G. Behavioral, neuroendocrine and physiological indicators of the circadian biology of male and female rabbits. Eur J Neurosci. 2020;51:429–53. https://doi.org/10.1111/ejn.14265.

Aguirre J, Meza E, Caba M. Dopaminergic activation anticipates daily nursing in the rabbit. Eur J Neurosci. 2017;45:1396–409. https://doi.org/10.1111/ejn.13571.

Alvariño JMR, Del Arco JA, Bueno A. Effect of mother–litter separation on reproductive performance of lactating rabbit females inseminated on day 4 or 11 post-partum. World Rabbit Sci. 1998;6:191–4.

Anonymous. https://www.newfoodmagazine.com/news/85045/global-rabbit-meat-market-grow/. Accessed 7 June 2021.

Askar AA, Ismail IE. Impact of heat stress exposure on some reproductive and physiological traits of rabbit does. Egypt J Anim Prod. 2012;49:151–9.

Bakr MH, Tusell L, Rafel O, Terré M, Sánchez JP, Piles M. Lactating performance, water and feed consumption of rabbit does reared under a Mediterranean summer circadian cycle of temperature v. comfort temperature conditions. Animal. 2015;9:1203–9.

Basurto E, Hoffman KL, Lemus AC, González-Mariscal G. Electrolytic lesions to the anterior hypothalamus-preoptic area disrupt maternal nest-building in intact and ovariectomized, steroid-treated rabbits. Horm Behav. 2018;102:48–54.

Bautista A, Mendoza-Degante M, Coureaud G, Martínez-Gómez M, Hudson R. Scramble competition in newborn domestic rabbits for an unusually restricted milk supply. Anim Behav. 2005;70:1011–21.

Bautista A, García-Torres E, Martínez-Gómez M, Hudson R. Do newborn domestic rabbits Oryctolagus cuniculus compete for thermally advantageous positions in the litter huddle? Behav Ecol Sociobiol. 2008;62:331–9.

Beyer C, Rivaud N. Sexual behaviour in pregnant and lactating domestic rabbits. Physiol Behav. 1969;4:753–7.

Blanchard RJ, Blanchard DC. Attack and defense in rodents as ethoexperimental models for the study of emotion. Prog Neuro-Psychopharmacol Biol Psychiatry. 1989;13(Suppl):S3–14. https://doi.org/10.1016/0278-5846(89)90105-x.

Brambell FWR. The reproduction of the wild rabbit, *Oryctolagus cuniculus (L.)*. J Zool. 1944;114:1–45.

Brecchia G, Bonanno A, Dall'Aglio C, Mercati F, Zerani M, Di Grigoli A, Boiti C. Neuroendocrine responses in neonatal mother deprived rabbits. Brain Res. 2009;1304:105–12.

Broekhuizen S, Mulder JL. Differences and similarities in nursing behaviour of hares and rabbits. Acta Zool Fenn. 1983;174:61–3.

Broekhuizen S, Bouman E, Went W. Variations in time of nursing in the brown hare (*Lepus europaeus*) and the European rabbit (*Oryctolagus cuniculus*). Mamm Rev. 1986;16:139–44. https://doi.org/10.1111/j.1365-2907.1986.tb00034.x.

Caba M, González-Mariscal G. The rabbit pup, a natural model of nursing anticipatory activity. Eur J Neurosci. 2009;30:1697–706.

Caba M, Silver R, González-Mariscal G, Jiménez A, Beyer C. Oxytocin and vasopressin immunoreactivity in rabbit hypothalamus during estrus, late pregnancy, and postpartum. Brain Res. 1996;720:7–16.

Caba M, Beyer C, González-Mariscal G, Morrell JI. Immunocytochemical detection of estrogen receptor alpha in the female rabbit forebrain: topography and regulation by estradiol. Neuroendocrinology. 2003a;77:208–22.

Caba M, Rovirosa MJ, Beyer C, González-Mariscal G. Immunocytochemical detection of progesterone receptor in the female rabbit forebrain: distribution and regulation by oestradiol and progesterone. J Neuroendocrinol. 2003b;15:855–64.

Caba M, Tovar A, Silver R, Morgado E, Meza E, Zavaleta Y, Juárez C. Nature's food anticipatory experiment: entrainment of locomotor behavior, suprachiasmatic and dorsomedial hypothalamic nuclei by suckling in rabbit pups. Eur J Neurosci. 2008;27:432–43.

Cano-Ramírez H, Hoffman KL. Activation of the orbitofrontal and anterior cingulate cortices during the expression of a naturalistic compulsive-like behavior in the rabbit. Behav Brain Res. 2017;320:67–74. https://doi.org/10.1016/j.bbr.2016.11.022.

Cano-Ramírez H, Hoffman KL. Activation of cortical and striatal regions during the expression of a naturalistic compulsive-like behavior in the rabbit. Behav Brain Res. 2018;351:168–77. https://doi.org/10.1016/j.bbr.2018.05.034.

Cano-Ramírez H, Pérez-Martínez LP, Hoffman KL. Dataset on c-Fos expression within components of corticostriatal thalamocortical circuits during the expression of a compulsive-like behavior in the female rabbit: brain-behavior relationships. Data Brief. 2020;34:106696. https://doi.org/10.1016/j.dib.2020.106696.

Castellini C. Reproductive activity and welfare of rabbit does. Ital J Anim Sci. 2007;6:743–7.

Castellini C, Canali C, Boiti C. Effect of mother–litter separation for 24 h, by closing the nest box or change of cage, on rabbit doe reproduction performance. World Rabbit Sci. 1998;6:199–203.

Chirino R, Beyer C, González-Mariscal G. Lesion to the main olfactory epithelium facilitates maternal behavior in virgin rabbits. Behav Brain Res. 2007;180:127–32.

Curley JP, Champagne FA. Influence of maternal care on the developing brain: mechanisms, temporal dynamics and sensitive periods. Front Neuroendocrinol. 2016;40:52–66. https://doi.org/10.1016/j.yfrne.2015.11.001.

Denenberg VH, Sawin PB, Fommer GP, Ross S. Genetic, physiological and behavioral background of reproduction in the rabbit: IV. An analysis of maternal behavior at successive parturitions. Behaviour. 1958;13:131–42.

Denenberg VH, Petropolus SF, Sawin PB, Ross S. Genetic, physiological and behavioral background of reproduction in the rabbit. VI. Maternal behavior with reference to scattered and cannibalized newborn and mortality. Behaviour. 1959;15:71–6.

Denenberg VH, Huff RL, Ross S, Sawin PB, Zarrow MX. Maternal behavior in the rabbit: the quantification of nest building. Anim Behav. 1963;11:494–9.

Deutsch JA. Nest building behaviour of domestic rabbits under semi-natural conditions. Br J Anim Behav. 1957;5:53–4.

Domínguez M, Aguilar-Roblero R, González-Mariscal G. Bilateral lesions of the paraventricular hypothalamic nucleus disrupt nursing behavior in rabbits. Eur J Neurosci. 2017;46:2133–40.

Drewett RF, Kendrick KM, Sanders DJ, Trew AM. A quantitative analysis of the feeding behavior of suckling rabbits. Dev Psychobiol. 1982;15:25–32.

El Nagar AG, Sánchez JP, Ragab M, Mínguez C, Baselga M. Genetic comparison of milk production and composition in three maternal rabbit lines. World Rabbit Sci. 2014;22:261–8.

Farkas TP, Szendrö ZS, Matics ZS, Radnai I, Mayer A, Gerencser ZS. Preference of rabbit does among different nest materials. In: Proceedings of the 11th World Rabbit Congress; 2016a June 15–18; Qingdao, China. p. 667–70.

Farkas TP, Szendrö ZS, Matics ZS, Radnai I, Mayer A, Gerencser ZS. Effect of different nest materials on performance of rabbit does. In: Proceedings of the 11th World Rabbit Congress; 2016b June 15–18; Qingdao, China. p. 197–200.

Fernández-Carmona J, Alqedra I, Cervera C, Moya J, Pascual JJ. Effect of lucerne-based diets on performance of reproductive rabbit does at two temperatures. Animal Sci. 2003;76:283–95.
Findlay A, Roth LR. Long-term dissociation of nursing behavior and condition of the mammary gland in the rabbit. J Comp Physiol Psychol. 1970;72:341–4.
Findlay A, Tallal P. Effect of reduced suckling stimulation on the duration of nursing in the rabbit. J Comp Physiol Psychol. 1971;76:236–41.
Forray A, Focseneanu M, Pittman B, McDougle CJ, Epperson CN. Onset and exacerbation of obsessive-compulsive disorder in pregnancy and the postpartum period. J Clin Psychiatry. 2010;71:1061–8. https://doi.org/10.4088/JCP.09m05381blu.
Fortun-Lamothe L, Lebas F. Effects of dietary energy level and source on foetal development and energy balance in concurrently pregnant and lactating primiparous rabbit does. Anim Sci. 1996;62:615–20.
Fuchs AR, Wagner G. Quantitative aspects of the release of oxytocin by suckling in unanaesthetized rabbits. Acta Endocrinol. 1963;44:581–92.
Fuchs AR, Cubile L, Dawood MY, Jorgensen FS. Release of oxytocin and prolactin by suckling in rabbits throughout lactation. Endocrinology. 1984;114:462–9.
García-Fernández HL, Chavira-Ramírez DR, González-Mariscal G. Long-lasting behavioral and neuroendocrine changes provoked in rabbits by cancelling a single nursing bout in early lactation. Dev Psychobiol. 2019;61:988–98. https://doi.org/10.1002/dev.21882.
García-Dalmán C, González-Mariscal G. Major role of suckling stimulation for inhibition of estrous behaviors in lactating rabbits: acute and chronic effects. Horm Behav. 2012;61:108–13.
Gibb JA. Sociality time and space in a sparse population of rabbits. J Zool Lond. 1993;229:581–607.
Gillan CM, Robbins TW. Goal-directed learning and obsessive-compulsive disorder. Philos Trans R Soc Lond Ser B Biol Sci. 2014;369:20130475. https://doi.org/10.1098/rstb.2013.0475.
González-Mariscal G. Mother rabbits and their offspring: timing is everything. Dev Psychobiol. 2007;49:71–6.
González-Mariscal G, Gallegos JA. New Zealand white rabbits show non-selective nursing in various types of nests. World Rabbit Sci. 2007;15:167–72.
González-Mariscal G, Gallegos JA. The maintenance and termination of maternal behavior in rabbits: involvement of suckling and progesterone. Physiol Behav. 2014;124:72–6.
González-Mariscal G, Melo AI. Bidirectional effects of mother-young contact on the maternal and neonatal brains. In: Von Bernhardi R, Eugenín JJ, editors. The plastic brain, Advances in experimental medicine and biology. Cham: Springer; 2017. p. 97–116.
González-Mariscal G, Poindron P. Parental care in mammals: immediate internal and sensory factors of control. In: Pfaff DW, Arnold AP, Etgen AM, Fahrbach SE, Rubin RT, editors. Hormones, brain and behavior. San Diego: Academic Press; 2002. p. 215–98.
González-Mariscal G, Rosenblatt JS. Maternal behavior in rabbits: a historical and multidisciplinary perspective. In: Rosenblatt JS, Snowdon CT, editors. Parental care: evolution, mechanisms and adaptive significance. New York: Academic Press; 1996. p. 333–60.
González-Mariscal G, Díaz-Sánchez V, Melo AI, Beyer C, Rosenblatt JS. Maternal behavior in New Zealand white rabbits: quantification of somatic events, motor patterns and steroid plasma levels. Physiol Behav. 1994;55:1081–9.
González-Mariscal G, Melo AI, Jiménez P, Beyer C, Rosenblatt JS. Estradiol, progesterone, and prolactin regulate maternal nest-building in rabbits. J Neuroendocrinol. 1996;8:901–7.
González-Mariscal G, Melo AI, Chirino R, Jiménez P, Beyer C, Rosenblatt JS. Importance of mother/young contact at parturition and across lactation for the expression of maternal behavior in rabbits. Dev Psychobiol. 1998;32:101–11.
González-Mariscal G, Melo AI, Parlow AF, Beyer C, Rosenblatt JS. Pharmacological evidence that prolactin acts from late gestation to promote maternal behavior in rabbits. J Neuroendocrinol. 2000;12:983–92.
González-Mariscal G, Jiménez P, Beyer C, Rosenblatt JS. Androgens stimulate specific aspects of maternal nest-building and reduce food intake in rabbits. Horm Behav. 2003;43:312–7.

González-Mariscal G, Chirino R, Beyer C, Rosenblatt JS. Removal of the accessory olfactory bulbs facilitates maternal behavior in virgin rabbits. Behav Brain Res. 2004a;152:89–95.

González-Mariscal G, Chirino R, Flores-Alonso JC, Rosenblatt JS, Beyer C. Intracerebroventricular injections of prolactin counteract the antagonistic effect of bromocriptine on rabbit maternal behaviour. J Neuroendocrinol. 2004b;16:949–55.

González-Mariscal G, Chirino R, Rosenblatt JS, Beyer C. Forebrain implants of estradiol stimulate maternal nest-building in ovariectomized rabbits. Horm Behav. 2005;47:272–9.

González-Mariscal G, Gallegos JA, Sierra-Ramírez A, Garza Flores J. Impact of concurrent pregnancy and lactation on maternal nest-building, estradiol and progesterone concentrations in rabbits. World Rabbit Sci. 2009a;17:145–52.

González-Mariscal G, Jiménez A, Chirino R, Beyer C. Motherhood and nursing stimulate c-FOS expression in the rabbit forebrain. Behav Neurosci. 2009b;123:731–9.

González-Mariscal G, Lemus AC, Vega-González A, Aguilar-Roblero A. Litter size determines circadian periodicity of nursing in rabbits. Chronobiol Int. 2013a;30:711–8.

González-Mariscal G, Toribio A, Gallegos-Huicochea JA, Serrano-Meneses MA. The characteristics of suckling stimulation determine the daily duration of mother-young contact and milk output in rabbits. Dev Psychobiol. 2013b;55:809–17.

González-Mariscal G, García-Dalmán C, Jiménez A. Biostimulation and nursing modify mating-induced c-FOS immunoreactivity in the female rabbit forebrain. Brain Res. 2015;1608:66–74.

González-Mariscal G, Caba M, Martínez-Gómez M, Bautista A, Hudson R. Mothers and offspring: the rabbit as a model system in the study of mammalian maternal behavior and sibling interactions. Horm Behav. 2016;77:30–41.

Gregerson KA. Prolactin: structure, function, and regulation of secretion. In: Neill JD, editor. Knobil and Neill's physiology of reproduction. San Diego: Elsevier; 2006. p. 1703–26. https://doi.org/10.1016/B978-012515400-0/50037-3.

Guglielmi V, Vulink NC, Denys D, Wang Y, Samuels JF, Nestadt G. Obsessive-compulsive disorder and female reproductive cycle events: results from the OCD and reproduction collaborative study. Depress Anxiety. 2014;31:979–87. https://doi.org/10.1002/da.22234.

Haber SN, Fudge JL, McFarland NR. Striatonigrostriatal pathways in primates form an ascending spiral from the shell to the dorsolateral striatum. J Neurosci. 2000;20:2369–82. https://doi.org/10.1523/JNEUROSCI.20-06-02369.2000.

Hammond J. Reproduction in the rabbit. London: Oliver and Boyd; 1925.

Hernández P, Gondret F. Rabbit meat quality. In: Maertens L, Coudert P, editors. Recent advances in rabbit sciences. Melle: ILVO; 2006. p. 269–90.

Hoffman KL, Rueda Morales RI. Toward an understanding of the neurobiology of "just right" perceptions: nest building in the female rabbit as a possible model for compulsive behavior and the perception of task completion. Behav Brain Res. 2009;204:182–91. https://doi.org/10.1016/j.bbr.2009.06.010.

Hoffman KL, Rueda Morales RI. D1 and D2 dopamine receptor antagonists decrease behavioral bout duration, without altering the bout's repeated behavioral components, in a naturalistic model of repetitive and compulsive behavior. Behav Brain Res. 2012;230:1–10. https://doi.org/10.1016/j.bbr.2012.01.041.

Hofmann M, Hoy S. Use of Avisoft SASLab Pro in investigation of vocalisation in rabbits. In: Proceedings of the 15th international conference on housing and diseases of rabbits, fur animals and pet animals; 2007 May 9–10; Celle, Germany. p. 181–5.

House SJ, Tripathi SP, Knight BT, Morris N, Newport DJ, Stowe ZN. Obsessive-compulsive disorder in pregnancy and the postpartum period: course of illness and obstetrical outcome. Arch Womens Ment Health. 2016;191:3–10. https://doi.org/10.1007/s00737-015-0542-z.

Hoy S, Schuh D. Sociometric investigations in groups of wild and domestic rabbits with one buck and two or three does. In: Proceedings of the 8th World Rabbit Congress; 2004 Sept 7–10; Puebla, Mexico. p. 1235–40.

Hoy S. Nursing behavior of wild and domestic rabbits. In: Maertens L, Coudert P, editors. Recent advances in rabbit sciences. Melle: ILVO; 2006. p. 75–7.

Hoy S. Behavior of rabbits. In: Hoy S, editor. Farm animal ethology. Ulmer; 2009. p. 192–3.
Hoy S, Selzer D. Frequency and time of nursing in wild and domestic rabbits housed outdoors in free range. World Rabbit Sci. 2002;10:77–83.
Hudson R, Distel H. The pattern of behaviour of rabbit kits in the nest. Behaviour. 1982;79:255–71.
Hudson R, Distel H. Nipple location by newborn rabbits: behavioural evidence for pheromonal guidance. Behaviour. 1983;85:260–75.
Hudson R, Distel H. The contribution of the olfactory and tactile modalities to the nipple-search behavior of newborn rabbits. J Comp Physiol A. 1985;157:599–605.
Hudson R, Distel H. Temporal pattern of suckling in rabbit pups: a model of circadian synchrony between mother and young. In: Reppert SM, editor. Development of circadian rhythmicity and photoperiodism in mammals. Ithaca: Perinatology Press; 1989. p. 83–102.
Hudson R, Bilkö A, Altbäcker V. Nursing, weaning and the development of independent feeding in the rabbit (Oryctolagus cuniculus). Z Säugetierk. 1996;61:39–48.
Hudson R, Cruz Y, Lucio RA, Ninomiya J, Martínez-Gómez M. Temporal and behavioral patterning of parturition in rabbits and rats. Physiol Behav. 1999;66:599–604.
Hudson R, Schaal B, Martínez-Gómez M, Distel H. Mother-young relations in the European rabbit: physiological and behavioral locks and keys. World Rabbit Sci. 2000;8:85–90.
Iraqi MM, Shenana ME, Baselga M. Some factors affecting production and milk composition characters in a crossbreeding experiment involving Gaballi and v-line rabbits in Egypt. World Rabbit Sci. 2007;15:151–9.
Jaspers E, Balsters JH, Kassraian Fard P, Mantini D, Wenderoth N. Corticostriatal connectivity fingerprints: probability maps based on resting-state functional connectivity. Hum Brain Mapp. 2017 Mar;38:1478–91.
Jilge B. The ontogeny of circadian rhythms in the rabbit. J Biol Rhythm. 1993;8:247–60.
Jilge B. Ontogeny of the rabbit's circadian rhythms without an external zeitgeber. Physiol Behav. 1995;58:131–40.
Jiménez A, González-Mariscal G. Maternal responsiveness to suckling is modulated by time post-nursing: a behavioral and c-FOS/oxytocin immunocytochemistry study in rabbits. J Neuroendocrinol. 2019; https://doi.org/10.1111/jne.12788.
Jiménez A, Young LJ, Triana-Del Río R, La Prairie JL, González-Mariscal G. Neuroanatomical distribution of oxytocin receptor binding in the female rabbit forebrain: variations across the reproductive cycle. Brain Res. 2015;1629:329–39.
Kahn L, Sutton B, Winston HR, Abosch A, Thompson JA, Davis RA. Deep brain stimulation for obsessive-compulsive disorder: real world experience post-FDA-humanitarian use device approval. Front Psych. 2021;12:568932. https://doi.org/10.3389/fpsyt.2021.568932.
Karpati K, Cheritah L, Bilko A. Nipple-switch and nipple-search behavior: factors contributing to the efficiency of suckling. In: Proceedings of the 7th World Rabbit Congress; July 4–7; Valencia, Spain. World Rabbit Science. 2000;8(Suppl) 1:1–5.
Kraft R. Comparative behavioral studies in wild and domestic rabbits. Thesis; 1976; University of Erlangen Germany. German.
Lebas F, Cousin M, Sardi G. Effet de la simultanéité de la lactation et de la gestation sur les performances laitières chez la lapine. Ann Zootech. 1972;21:129–31.
Leckman JF, Mayes LC, Feldman R, Evans DW, King RA, Cohen DJ. Early parental preoccupations and behaviors and their possible relationship to the symptoms of obsessive-compulsive disorder. Acta Psychiatr Scand Suppl. 1999;396:1–26. https://doi.org/10.1111/j.1600-0447.1999.tb10951.x.
Lincoln DW. Suckling: a time constant in the nursing behavior of the rabbit. Physiol Behav. 1974;13:711–4.
Lloyd HG, McCowan D. Some observations on the breeding burrows of the wild rabbit, Oryctolagus cuniculus, on the island of Skokholm. J Zool (Lond). 1968;156:540–9.
Lorberbaum JP, Newman JD, Horwitz AR, Dubno JR, Lydiard RB, Hamner MB, Bohning DE, George MS. A potential role for thalamocingulate circuitry in human maternal behavior. Biol Psychiatry. 2002;51:431–45. https://doi.org/10.1016/s0006-3223(01)01284-7.

Ludwiczak A, Składanowska-Baryza J, Kuczynska B, Stanisz M. Hycole doe milk properties and kit growth. Animals. 2020;10:214–30.

Lukefahr SD, McNitt JI, Cheeke PR, Patton NM. Rabbit production. 2nd ed. Oxfordshire: CABI; 2021.

Maertens L, De Groote G. Comparison of feed intake and milk yield of does under normal and high ambient temperature. J Appl Rabbit Res. 1990;13:159–62.

Maertens L, Lebas F, Szendrö Z. Rabbits milk: a review of quality, quality and non-dietary affecting factors. World Rabbit Sci. 2006;14:205–30.

Mahmoud EAA. A study on some factors affecting milk yield in New Zealand white rabbits under Egyptian conditions. Benha Vet Med J. 2013;25:13–22.

Martínez-Gómez M, Juárez M, Distel H, Hudson R. Overlapping litters and reproductive performance in the domestic rabbit. Physiol Behav. 2004;82:629–36.

Matics Z, Szendrö Z, Hoy S, Nagy I, Radnai I, Biró-Németh E, Gyovai M. Effect of different management methods on the nursing behaviour of rabbits. World Rabbit Sci. 2004;12:95–108.

McGuinness M, Blissett J, Jones C. OCD in the perinatal period: is postpartum OCD (ppOCD) a distinct subtype? A review of the literature. Behav Cogn Psychother. 2011;39:285–310. https://doi.org/10.1017/S1352465810000718.

McNeilly AS, Friesen HG. Prolactin during pregnancy and lactation in the rabbit. Endocrinology. 1978;102:1548–54.

McNitt J, Lukefahr S. Effects of breed, parity, day of lactation and number of kits on milk production of rabbits. J Anim Sci. 1990;68:1505–12.

Melo AI, González-Mariscal G. Placentophagia in rabbits: incidence across the reproductive cycle. Dev Psychobiol. 2003;43:37–43.

Mena F, Clapp C, Aguayo D, Martínez de la Escalera G. Prolactin and propranolol prevent the suckling-induced inhibition of lactation in rabbits. Physiol Behav. 1990a;48:311–5.

Mena F, Clapp C, Martínez de la Escalera G. Age-related stimulatory and inhibitory effects of suckling regulate lactation in rabbits. Physiol Behav. 1990b;48:307–10.

Meza E, Juárez C, Morgado E, Zavaleta Y, Caba M. Brief daily suckling shifts locomotor behavior and induces PER1 protein in paraventricular and supraoptic nuclei, but not in the suprachiasmatic nucleus, of rabbit does. Eur J Neurosci. 2008;28:1394–403. https://doi.org/10.1111/j.1460-9568.2008.06408.x.

Meza E, Waliszewski SM, Caba M. Circadian nursing induces PER1 protein in neuroendocrine tyrosine hydroxylase neurons in the rabbit doe. J Neuroendocrinol. 2011;23:472–80. https://doi.org/10.1111/j.1365-2826.2011.02138.x.

Meza E, Aguirre J, Waliszewski S, Caba M. Suckling induces a daily rhythm in the preoptic area and lateral septum but not in the bed nucleus of the stria terminalis in lactating rabbit does. Eur J Neurosci. 2015;41:196–204. https://doi.org/10.1111/ejn.12776.

Miguel EC, Lopes AC, McLaughlin NCR, Norén G, Gentil AF, Hamani C, Shavitt RG, Batistuzzo MC, Vattimo EFQ, Canteras M, De Salles A, Gorgulho A, Salvajoli JV, Fonoff ET, Paddick I, Hoexter MQ, Lindquist C, Haber SN, Greenberg BD, Sheth SA. Evolution of gamma knife capsulotomy for intractable obsessive-compulsive disorder. Mol Psychiatry. 2019;24:218–40. https://doi.org/10.1038/s41380-018-0054-0.

Mohamed M, Szendrö ZS. Studies on nursing and milk production in does and milk intake and suckling behavior of their kits. In: Proceedings of the 5th World Rabbit Congress; 1992 July 25–30; Corvallis, USA. p. 708–16.

Mowrer OH, Lamoreaux RR. Fear as an intervening variable in avoidance conditioning. J Comp Psychol. 1946;39:29–50. https://doi.org/10.1037/h0060150.

Myers K, Poole W. The effects of season and population increase on behaviour. CSIRO Wildl Res. 1961;6:1–41.

Neville MC. Lactation and its hormonal control. In: Neill JD, editor. Knobil and Neill's physiology of reproduction. San Diego: Elsevier; 2006. p. 2993–3054.

Nitschke JB, Nelson EE, Rusch BD, Fox AS, Oakes TR, Davidson RJ. Orbitofrontal cortex tracks positive mood in mothers viewing pictures of their newborn infants. NeuroImage. 2004;21:583–92. https://doi.org/10.1016/j.neuroimage.2003.10.005.

Numan M, Insel TR. The neurobiology of parental behavior. New York: Springer; 2003.

Numan M, Stolzenberg DS. Medial preoptic area interactions with dopamine neural systems in the control of the onset and maintenance of maternal behavior in rats. Front Neuroendocrinol. 2009;30:46–64. https://doi.org/10.1016/j.yfrne.2008.10.002.

Numan M, Fleming AS, Lévy F. Maternal behavior. In: Neill JD, editor. Knobil and Neill's physiology of reproduction. San Diego: Elsevier; 2006. p. 1921–93.

Olazábal DE, Pereira M, Agrati D, Ferreira A, Fleming AS, González-Mariscal G, Lévy F, Lucion AB, Morrell JI, Numan M, Uriarte N. New theoretical and experimental approaches on maternal motivation in mammals. Neurosci Biobehav Rev. 2013;37:1860–74. https://doi.org/10.1016/j.neubiorev.2013.04.003.

Pałka S, Kmiecik M, Kozioł K, Otwinowska-Mindur A, Migdał L, Bieniek J. The effect of breed on litter size. Sci Ann Pol Soc Anim Prod. 2017;13:25–9.

Parer I, Fullaghar PI, Malafant KW. The history and structure of a large warren of the rabbit in Canberra, A.C.T. Aust Wildl Res. 1987;14:505–13.

Partridge GG, Lobley GE, Fordyce RA. Energy and nitrogen metabolism of rabbits during pregnancy, lactation, and concurrent pregnancy and lactation. Br J Nutr. 1986;56:199–207.

Pascual JJ, Cervera C, Blas E, Fernández-Carmona J. Milk yield and composition in rabbit does using high fat diets. In: Proceedings of the 6th World Rabbit Congress; 1996 Toulouse, France; 1:259–61.

Pascual J, Cervera C, Blas E, Fernández-Carmona J. High-energy diets for reproductive rabbit does: effect of energy source. Nutr Abstr Rev. 2003;73:27R–39R.

Rafai P, Papp Z. Temperature requirement of does for optimal performance. Arch Exp Vet Med Leipzig. 1984;38:450–7.

Ramírez VD, Beyer C. The ovarian cycle of the rabbit: its neuroendocrine control. In: Knobil E, Neill J, editors. The physiology of reproduction. New York: Raven Press; 1988. p. 1873–92.

Ranote S, Elliott R, Abel KM, Mitchell R, Deakin JF, Appleby L. The neural basis of maternal responsiveness to infants: an fMRI study. Neuroreport. 2004;15:1825–9. https://doi.org/10.1097/01.wnr.0000137078.64128.6a.

Rödel HG, Dausmann KH, Starkloff A, Schubert M, von Holst D, Hudson R. Diurnal nursing pattern of wild-type European rabbits under natural breeding conditions. Mamm Biol. 2012;77:441–6. https://doi.org/10.1016/j.mambio.2012.04.002.

Ross S, Denenberg VH, Sawin PB, Meyer P. Changes in nest building behavior in multiparous rabbits. Br J Anim Behav. 1956;4:69–74.

Ross S, Denenberg VH, Frommer GP, Sawin PB. Genetic, physiological and behavioral background of reproduction in the rabbit. V. Non-retrieving of neonates. J Mammal. 1959;40:91–6.

Ross S, Zarrow MX, Sawin PB, Denenberg VH. Maternal behavior in the rabbit under semi-natural conditions. Anim Behav. 1963;11:283–5.

Savietto D, Cervera C, Blas E, Baselga M, Larsen T, Friggens NC, et al. Environmental sensitivity differs between rabbit lines selected for reproductive intensity and longevity. Animal. 2013;7:1969–77.

Savietto D, Cervera C, Ródenas L, Martínez-Paredes E, Baselga M, García-Diego FJ, et al. Different resource allocation strategies result from selection for litter size at weaning in rabbit does. Animal. 2014;8:618–28.

Schaal B, Coureaud G, Langlois D, Giniès C, Sémon E, Perrier G. Chemical and behavioural characterization of the rabbit mammary pheromone. Nature. 2003;424:68–72.

Schaal B, Coureaud G, Moncomble A, Langlois D, Perrier G. Odour cues and pheromones in the mediation of rabbit female-offspring relations. In: Maertens L, Coudert P, editors. Recent advances in rabbit science. Melle: ILVO; 2006. p. 79–81.

Schley P. Rabbits. Stuttgart: Eugen Ulmer; 1985. German.

Schlolaut W, editor. The big book of the rabbit. 2nd ed. Frankfurt on Main: DLG; 1998. German.

Schulte I. Investigations on nursing and suckling behavior and on mother-child-relationship in rabbits of New Zealand white breed under the use of infrared video technique. Thesis; 1998; University of Leipzig. German.

Schulte I, Hoy St. Investigations on suckling behavior in rabbits using infrared video technique. Aktuelle Arbeiten zur artgemäßen Tierhaltung. KTBL-Schrift 1996;373, p. 210–7. German.

Seitz K. Investigations on nursing behavior of domestic rabbit does and on milk intake, liveweight development and losses of kits. Thesis 1997; University of Giessen. German.

Selzer D. Comparative investigations on behavior of wild and domestic rabbits kept under different housing conditions. Thesis 2000; University of Gießen. German.

Selzer D, Lange K, Hoy ST. Investigations on mother-child-relationship in domestic rabbits considering different housing conditions. In: Proceedings of the 12. Arbeitstagung über Haltung und Krankheiten der Kaninchen, Pelztiere und Heimtiere. 2001; October 9–10; Celle, Germany. p. 106–14. German.

Sofroniew M. Vasopressin, oxytocin and their related neurophysins. In: Björklund A, Hökfelt K, editors. Handbook of chemical neuroanatomy, Vol 4, part 1. Amsterdam: Elsevier; 1985. p. 93–165.

Soriano-Mas C. Functional brain imaging and OCD. Curr Top Behav Neurosci. 2021; https://doi.org/10.1007/7854_2020_203.

Strathearn L, Li J, Fonagy P, Montague PR. What's in a smile? Maternal brain responses to infant facial cues. Pediatrics. 2008;122:40–51. https://doi.org/10.1542/peds.2007-1566.

Szechtman H, Harvey BH, Woody EZ, Hoffman KL. The psychopharmacology of obsessive-compulsive disorder: a preclinical roadmap. Pharmacol Rev. 2020;72:80–151. https://doi.org/10.1124/pr.119.017772.

Szendrö Z, Papp Z, Kustos K. Effect of environmental temperature and restricted feeding on production of rabbit does. In: Proceedings of the 2nd international conference on rabbit production in hot climates; 1998 Sep 7–9; Adana, Turkey, 41, p. 11–7.

Szendrö Z, Gerencsér Z, McNitt JI, Matics Z. Effect of lighting on rabbits and its role in rabbit production: a review. Lives Sci. 2016;183:12–8.

Szendrö Z, Papp Z, Kustos K. Effect of ambient temperature and restricted feeding on the production of rabbit does and their kits. Acta Agraria Kaposvariensis. 2018;22:1–17.

Theau-Clément M, Mercier P. Effect of a doe–litter separation on rabbit doe reproductive performance and growth of the young. World Rabbit Sci. 1999;7:177–9.

Theau-Clèment M, Boiti C, Bonnano A, Eiben C, Maertens L, Szendrö ZS. Alternative methods for the synchronisation of oestrus in lactating does. In: Maertens L, Coudert P, editors. Recent advances in rabbit science. Melle: ILVO; 2006. p. 21–38.

Ubilla E, Rebollar PG, Pazo D, Esquifino A, Alvariño JMR. Effects of doe–litter separation on endocrinological and productivity variables in lactating rabbits. Lives Prod Sci. 2000;67:67–74.

Val-Laillet D, Nowak R. Early discrimination of the mother by rabbit pups. Appl Anim Behav Sci. 2008;111:173–82.

Verga M, Luzi F. Behavior of kits. In: Maertens L, Coudert P, editors. Recent advances in rabbit science. Melle: ILVO; 2006. p. 83–6.

Von Holst D. Living in the group: impact on behavior, fertility, health and Lebenserwartung European wild rabbits. In: Proceedings Aktuelle Arbeit zur artgemäßen Tierhaltung, 2001; November 20–22; Freiburg, Germany, KTBL-Schrift 407, p. 51–63.

Von Holst D, Hutzelmeyer H, Kaetzke P, Khaschei M, Rödel HG, Schrutka H. Social rank, fecundity and lifetime reproductive success in wild European rabbits (*Oryctolagus cuniculus*). Behav Ecol Sociobiol. 2002;51:245–54.

Wakerley JB. Milk ejection and its control. In: Neill JD, editor. Knobil and Neill's physiology of reproduction. San Diego: Elsevier; 2006. p. 3129–90.

Woody EZ, Hoffman KL, Szechtman H. Obsessive compulsive disorder (OCD): current treatments and a framework for neurotherapeutic research. Adv Pharmacol. 2019;86:237–71. https://doi.org/10.1016/bs.apha.2019.04.003.

Wu Y, Zhao A, Qin Y. Effect of lighting schedule, intensity and colour on reproductive performance of rabbit does. World Rabbit Sci. 2021;29:59–64.

Xiccato G, Parigi-Bini R, Dalle Zotte A, Carazzolo A, Cossu ME. Effect of dietary energy level, addition of fat and physiological state on performance and energy balance of lactating and pregnant rabbit does. Anim Sci. 1995;61:387–98.

Zarrow MX, Sawin PB, Ross S, Denenberg VH. Maternal behavior in the rabbit: evidence for an endocrine basis of maternal nest building and additional data on maternal nest building in the Dutch-belted race. J Reprod Fertil. 1961;2:152–62.

Zarrow MX, Farooq A, Denenberg VH, Sawin PB, Ross S. Maternal behavior in the rabbit: endocrine control of maternal nest building. J Reprod Fertil. 1963;6:375–83.

Zarrow MX, Denenberg VH, Kalberer WD. Strain differences in the endocrine basis of maternal nest-building in the rabbit. J Reprod Fertil. 1965;10:397–401.

Zimmermann E. Weight gain in young rabbits in dependence on parity of doe. In: Proceedings of the 7th Congress on housing and diseases of rabbits, fur providing animals and pet animals; 1990; Celle, Germany, p. 46–9.

Parental Behavior in Bovines

Lena Lidfors

Abstract After approximately 9 months, cows give birth to one calf. They usually try to separate from the herd if kept outdoors, but this depends on the possibilities to hide. When the calf is born, the cow immediately stands up and starts licking the calf, and this is most intense for the first hour. During this time, the cow is very protective of her calf and may attack anyone trying to come close to it, especially if the cow is not so used to human handling. The calf tries to stand up and falls in the beginning, but it usually stands about half an hour later. Once standing, the calf searches for the udder, and after approximately 2 h, the calf is suckling the colostrum from the cow's teats. The calf is born without an immune defense, and it, therefore, needs to suckle enough colostrum from the cow to get protection against infections. If the calf does not succeed in suckling within 4–6 h after birth, the farmer must milk the cow and give the calf at least 2–4 l of colostrum in a teat bottle. If the farmer has frozen colostrum of good quality in the freezer, and it is either difficult to milk the cow or her colostrum is of too low quality, the frozen colostrum can be warmed up and given to the calf. Beef calves are usually allowed to stay with their mother until they are 6–8 months old, when they are separated from the cow and then weaned. Dairy calves are usually separated from their mother immediately or within a few days after birth and raised by artificial milk feeding. In the beginning or during the whole milk period (6–12 weeks), they may be kept individually indoors in crates or outdoors in hutches. After weaning, they are usually group housed.

Keywords Parturition · Isolation · Maternal licking · Vocalization · Calf behavior · Suckling · Mother-young attachment · Weaning · Cow-calf separation

---

L. Lidfors (✉)
Department of Animal Environment and Health, Swedish University of Agricultural Sciences, Skara, Sweden
e-mail: Lena.Lidfors@slu.se

G. González-Mariscal (ed.), *Patterns of Parental Behavior*, Advances in Neurobiology 27, https://doi.org/10.1007/978-3-030-97762-7_6

1 Introduction

Domestic bovines are mainly of two types, those without a hump (*Bos taurus*) and those with a hump called zebu cattle (*Bos indicus*). *Bos taurus* are mainly found in the temperate regions of the world and are split into three types: (1) dairy cows, where the calf by tradition is removed from the cow shortly after birth in order to get as much milk from the cow as possible; (2) beef cattle, where the calf stays with the mother until 6–8 months and thereafter is fed to develop muscle used for human meat consumption; and (3) dual-purpose cattle, where the cow is both kept for milk and meat production. Through genetics, these different types have been bred for increased production over a long time. *Bos indicus* have also been split into dairy and beef breeds in the hot regions of the world where they are the main cattle breeds. Attempts have been made to keep dairy cows of *Bos taurus* in the hot regions, but they cannot cope with the high temperatures. Therefore, dairy cows of *Bos indicus* have been crossbred with *Bos taurus* in order to increase the milk production in cattle kept in these hot regions. Despite these different cattle types and breeds, their parental behavior does not seem to differ much. Therefore, this chapter will present results from studies on both *Bos taurus* and *Bos indicus* cattle, assuming that there are no major differences in their behavior.

To understand the basic parental behavior in bovines, it is important to study them in natural or seminatural environments where they can show their natural behavior as much as possible. This has been done on Camargue cattle roaming free in the south of France (Schloeth 1958, 1961), Chillingham cattle kept in large areas in the south of Scotland (Hall 1989), feral cattle on the Orkney island of Swona (Hall and Moore 1986), feral cattle on Amsterdam Island (Berteaux and Micol 1992) and semi-wild free-ranging Maremma calves in Italy (Vitale et al. 1986). Our own studies have been done on crossbred beef cattle kept on a ranch in the south of Sweden during 1986–1988 (Lidfors 1994). Studies of natural parental behavior in *Bos indicus* have been carried out on Maasai cattle in Kenya (Reinhardt et al. 1978). Some studies have also observed dairy cows around calving to try to understand their maternal behavior (Edwards and Broom 1982) and determine if it is different from the free-ranging semi-wild cattle (Lidfors 1996). There are also studies on how to keep and manage cows before and during calving in order to follow their natural behaviors. This will be presented in a second section.

Removing the calf from the mother at birth, a practice in dairy cattle, has been criticized during the last couple of years, and there are ongoing studies on how to manage milking the cow and still keep the calf with the mother. In the third section of this chapter, I will present studies on different methods to keep the calf with its mother or raise it by a foster cow.

One important aspect of bovine behavior is that the natural group composition consists of a maternal group with cows of different ages and their offspring. Male calves leave the maternal group at around 1.5–2 years and form bachelor groups. When the young bulls reach an age of around 4–5 years, they become more solitary and compete to mate with the cows in the maternal groups. Hall (1989) found that the older bulls kept territories to which they tried to attract cows for mating. As the bulls are not involved in any parental behavior, this chapter will mainly use the word maternal behavior or mother-young behavior.

2 Origin and Distribution of Cattle

Cattle have their origin in the aurochs (*Bos primigenius*) which evolved in Asia and then spread all over Europe. Aurochs were large animals and bulls weighed more than 1 ton. They were heavily hunted for their meat and slowly declined in number until the last animal died in an animal park in Poland in 1627. Since then, the aurochs became extinct. The Hannover Zoo has tried to rebreed the aurochs from up to ten different old cattle breeds, and the offspring of those can be found in different zoos around the world.

Cattle were domesticated by humans already 9000 years ago (Clutton-Brock 1999). The main reasons for domestication were meat, milk, blood, power, leather, and feces for human activities (e.g., fertilizers for agriculture). Archeological remains show that already on 2050 BC humans were milking cows (Ekesbo and Gunnarsson 2018). From the start of domestication until the 1920s, dairy cows were hand milked three times per day mostly by women (Ekesbo and Gunnarsson 2018). Around the 1950s, milking machines became more common, which increased the number of cows that could be milked at the same time and reduced the number of milking to twice a day; mainly men took over the work (Ekesbo and Gunnarsson 2018). Today, dairy cattle herds are larger, and carrousel milking or automatic milking systems are common.

The world population of cattle during the inventory in 2021 was found to be 1,000,967,000 heads (Cook 2021). India has the largest number of cattle (30.52%, including water buffalo), followed by Brazil (25.25%), China (9.55%), the USA (9.35%), and the European Union (8.55%) (Cook 2021).

3 Breeds of Cattle

Cattle are divided into *Bos taurus* (humpless cattle) and *Bos indicus* (zebus, humped cattle, Brahman cattle). *Bos taurus* are the typical cattle of Europe, northeastern Asia, and parts of Africa, and most of the breeds are adapted to cooler climates. *Bos taurus* are used for milk production, beef production, to keep the landscape open, and their by-products such as hides are used in the leather industry, whereas manure is used to fertilize growing crops. *Bos indicus* originated in South Asia, and it is characterized by a fatty hump on their shoulders, drooping ears, and a large dewlap. They are adapted to high temperatures and are used as draught oxen, dairy cattle, beef cattle, and their by-products, such as hides and dung, for fuel, and manure is used for fertilizing crops. *Bos indicus* are bred with *Bos taurus* to increase the milk production in several countries where the climate is too hot to keep pure *Bos taurus*.

There are more than 1000 breeds of cattle around the world (Hall 2002). Cook (2021) presents cattle breeds based on their origin of country. To mention some common *Bos taurus* dairy breeds, Holstein is the most common breed in North America and Europe today. Other breeds are the Ayrshire or Red breeds, from which others like, Swedish Red, etc. are derived. Jersey cows are another dairy breed that is common in some countries. Common beef breeds of *Bos taurus* are Charolais, Limousin, Hereford, and Aberdeen Angus (Black Angus vs. Red Angus). Dual-purpose cattle breeds are bred and kept to produce both milk and meat. The females are used for milk production and breeding, whereas unwanted bulls are kept for beef production. These breeds are ideal for smaller farms with limited specialization. The second most common cattle breed in the world is the dual-purpose *Bos taurus* Fleckvieh/Simmental with an estimated 41 million cattle. Other breeds are Shorthorn and Red Poll cattle both developed in England. Within *Bos indicus*, Sahiwal from Pakistan and Gir from Gujarat are the most common dual-purpose cattle breeds which have spread to many countries around the world.

Adult bovines of *Bos taurus* are large animals where cows weigh from 600 to 700 kg and bulls weigh from 900 to 1200 kg. However, there are large differences depending on the breed. The old dairy breeds that used to graze on the lower mountains and forests during summer times in the old farm landscape are one of the smaller cattle, weighing from 350 to 600 kg. These old dairy mountain breeds are polled (i.e., they are inherently born without the possibility to grow horns). Today, there are some beef breeds where the horns have been bred away, for example, Aberdeen Angus, Hereford, and Charolais.

For more cattle breeds, please visit web pages such as http://afs.okstate.edu/breeds/cattle/, https://www.britannica.com/animal/cattle-livestock, and http://www.thecattlesite.com/breeds/beef/.

4 Cow-Calf Behavior When Kept in the Wild or Under Seminatural Conditions

Behavior Around Calving

After approximately 9 months (± 14 days), the cow gives birth to usually one calf. Twins are reported to be born in 3–5% of Holstein Friesian dairy cattle (Gáspárdy et al. 2018), in 4.2% of Holstein calvings in Minnesota (Del Río et al. 2007), and in 3.4% of Swedish Red and 4.5% in Swedish Holstein (Swedish Dairy Association 2007). The frequency of twins increases with increasing age and milk yield (Ekesbo and Gunnarsson 2018). The cow can also get triplets, but that is rare. The calf of Camargue cattle is generally born during early spring (Schloeth 1961). In the feral cattle on Amsterdam Island, reproduction appears to be seasonal, as 78% of births occurred within 4 months (Berteaux and Micol 1992). Reinhardt et al. (1986) found that breeding in semi-wild Scottish Highland cattle was synchronized in May and

June and calving coincided with the spring flush of grass. The age at first calving ranged from 2 to 4 years in feral cattle on Amsterdam Island (Berteaux and Micol 1992) and 2–3 years in semi-wild Scottish Highland cattle (Reinhardt et al. 1986). Semi-wild Boran cattle had their first calves at about 34 months (Reinhardt 1982). The fertility (rate of conception) was estimated at 0–73 fetuses per cow per year (Berteaux and Micol 1992). The calving interval was on average 391 days in a semi-wild Scottish Highland cattle herd (Reinhardt et al. 1986) and 430 days with a range from 304 to 801 days in a semi-wild Boran cattle herd (Reinhardt 1982). The calving interval of managed Boran cattle living on the same ranch as the semi-wild herd averaged 494 days with a range of 306–798 days.

During the first year, calving mortality was 5% in the semi-wild Scottish Highland cattle herd (Reinhardt et al. 1986). Calf mortality from birth until weaning in extensively kept beef cattle was reported by Portuguese farmers to be 5.7% (Santos et al. 2019). Holstein cows were found to have a calf mortality of 7.2% with singleton births and 28.2% with twins (Del Río et al. 2007). Further calf mortality for singletons was 5.0% for multiparous cows and 10.7% for first calving primiparous cows (Del Río et al. 2007). There are different reasons for calf mortality in free-ranging cattle, such as difficult calving, weak calves, weather conditions during the calving season, accidents, and predation. In dairy cattle, many studies have analyzed the reasons for calf mortality, but this is outside the scope of this chapter.

The time of birth has been varying in different studies; in zebu cattle (*Bos indicus*), 55% of the births were during the day and 45% during the night (Reinhardt 1982), whereas in Hereford (*Bos taurus*), 82% of the calves were born between midday and midnight (George and Barger 1974). In a herd of mixed beef and dairy breeds, calving was equally distributed throughout the day and night hours, although significantly more cows calved from 11 to 13 h (Owens et al. 1985). In dairy cows, more calves were born at night than during the day (Edwards 1982).

One to several days before calving, the cow moves away from the herd in short trips (Brownlee 1950; Reinhardt et al. 1977; Edwards and Broom 1982). However, Lidfors et al. (1994a) found that cows of an old dairy breed kept in a forest area did not leave the herd until a few hours before their calving in a secluded place. If the cow stays in the herd, she may approach and attempt to sniff or lick alien calves (Edwards and Broom 1982; Lidfors et al. 1994a). This was seen by nearly half of the cows within 24 h of their own calving by Finger and Brummer (1969), which lead them to propose a hormonal basis for maternal behavior. This behavior is rarer in primiparous than multiparous cows, suggesting that experience also may have some effect on the elicitation of maternal behavior (Finger and Brummer 1969). Owens et al. (1985) observed a few cases of calf-stealing by pre-parturient cows. In my own research of mother-young behavior in beef cows kept outdoors all year round on a ranch, I have also observed attempts of calf-stealing in a primiparous cow after having rejected her own calf and in multiparous cows that have lost their own calf (Lidfors 1994).

There are different reports about where cows give birth. The Maasai cattle (*Bos indicus*) cows were found to give birth in a concealed place (Reinhardt et al. 1977).

Fig. 1 Cow with young calf in isolation from other cows in the Finnish forest. (Photo: Lena Lidfors)

Chillingham cattle cows were found to isolate themselves at calving (Hall 1979), as did cows kept free in a Russian study (Baskin and Stepanov 1993).

However, free-ranging beef cows on a ranch were only rarely isolated from the herd at parturition (Lidfors et al. 1994a). Similar results were found in beef and dairy cows (Edwards 1982; Kiley-Worthington and de la Plain 1983; Wood-Gush et al. 1984). Although most cows calved in the cow groups, some of them were observed to give birth completely separated from the rest (Edwards and Broom 1982). Kiley-Worthington and de la Plain (1983) found that a purposeful separation from the herd was rare among beef cows, because during the course of their "restless" behavior, pre-parturient cows tended not to synchronize their activities with the herd to the same degree as usual resulting in that they were left behind as the herd grazed on. In the free-ranging beef cows, the calving cows were found in the shelters with three walls, a roof, and a deep straw bed more often than the pregnant cows were; thus, a slight separation from the herd was achieved (Lidfors et al. 1994a). Wagnon (1963) found that especially nervous older cows and heifers would seek an isolated spot before calving. This was also found in our own studies, where the primiparous cows of free-ranging Finn cattle were more prone to isolate themselves in the forest at calving than the multiparous cows (Lidfors et al. 1994a; Fig. 1).

Lidfors (1994) presented three main reasons for why domesticated cows do not seek as much isolation at calving as their ancestors:

1. The geography of the calving areas did not stimulate the isolation-seeking behavior at calving.
2. The domestication of beef and dairy cows has reduced their motivation to seek isolation at calving.

3. Cows are flexible in their behavioral repertoire so that when the calving area is grassland without any trees or bushes, they calve relatively close to other cows, and when the area is dense with trees and bushes, they seek a secluded place for calving. This is supported by the results on free-ranging beef cattle on an open ranch compared to Finn cattle in the forest (Lidfors et al. 1994a), but also from studies on American bison cows (Lott and Galland 1985) and reindeer (Espmark 1971).

During calving, the cow is lying down, but she stands up and changes the lying side several times before the calf is finally born. Selman et al. (1970a) found that during 20 out of 30 calvings, the cow was lying down and giving birth until the hips of the calf were out, after which the cow stood up and dropped the calf to the ground. Nine cows completed calving while lying down, and only one stood up during the complete calving (Selman et al. 1970a). Lidfors and Jensen (1988) found that immediately after parturition, all 19 out of the 20 cows stood up and started to lick the calf. Twelve of these cows licked their calf and seven performed other behaviors such as eating amniotic membranes or sniffing the calf (Lidfors and Jensen 1988).

The Behavior of the Calf Immediately After Birth

The first sign of life in the calf is often the blinking of the eyes as it dangles from its mother's vulva (Kiley-Worthington and de la Plain 1983). The first obvious movements consist of head shaking, snuffing, and sneezing, which often is as soon as the calf's shoulders are passed (Selman et al. 1970b; Kiley-Worthington and de la Plain 1983). This is probably to help clear the nostrils and ears of birth fluids (Kiley-Worthington and de la Plain 1983). After being expelled or upon falling on the ground, the next sign is the beginning of regular breathing, often preceded by a coughing splutter (Kiley-Worthington and de la Plain 1983). For the next few minutes, the calf lies stretched out on one side (Kiley-Worthington and de la Plain 1983; Hermann and Stenum 1982). The calf's first head lifting has been reported to be 3.3 min after birth (Le Neindre 1982). From the sidewise position, it moves its legs and pushes itself into a sternal lying position (Kiley-Worthington and de la Plain 1983; Hermann and Stenum 1982). Crossbred beef calves were found to take a shorter time than dairy calves to rise to sternal recumbence (Hermann and Stenum 1982). Lidfors et al. (1994a) found that to reach sternal recumbence, it took 7.3 min for dairy calves born in group housing and 6.6 min in dairy calves born in an individual calving pen (Lidfors 1996).

The next behavior of the calf is trying to rise to its feet (Hermann and Stenum 1982; Kiley-Worthington and de la Plain 1983), in the typical bovine way – hind legs first (Brownlee 1950; Kiley-Worthington and de la Plain 1983). These attempts are clumsy and usually soon result in collapse, which often is helped along by the mother's rough licking (Kiley-Worthington and de la Plain 1983). The newborn calf, though it has an innate tendency to perform the appropriate movements in the appropriate order, requires practice before perfection is attained (Brownlee 1950). Time from birth until first attempt at standing has been reported to be from 9 to 32 min in different studies and type of cattle (Table 1). The interval between birth and first successful standing has been found to be from 35 to 66 min in beef cattle calves, 36–67 min in calves born to multiparous dairy cows, and 67–105 min in calves born to primiparous dairy cows (Table 1). Both breed and environment can modify the activity of newborn calves (Le Neindre 1982).

Having got to its feet, the calf will then stand rather unsteadily for about 30 s to a minute, before taking its first steps (Kiley-Worthington and de la Plain 1983). These are the typical four-step form of ambulation or tentative walking (Frasers 1989), and it may quickly end in a fall (Kiley-Worthington and de la Plain 1983). After a fall, the calf will again struggle to its feet and begin staggering over to its mother (Kiley-Worthington and de la Plain 1983). Some calves are repeatedly knocked over by their dams during the grooming phase (Selman et al. 1970b). The fetal digital pads make the calf unsteady, but they rapidly become shredded and removed from the soles when the calf starts walking (Frasers 1989). Within a few steps, walking becomes more coordinated, and some calves were observed to leap around within 3 h after birth (Kiley-Worthington and de la Plain 1983). Leaping involves the rapid forward movement of the forequarters, followed by a bucking of the hind, often accompanied by the sideways kicks of one hind leg (Kiley-Worthington and de la Plain 1983).

Table 1 The mean time in minutes (± SD) from birth until calves were attempting to stand up but falling and first successful standing in beef and dairy cattle from different studies

References	Type of cattle	Number of calves	Attempt to stand	Successful standing
Selman et al. (1970b)	Beef	10	–	35.4 ± 14.8
George and Barger (1974)	Beef	35	~30	66 ± 9
Lidfors and Jensen (1988)	Beef	20	9.6 ± 6.5	45.8 ± 38.2
Le Neindre (1982)	Beef + dairy	7 + 6	9.4 (4–19)	51.1 (32–83)
Hermann and Stenum (1982)	Beef x dairy cross	8	14 ± 9.7	40.1 ± 18.5
Selman et al. (1970b)	Dairy multiparous	10	–	58.1 ± 20.6
Hermann and Stenum (1982)	Dairy multiparous	6	11.7 ± 7.1	67.5 ± 34.6
Selman et al. (1970b)	Dairy primiparous	10	–	72.7 ± 71.6
Hermann and Stenum (1982)	Dairy primiparous	4	32.8 ± 23.0	104.8 ± – 46.2
Lidfors et al. (1994a)	Dairy (primiparous + multiparous)	5 + 6	16.5 ± 13.4	36.3 ± 27.0
Lidfors (1996)	Dairy	22–24	23.9 ± 17.2	60.5 ± 31.9

The calf's first steps are usually directed toward its mother, and, upon reaching her, it begins to smell, rub, and lick her (Walker 1950; Kiley-Worthington and de la Plain 1983). Kiley-Worthington and de la Plain (1983) found that any other calves or cows in the vicinity will also be sniffed and even followed and the calf may search for a teat and suck another cow. The calf, as soon as it is able to, begins to investigate the environment by smelling and touching objects (Walker 1950; Kiley-Worthington and de la Plain 1983).

Selman et al. (1970b), who compared calves of beef breeds with those of dairy breeds, found that the former calves stood up significantly earlier after birth than the latter, and this was explained by the beef cows carrying out their grooming activities more rigorously and for a longer period than most of the other dams (Selman et al. 1970a, b). In other studies, no correlation was found between maternal behavior, such as licking and vocalizing, and the amount of time before first standing of the calf (Edwards and Broom 1982; Hermann and Stenum 1982; Kiley-Worthington and de la Plain 1983).

The rigor and speed with which a calf becomes active after birth often depend on the ease with which parturition takes place. Male calves are usually heavier at birth than female calves (Witt 1963; Dufty 1973), they have a longer gestation length (Witt 1963), and they are presented longer at the vulva than female calves (Owens et al. 1985). Since rapid growth of the fetus occurs during the final stages of gestation, a prolonged gestation results in a bigger calf at birth (Owens et al. 1985). George and Barger (1974) found no significant differences between male and female calves in birth weight, duration of birth, time to stand, or time to first suckling. However, those cows giving birth to male calves retained their placenta longer than did those having female calves.

Maternal Licking of the Calf

Licking the newborn calf is the predominant behavior shown by the cow immediately after calving, occupying 30–50% of the first hour postpartum (Edwards and Broom 1982; Lidfors and Jensen 1988). Licking the calf decreases markedly with time after calving (Finger and Brummer 1969; Edwards and Broom 1982; Lidfors and Jensen 1988), but it often continues at intervals during the first few hours postpartum (Leuthold 1977). The stimulus characteristic of the newborn calves contributing most to their special attractiveness appears to be the impregnation of their coat with amniotic fluid (Finger and Brummer 1969). Cows have been observed to stay at the place where amniotic fluids were expelled, licking, eating the litter, and vocalizing toward it (Lidfors 1994), and they tend to stay and give birth on the place where the first amniotic fluid was spilt (George and Barger 1974; Owens et al. 1985). The decrease in attractiveness of a calf with time after birth may be related to the progressive removal of amniotic fluids from the coat because of licking and evaporation (Finger and Brummer 1969).

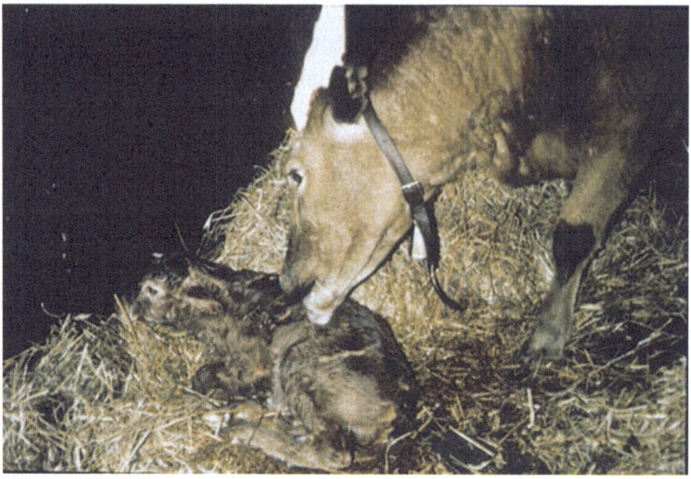

Fig. 2 Cow licking its newborn calf after being born in a shelter in the Finnish forest. (Photo: Aunemaria Yllipekkala)

Suggested functions of licking of the neonate include:

- Removal of amniotic fluid and fetal membranes (Lidfors 1994; Leuthold 1977).
- Stimulation of activity (Metz and Metz 1986; Lidfors 1994).
- Stimulation of breathing (Metz and Metz 1986).
- Stimulation of circulation (Metz and Metz 1986).
- Stimulation of urination and defecation (Metz and Metz 1986).
- Drying of the coat and associated reduction of heat loss (Lidfors 1994).
- Improved hygiene at birth place and calf body surface, reducing the risk of both infection and predation (Sambraus 1973).
- Mother learning to recognize its calf (Hudson and Mullord 1977) (Fig. 2).

Vocalization in Calves and Cows

The newborn calf rarely vocalizes, and the calls are of the "men" type with the mouth open (Kiley-Worthington and de la Plain 1983). When the calf vocalizes, it appears to be a response to an "uncomfortable" situation (Kiley-Worthington and de la Plain 1983). Occasionally, it vocalizes in response to its mother's calling or when it has sniffed at the mother or another cow, calf, or a strange object (Kiley-Worthington and de la Plain 1983). Crossbred beef-dairy calves vocalized significantly later than calves from both primiparous and multiparous dairy cows (Hermann and Stenum 1982).

Cows perform a low "mm" vocalization (Kiley 1972) or closed-mouth calls (Green et al. 2021) after birth. This vocalization is most frequent during the first hour after calving (Hermann and Stenum 1982; Metz and Metz 1985; Lidfors 1996). Green et al. (2021) found that the most frequently occurring behavior expressed by dairy cows kept with their newborn calf was closed-mouth vocalizations (53.6%), followed by touching the calf (including licking) (19.2%), alert behavior to the calf (16.9%), and then open-mouth (5.6%) and mixed calls (4.7%). Hall et al. (1988) showed that free-roaming Chillingham cattle have individual calls, and this has also been shown for dairy cattle by Green et al. (2019, 2020). If cow and calf are separated, recognizing the dam's call may be crucial for the calf's survival (Fig. 3).

The function of the closed-mouth vocalization when the cow and calf are isolated is probably to help in the attachment so that the calf learns to recognize its mother's calls when it is hiding (first 4–7 days) and the cow comes to nurse it (Lidfors 1994). Own observations in the forest of Finn cows approaching their calf when it was hiding under a bush were that the cow stopped about 20 m from the calf and vocalized. Then the calf stood up and ran to its mother, suckled for some time, and then returned to lie down under the bush and the cow left. According to Sambraus (1971), during the first 3 days after birth, the calves do not react to their mother's vocalizations. From the fourth to the seventh day, the calves react to their mother's vocalizations, but they do not seem to recognize her visually (Sambraus 1971). When the

Fig. 3 Cow searching for and calling after calf and calf calling on its dam from studies of free-ranging beef cattle on a ranch in southern Sweden. (Photo: Lena Lidfors)

dam vocalizes, the calf runs in her direction, but it stops if she ceases calling and only continues walking in her direction if she calls again (Sambraus 1971). Not until calves are 8 days old do they manage to recognize their mother by vision (Sambraus 1971). This is around the time when the calf leaves its hiding and is integrated into the maternal herd.

To find out if domestic cattle have unidirectional recognition of the mother by offspring, as occurs in hider species, or bidirectional recognition which occurs in "follower" species, de la Torre et al. (2016) conducted a playback experiment in two free-ranging beef cattle herds. They found that cows were more likely to orient themselves and move toward a loudspeaker with calls from their own calf than from calls of other calves. Calves were also more likely to orient themselves and move toward a loudspeaker and to call back to calls from their own mother than to calls from other females. This shows that cattle have bidirectional recognition, which is typical of follower species. Younger calves had mothers that tended to respond more strongly to playback calls than mothers of older calves (De La Torre et al. 2016).

Cow-Calf Attachment

The stimulus characteristics of the calf play a role in eliciting maternal behavior (Finger and Brummer 1969). The different types of behavior the cow exhibits after parturition, such as licking the newborn and vocalizing, function in different ways as stimuli for the calf and influence its condition (Hermann and Stenum 1982). In species with precocial young, rapid formation of a specific mother-infant bond is essential, and licking of the neonate plays a very important role in this process (Klopfer et al. 1964). Performance of licking provides the dam with much olfactory and gustatory input and leads to what has been described as "olfactory imprinting" of the infant on its dam (Klopfer et al. 1964; Lent 1974).

The first suckling marks the successful establishment of a mother-young relationship that, however, requires more time to develop fully (Leuthold 1977). This relationship is individualized and exclusive, probably because of a two-way learning process partly corresponding to the "imprinting" (Leuthold 1977). During a "critical period," both mother and young learn to recognize each other's individual characteristics, which normally takes longer in the young than in the mother (Leuthold 1977). Owens et al. (1985) did not find any significant effect of the cow's pre- and postpartum behavior traits on calf survival.

Hudson and Mullord (1977) found that a 5 min contact with a calf immediately postpartum is sufficient for the formation of a strong, specific maternal bond with that calf. This bond lasts even if the calf is removed for 12 h and then returned (Hudson and Mullord 1977). After a 24 h separation, the cow still shows signs of distress, but it can no longer recognize her own calf (Hudson and Mullord 1977).

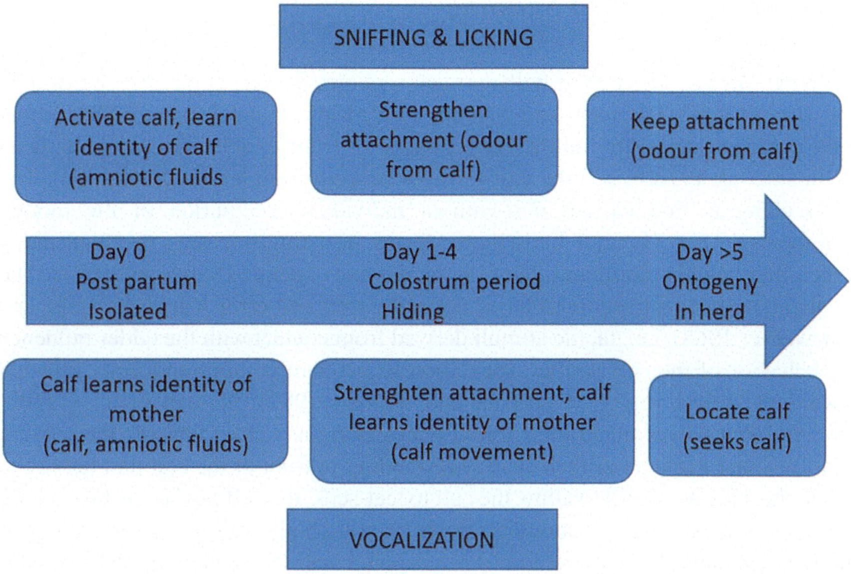

Fig. 4 Model to explain how the cow's sniffing and licking and vocalization create an attachment between mother and young during postpartum isolation and how this attachment is maintained during calf development. (After Lidfors 1994)

Once the mother-young bond is established, it is difficult, if not impossible, to substitute either of the partners (Leuthold 1977). The exclusiveness of the mother-young bond is often reinforced by the aggressiveness of the females toward strange young (Leuthold 1977).

During the 2–3 h after birth, most cows spend 90% of their time within 2 m of their calves and sniff and lick them frequently (Edwards and Broom 1982). After this period, the calf's activity level increases and suckling is more likely to occur (Edwards and Broom 1982). The distance between cow and calf increases markedly between the first and second hour postpartum, and then it continues to increase steadily with time after calving (Edwards and Broom 1982). Cows seldom remain lying while their calves are active; rather, they frequently follow them when they walk away (Edwards and Broom 1982). They respond to the calf's activity by turning or walking toward it, licking it occasionally, and vocalizing (Edwards and Broom 1982). The activity bouts of the cow and calf are synchronized, but the degree of synchronization decreases with parity (Edwards and Broom 1982). Based on the results in a PhD thesis, Lidfors (1994) suggested a model for the cow-calf attachment (Fig. 4).

Teat-Seeking and First Suckling

Once the calf has stood up, it will start teat-seeking (Selman et al. 1970b; Edwards and Broom 1982; Hermann and Stenum 1982; Lidfors and Jensen 1988; Lidfors 1996). At this point, the mother is generally in front of its calf licking its head (Selman et al. 1970b), and the calf receives no help from her (Lidfors 1994). The teat-seeking is first carried out without individual recognition of the mother (Sambraus 1971). The calf finds the udder by initiating releasing mechanisms at which first visual stimuli and, later on, tactile ones operate (Sambraus 1971). The localization of a teat seems to be achieved by trial and error learning (Hafez and Lineweaver 1968), i.e., tactile stimuli derived from contact with the udder influence the behavior of the calf so that orientation to the dam is more and more directed toward the udder region (Hermann and Stenum 1982). Selman et al. (1970b) observed that calves made their initial teat-seeking advances towards the calving pen walls and mangers as their dam was standing in front of the calf and licking it. When the dam stood still to allow the calf to teat-seek, the calf pushed its head along her body and explored the shoulder, brisket, axilla, body wall, udder or groin, and thigh and sometimes continued until it reached a wall (Selman et al. 1970b). With time, teat-seeking became increasingly oriented toward the dam, but was largely directed at the dam's fore or hindquarter (Selman et al. 1970b). The shape of the dam's underbelly affected where the calf concentrated its teat-seeking; when the udder and teats were placed at the highest part of the underbelly, pushing was mainly concentrated on the udder area (beef cows and primiparous dairy cows) (Selman et al. 1970b). However, when the abdomen and udder were large and the highest part of the underbelly was the xiphoid-axillary region, teat-seeking was often made around the forelegs (multiparous dairy cows). In free-ranging beef cows, the dam was often the only object calves would orient toward and multiparous cows were observed to orient themselves so that the udder was in front of the calves' head; thus, they appeared to try to help the calves (Lidfors 1994). Primiparous dairy cows in individual calving pens were sometimes observed standing in front of the calf, thus making it more difficult for the calf to find the teat (Lidfors 1996) (Fig. 5).

Calves are born without an immunoglobulin protection, and it is, therefore, crucial that they get colostrum within some hours after birth. Lidfors et al. (1994a) found that 63.6% (7 out of 11) of calves born in a group housing system for dairy cows did not suckle within the first 3 h postpartum and had to be helped to find the teat or were given colostrum in a teat-bottle. In another study where dairy cows were individually housed, 31.8% (7 out of 22) of calves did not suckle within the first 4 h after calving (Lidfors 1996). In earlier studies, it has been found that 13–45% of calves did not suckle within the first 6–8 h after calving (Selman et al. 1970b; Edwards 1982; Illmann and Špinka 1993). In all these studies, the cows were from a dairy breed, and there are no reports from beef cattle having these problems with suckling. The higher the dairy cows' lactation number, i.e., the number of times she has given birth and been milked for 8 months, the longer time it took to the first suckling (Edwards 1982; Lidfors 1996). Older dairy cows tend to have udders that are more pendulous and larger teats, due to being milked by milk machines, and several calves spend a long time mouthing and nibbling teats that are

Fig. 5 Older calf suckling on its dam. (Photo: Lena Lidfors)

too large (Selman et al. 1970b; Edwards and Broom 1982). Ventorp and Michanek (1992) studied teat-seeking in 42 calves kept with their mother in individual pens and found that a smaller distance from udder to floor led to increased variation and a significant increase in the time spent teat-seeking. The distance between the udder and the floor also had a significant effect on the time to the first suckling.

Edwards et al. (1982) analyzed the serum of calves before and after their first suckling and found that before suckling the levels of IgG1, IgG2, and IgM were low and IgA was not detectable. Calves observed to have suckled during the first 6 h postpartum had adequate levels of Ig 48 h postpartum. The levels of all classes of Ig were associated to the latency to first suckling, and they decreased in relation to parity of the dam and month of calving.

The first suckling after birth has been reported to occur at 72–115 min after birth for beef cattle and at 107–261 min for dairy cattle (Table 2). Ventorp and Michanek (1992) found that newborn dairy calves kept in individual calving pens with their mother suckled at a median of 4 h and 9 min postpartum (n = 21). Calves that were active early usually suckled earlier. The time to first suckling correlates with the time spent teat-seeking and with the total time suckling (Kiley-Worthington and de la Plain 1983; Ventorp and Michanek 1992). The larger the calf, the longer time it takes until its first suckling (Kiley-Worthington and de la Plain 1983).

Within the first 3 h after birth, the average time spent actually sucking is low, and 73% of the bouts are 1 min or less in duration (Kiley-Worthington and de la Plain 1983). The suckling time of calves in the first suckling bout was 21.6 min, and those that had a second suckling bout had a suckling time of 1 min (Hermann and Stenum 1982). The number of suckling bouts during the first 8 h postpartum varied from none to four (Selman et al. 1970b). Walker (1950) reports that during the first day, calves feed five times.

At the initiation of suckling, the calf tends to move its body closer to the dam until their bodies are actually touching (Selman et al. 1970b). The tendency for

Table 2 Time (min. ± SD) from birth until first udder contact and first suckling and duration of first suckling in calves from different studies of beef and dairy breeds

References	Type of cattle	Number of calves	Udder contact	Suckling	Duration of first suckling
Selman et al. (1970b)	Beef	8	–	81.4 ± 52.2	16.7 ± 9.2[a]
George and Barger (1974)	Beef	33	–	115 ± 11	–
Lidfors and Jensen (1988)	Beef	20	64.1 ± 49.5	97.3 ± 44.2	13.2
Le Neindre (1982)	Beef + dairy	7 + 6	–	72.6 (41–132)	25.7 (12–44)
Hermann and Stenum (1982)	Beef x dairy	7	43.1 ± 19.6	60.9 ± 18.0	31.1 ± 12.4
Selman et al. (1970b)	Dairy multip.	7	–	261.1 ± 129.1	16.7 ± 9.2[a]
Hermann and Stenum (1982)	Dairy multip.	4	78.0 ± 32.7	203.3 ± 60.6	37.0 ± 14.7[b]
Hermann and Stenum (1982)	Dairy primip.	4	131.8 ± 43.4	221.5 ± 99.5	29.0 ± 13.7[b]
Selman et al. (1970b)	Dairy primip.	8	–	218.3 ± 113.8	16.7 ± 9.2[a]
Lidfors et al. (1994a)	Dairy	4	52.6 ± 34.7	107.0	–
Lidfors (1996)	Dairy	15	92.2	120.9 ± 55.1	19.9 ± 9.7

[a]Authors did not separate between beef and dairy calves (n = 23 calves)
[b]Only three calves from each group could be observed completely

dams is to lick the hindquarters of suckling calves to favor their closeness (Selman et al. 1970a; Hermann and Stenum 1982). Most calves suckle from a front teat the first time (Selman et al. 1970b; Edwards and Broom 1982). If the rear teats are lower than the front, as frequently happens in the older cows, they may not be suckled for up to a week postpartum (Kiley-Worthington and de la Plain 1983). The first time the calf suckles it tends to do it from the same side of the dam as previously (Selman et al. 1970b) and to use only one or two teats (Kiley-Worthington and de la Plain 1983; Walker 1950). Walker (1950) reports that it was not until the third or fourth day that all four teats were suckled at one feed. Calves of primiparous cows directed a smaller percentage of total suckling to a favored teat and changed teats more frequently during a bout of suckling than did calves of older cows (Edwards and Broom 1982). When the cow's udder is small, the calf suckles all four teats from one side (Walker 1950).

5 Cow-Calf Behaviors During Development

Suckling During Growth

Suckling frequency decreases with age (Walker 1962; Wagnon 1963; Nicol and Sharafeldin 1975; Papini et al. 1983; Kiley-Worthington and de la Plain 1983; da Costa et al. 2006), while suckling duration increases with age (Nicol and Sharafeldin

1975; Papini et al. 1983; Kiley-Worthington and de la Plain 1983). This combined results in a relatively constant total suckling time of 30–35 min at all ages (Nicol and Sharafeldin 1975). The total suckling time per 24 h has been reported to be 37–57 min (Wagnon 1963; Hafez and Lineweaver 1968), and in zebu cattle, it was 23.76 (± 0.47) min per 12 h (da Costa et al. 2006).

Across the first weeks of life, calves suckle frequently, probably six to ten times daily (Walker 1950). During growth, they suckle three to five times per 24 h (Walker 1962; Wagnon 1963). Calves of Hereford x Friesian cows and Blue Grey cows kept under grazing conditions had 5.5 suckling bouts per 24 h, whereas housed Blue Grey cows had 9.1 suckling bouts per 24 h (Somerville and Lowman 1979). Older calves tended to suckle less frequently, and milk yield was considered a possible factor involved in this. Da Costa et al. (2006) found that zebu calves suckled 2.57 times per 12 h.

A suckling session lasts 10.4–17.1 min (Wagnon 1963; Papini et al. 1983). Somerville and Lowman (1979) found that the mean duration of suckling was 6.9 min in Hereford x Friesian cows on pasture, 10.7 min in Blue Grey cows on pasture, and 10.6 min in housed Blue Grey cows. Da Costa et al. (2006) found the duration of a suckling session in zebu calves to be 9.25 min. The amount of suckling and the length of the bouts depend on many factors, such as how much milk the cow has (Kiley-Worthington and de la Plain 1983). Calves in lightly grazed pastures and in herds receiving supplements averaged 44 min a day nursing; in closely grazed pastures and unsupplemented animals, nursing lasted an average of 55 min (Wagnon 1963).

Lidfors et al. (1994b) showed that calves suckling primiparous beef cows (with a presumably low milk production) had a pre-stimulation of the teats and udder where the calf sucked less than three times on each teat before switching to another teat accompanied by butting with the muzzle on the udder. After about 1 min, the calves stopped and sucked on the same teat for a longer duration, while milk foam could be observed around the mouth. After some time, the calf changed to another teat and kept sucking that for some time, and then to the next, and so on. After around 5–6 min, the calves started changing teats more rapidly again and then also butted on the udder repeatedly (Lidfors et al. 1994b). This last phase of suckling was presumed to be a post-stimulation of the udder that may have led to some milk running down to the teats, but also a stimulation for future milk production in the cow (Lidfors et al. 1994b).

Mayntz and Costa (1998) also found that calves' suckling behavior consists of short bouts during pre-stimulation and a sudden increase followed by a gradual decrease in bout length during ejection. Post-stimulation was found to be carried out with short bouts again and to take about two-thirds of the mealtime, i.e., 12–15 min on average (Mayntz 1996; Mayntz and Costa 1998). Mayntz and Costa (1998) also showed that a calf changes teats 400–600 times during a suckling. Each sucking bout, i.e., duration of uninterrupted sucking on each teat, depends on the amount of milk available in that cistern (Mayntz and Costa 1998). Mayntz (2005) explains that the inflow into the udder cavern is slower than the outflow through the teat canal, and therefore it is an optimal strategy to abandon an empty teat and to suck the other ones while the first is refilled.

Rasmussen and Mayntz (1998) measured the pressure difference across the teat canal when calves are suckling and found that it is between 60 and 110 kPa. There are peaks of pressure difference that last for a couple of milliseconds, and they are applied with a frequency of 2.0–2.4 Hz (Rasmussen and Mayntz 1998). Between those peaks, the pressure difference is decreased to almost 0 kPa (Rasmussen and Mayntz 1998). The pressure difference consists of about 60% "under-pressure" beneath the teat tip and about 40% "over-pressure" in the teat cistern (Rasmussen and Mayntz 1998). The information from this research was aimed at developing better milking equipment for dairy cows.

Breed differences in suckling behavior were noted by Walker (1962), and Lewandrowski and Hurnik (1983) found that daily suckling frequency was greater for Hereford than for crossbred calves. Da Costa et al. (2006) found that the Nelore breed had an effect on the number of suckling meals and total suckling duration.

Nicol and Sharafeldin (1975) found that calves which suckled more often than the average did so consistently and had above-average total suckling time. Nicol and Sharafeldin (1975) also found a low positive correlation between calf daily live weight and suckling time. Walker (1962), on the other hand, did not find any evident relation between the numbers of suckling episodes or the time spent suckling and the weaning weight of the calf.

Female calves suckle slightly longer but less frequently than male calves (Papini et al. 1983). Nicol and Sharafeldin (1975), on the other hand, found no real differences in the suckling behavior between steer and heifer calves. In zebu cattle, male calves showed a larger number of suckling meals (2.60 vs. 2.12 meals/12 h) and a longer total suckling duration (25.05 vs. 21.51 min/12 h) than female calves (da Costa et al. 2006).

Nursing activities occur at all hours of day and night, but they happen more often between 5 and 6 a.m. with other peaks around midday, in the afternoon around dusk, and around midnight (Walker 1962; Wagnon 1963; Sambraus 1971; Papini et al. 1983). Only when the hours of darkness were longer than 5 h did the calf feed at night (Walker 1950). Somerville and Lowman (1979) found a relatively uniform distribution of suckling throughout the "day" in Blue Grey cows both on pasture and when housed, but with three peaks of suckling activity in Hereford × Friesian cows under grazing conditions. Da Costa et al. (2006) found that zebu calves suckled at any time during the daylight. At a young age, a distinct pattern of three main suckling periods during the day was clear, with the most marked peak soon after daybreak (Nicol and Sharafeldin 1975). There was some suggestion that the pattern is less marked by the time the calves are over 100 days old, and this may be due to the increase in grazing time (Nicol and Sharafeldin 1975).

The suckling procedure is initiated by the calf that calls for its mother or by the cow that visits her calf at its resting place (Sambraus 1971; Reinhardt et al. 1977; Nicol and Sharafeldin 1975). The mother sniffing the calf's nose or genitals always precedes suckling (Reinhardt et al. 1977), and strange calves are displaced with a stroke from the hind leg or with head butting (Sambraus 1971). This behavior was more common in primiparous cows than in older cows (Edwards and Broom 1982). The calves regularly seek their mother's teats when they are frightened (Reinhardt et al. 1977) or if she calls for reasons other than suckling (Sambraus 1971). Calves initiated 83–87% of the suckling bouts and terminated them spontaneously in 63–81% of cases during the first 2 months after birth (Wagnon 1963; Papini et al. 1983). The cow ended about 13% of the suckling bouts by walking away from the calf or by lying down, and 6% of the bouts by fighting for supplements, grazing, or walking to the water container or the salt lick (Wagnon 1963).

When the cow terminates the suckling, the calf may suckle from behind (Walker 1950; Schloeth 1961). When the calf first reaches the teats, it will often bump the udder vigorously with the head, presumably to stimulate the letdown of the milk (Walker 1950; Leuthold 1977). Once the milk appears, no further bumping occurs (Walker 1950), and the calf suckles quietly with the tail held slightly erect and waving gently from side to side (Selman et al. 1970b). When the milk supply decreases or fails, the calf will bump again, and if this has no effect, it will eventually change to another teat (Walker 1950). Rapid tail wagging occurred only during teat-seeking or when it seemed that the milk flow was reduced or had stopped (Selman et al. 1970b). During suckling, the calf generally stands in a parallel position (Sambraus 1971). As the calf gets older, the bumping of the udder becomes more vigorous, and it may cause obvious discomfort to the mother, which results in increased evasive action by her and may be a contributory factor to the weaning process (Leuthold 1977).

During the nursing period, the cows spent 41% of the time standing at ease, with some licking of their calves, and 34% of the day ruminating and a grazing (Wagnon 1963). During the licking and sniffing, the cow has an opportunity for identifying her offspring (Schloeth 1958; Sambraus 1971; Poindron and Le Neindre 1975). During suckling, the calf generally stands in an opposite parallel position (Sambraus 1971), which makes it possible for the cow to lick especially the anogenital region (Hermann and Stenum 1982). Licking of this region may also function as a stimulus for defecation and urination (Walker 1950; Hafez and Lineweaver 1968). During the first months postpartum, cow and calf often urinate during the suckling procedure (Reinhardt et al. 1977).

It is common for young ungulates to attempt to suckle from alien adults, but such attempts are usually rejected (Sambraus 1971; Lent 1974; Nicol and Sharafeldin 1975). Yet, dairy cows have been selected for weak maternal bonding and docility during milking, and in one study it was found that 80% of the cows allowed alien calves to suckle (Finger and Brummer 1969).

Das et al. (2000, 2001) observed the suckling behavior of calves during restricted suckling 30 min after milking. The total suckling duration was 11.8 min in zebu calves and 9.4 min in crossbred calves, and it decreased in both breeds from 1 to 6 months (Das et al. 2000). The separate suckling bouts were longer in the zebu cows (2.8 min) than in crossbred cows (2.3 min), and zebu cows also had more suckling bouts per session (3.8 and 3.2, respectively) than the crossbreds (Das et al. 2001). Zebu cows stayed in close contact with their calves for a longer time and directed more agonistic actions against alien calves and cows than the crossbred cows (Das et al. 2001). The number of suckling bouts decreased from 3.8 at 1 month of age to 1.1 at 6 months (Das et al. 2000). The duration of each suckling bout decreased from 3.5 min at 1 month of age to 1.6 min at 6 months. Play behavior occurred mainly after nursing, and the duration increased across age from 1 to 6 months (Das et al. 2000).

The First Weeks of Life

Cattle have been classified as a "hider" species, since the calf lies away from its dam (Scheurmann 1974). In such species, the neonate walks away from its mother at the end of an activity bout and selects its own lying place (Lent 1974). During the first 2–3 days of life, the calf hides (Schloeth 1961; Vitale et al. 1986) or remains close to where it was born (Wagnon 1963) or to its mother (Kiley-Worthington and de la Plain 1983). It has been noted that infant ungulates lie down near vertical objects rather than in the open (Lent 1974; Scheurmann 1974). If there is a lack of good hiding places for the calves, they may walk under a fence of the pasture and hide outside, at the risk of getting injured or loose contact with their dam (König et al. 1997). Therefore, it is recommended to provide suitable hiding places inside the pasture (König et al. 1997).

Wood-Gush et al. (1984) found little evidence of the cows trying to remain apart from their calves, which was suggested to be due to either a loss in need for isolation in breeds of *Bos taurus* or that the usual European farm does not have the necessary environmental features to stimulate the behavior. They, therefore, suggested that there is a graduation in the Bovidae from the true hiders – in which the calf lies alone and is visited by the dams – to the true follower, and cattle are an intermediate within the spectrum. Scheurmann (1974) found calves born on the pasture to be very passive between their second and fifth days of life and they only left their lying place for suckling and playing. Lidfors (1994) observed that calves born in a forest area were hiding for the first 4–7 days and only became active when the mother approached their hiding place and called them for suckling. When suckling was over, the calves walked back to their hiding place and the cow left. She may remain with the herd and just visit the calf for suckling (Schloeth 1961) or graze in the surroundings (Vitale et al. 1986). When the cow has to leave to eat or drink, the calf will remain secluded until she returns (Wagnon 1963). In such instances, the cow usually does not remain away for a long time (Wagnon 1963). If the calf is found and disturbed, the cow might move it to another area (Wagnon 1963). Reinhardt et al. (1977) report that the mother cow keeps on the very periphery of the herd, grazing and resting in close contact to her newborn for up to 2 weeks after birth. During this phase, many cows become hostile and vigorously threaten and/or charge animals of different species that come too close (Reinhardt et al. 1977).

When the calf is 3–4 days old (Schloeth 1961), 1–3 weeks old (Reinhardt et al. 1977), or around 1 week old (Lidfors 1994), the cow takes it to the herd where the calf joins a subgroup consisting of other young which do not yet regularly graze (Schloeth 1961; Reinhardt et al. 1977). There is always one cow or the bull that stays with the young calves (Wagnon 1963; Reinhardt et al. 1977), but when supplements are being fed, all cows will come to the troughs leaving a group of calves unattended (Wagnon 1963). While a "babysitting" cow usually butts other calves away when they come near, she usually comes running to defend any nearby calf that calls in fright (Wagnon 1963).

Distance Between Calf and Cow

The distance between cow and calf increases with age (Kiley-Worthington and de la Plain 1983; Papini et al. 1983), and by the end of the second week, the calf is between 5 and 50 m away from its mother (Kiley-Worthington and de la Plain 1983). Papini et al. (1983) found that cow and calf kept more than 15 m apart most of the time, whereas Kiley-Worthington and de la Plain (1983), using another scale of measurement, found that they remained over 50 m apart most of the time. Before 2 months, the dam usually seeks up the calf to nurse and interacts with it, but at about 2 months, the calf seeks out the mother and maintains the bond (Kiley-Worthington and de la Plain 1983). By about six months the cow/calf distance has changed to that more characteristic of adults with a peak of 30–40 m apart (Kiley-Worthington and de la Plain 1983). Near its mother, the calf just stands; however, at distances larger than 15 m lying down is the dominant activity (Papini et al. 1983). Most long distance cow-calf separations result from the cow moving to feed or to obtain water (Wagnon 1963; Price et al. 1985). Reunion of mother and young depends on auditory communication; the calf normally travels most of the intervening distance (Price et al. 1985).

When moving with the herd the calves keep in close vicinity to their mothers (Reinhardt et al. 1977; Reinhardt 1982). The primiparous cows are in general further away from their calves than the more experienced cows (Kiley-Worthington and de la Plain 1983). The time of year the calves are born and breed differences do not affect mother/calf distance (Kiley-Worthington and de la Plain 1983).

Contact Between Calf and Cow During Growth

Cow to calf contact includes auditory (calling), tactile (suckling, rubbing and licking), olfactory (smelling) and gustatory (tasting and licking) stimuli (Kiley-Worthington and de la Plain 1983).

Cow and calf seem to recognize each other's calls (Reinhardt et al. 1977; Kiley-Worthington and de la Plain 1983). When they are not together cow and calf keep regular acoustical contact (Wagnon 1963; Reinhardt et al. 1977). This consists of (1) the "muuh" of cows and elder calves when calling each other over greater distances, (2) the "eeh" or "oeh" of young calves calling for their mothers, and (3) the "hhm" of cows when they are reunited with their calves (Reinhardt et al. 1977).

After the first week of birth, the mother rarely produces the low "mm" contact call characteristic of the first week (Kiley-Worthington and de la Plain 1983). The calf calls very seldom to the mother, or anyone else, unless it is very cold, hungry or lonely (Kiley-Worthington and de la Plain 1983). Most recorded vocalizations by the calf occurred just before or during an interaction with the mother (Wood-Gush et al. 1984). Calves often ignore their mothers' calling, but there are also occasions when a cow might ignore the calling of its calf (Wagnon 1963). When a young calf calls in fright most nearby cows answer and come running to its defense (Wagnon 1963). This was also observed in my studies of free-ranging beef cattle, and once a calf that was playing made a sound that made all cows come rushing to its rescue.

Typical licking periods occur when the herd's activity changes; when the cows stop grazing, most of them visit the "kindergarten" not only for suckling but also for mere social licking (Reinhardt et al. 1977). This behavior presumably serves as physical contact gesture through which the amicable relationship is confirmed (Reinhardt et al. 1977). While on the average the time cows spend grooming their calves is very small, individual cows groom their calves for up to 15 min/day while some pay little attention to them other than to permit suckling (Nicol and Sharafeldin 1975). The cows show a highly significant decrease of licking in the second month (Papini et al. 1983).

Cows commonly sniff their calves when returning to them after grazing (Kiley-Worthington and de la Plain 1983; Lidfors 1994). The percentage of time spent touching the calf did not change very much during growth but it was about 10% of total daytime up to 2 months and 5% between 2 and 6 months (Kiley-Worthington and de la Plain 1983). Wood-Gush et al. (1984) found that the number of interactions with the mother ranged from an average of 1.45 per 30 min in the first week to 0.15 per 30 min in the last week. No aggression between mother and calf occurred (Wood-Gush et al. 1984).

Cow behavior in searching for the calf consists of standing looking about as she calls and listens for a reply, walking and calling, and even running a bit at times (Wagnon 1963). Sometimes, when cows move from one part of their pasture to another, one or more calves do not travel along, and the cow then stops and calls at her calf until it starts following her (Wagnon 1963). Calves identify their mothers largely by sight, but scent and sound also appear to play a part in some recognitions (Kiley-Worthington and de la Plain 1983).

There are individual differences in duration that the cows spend in contact with their calf and in the type of contact they use (Kiley-Worthington and de la Plain 1983). There is also variation in the type of contact made by the calves to their mother (Kiley-Worthington and de la Plain 1983). There is an indication that the more frequent the contact made by the cow toward her calf, the more frequently the calf will contact her (Kiley-Worthington and de la Plain 1983). Mothers and their calves contact each other more than other herd members do, and both cows and calves contact their peers more than other herd members (Kiley-Worthington and de la Plain 1983).

A high proportion of the interactions with the mother are unconnected with suckling, while the interactions with other adults involve threat by the adult and avoidance by the calf (Wood-Gush et al. 1984). The calves' interactions with other cows tend to remain constant over the observation period, but the proportion of threats from other adults increased from 0.03 per 30 min in the first week to an average of 0.13 per 30 min in the sixth to eighth weeks (Wood-Gush et al. 1984).

Mother cows associate with cows of other families, and calves engage in friendship relations with age-mates of other families (Reinhardt and Reinhardt 1981b). The resulting interfamily cohesions lead to the formation of what might be called clans (Reinhardt and Reinhardt 1981b). The calves preferentially not only graze, walk, and rest with their mothers, but they also prefer them as adult play partners, whom they attempt to mount and induce to push (Reinhardt et al. 1977; Kiley-Worthington and de la Plain 1983).

Weaning by the Cow or by Humans After Longer Suckling Periods

The calves of Camargue cattle and zebu cattle are weaned at 8–11 months of age (Schloeth 1961; Reinhardt and Reinhardt 1981a). In the zebu cattle, heifer calves were weaned in the ninth month, while bull calves were weaned in the 11th month (Reinhardt and Reinhardt 1981a). It is normally not the calf that stops sucking but the cow that no longer tolerates to be sucked; she kicks, butts, and threatens her calf, whenever it attempts to reach her teats (Reinhardt and Reinhardt 1981a). Cows suffering from dysfunction of the ovaries continue nursing for more than another 9 months, and then the calf voluntarily stops suckling (Reinhardt et al. 1977). The affiliative bond between cow and calf remains unimpaired following the birth of another calf (Reinhardt and Reinhardt 1981b). In the study of free-ranging Finn cattle in a forest area, all cows had weaned their calves at least 3 weeks before the birth of their next calf.

Beef calves are usually not kept with their mothers until natural weaning, and farmers usually separate the calf from the mother at around 5–9 months of age. Breaking the bond between the cow and the calf creates stress leading to cow and calf vocalizing for up to 3 days after the separation. One way to reduce the stress is to separate weaning off from milk and separation from the mother at different times. Researchers have managed to reduce the weaning stress by the use of so-called "fence-line weaning" where the cow and calf are placed in adjacent pastures so that they can sniff and lick each other, but cannot carry out suckling (Price et al. 2003). Another way of reducing the stress at weaning is to place a nose-flap in the muzzle of the calf that stops the calf from suckling. Haley et al. (2005) showed that calves were calmer after the separation from the cow 4 days after they had the nose-flap mounted compared to calves that did not get a nose-flap. Hötzel et al. (2010) placed nose-flaps on 6-month-old beef calves reared by cows of different milk yields for 11 days to wean them off milk. They found that the distance between the calves and their mother as well as the frequency of grazing and rumination decreased after nose-flaps were placed on the calves. They also found that vocalization, suckling attempts, walking, and standing increased from 3 days before weaning to day 5 after nose-flaps were placed on the calves (Hötzel et al. 2010). The calves' behavioral responses to prevention of suckling did not differ between those reared by cows with high milk yields and those raised by cows with low milk yields (Hötzel et al. 2010).

6 Reproductive Endocrinology Around Parturition

Most of the studies on the endocrinology of cows around parturition and postpartum have been conducted on dairy cows, and it needs to be held in mind that the breeding for high milk production may cause some differences in their endocrinology compared to beef cows.

The process of normal calving is presented on the Veterinary Disease Information Blog (https://vetstudentresearch.blogspot.com/2015/07/the-processes-of-normal-calving.html).

An overview of the hormones involved and how they influence normal calving processes in the calf and cow can be found in Fig. 6.

The normal postpartum reproductive endocrinology has been extensively studied and was reviewed by Leslie (1983): Approximately 6 weeks prior to parturition, the biosynthetic functions of the placenta around the calf result in gradually increasing plasma levels of estrogens (Convey 1974). The peripheral blood level of a metabolite of PGF2-alpha undergoes a dramatic increase shortly before parturition and early in the postpartum period (Thatcher et al. 1980). This was viewed as an

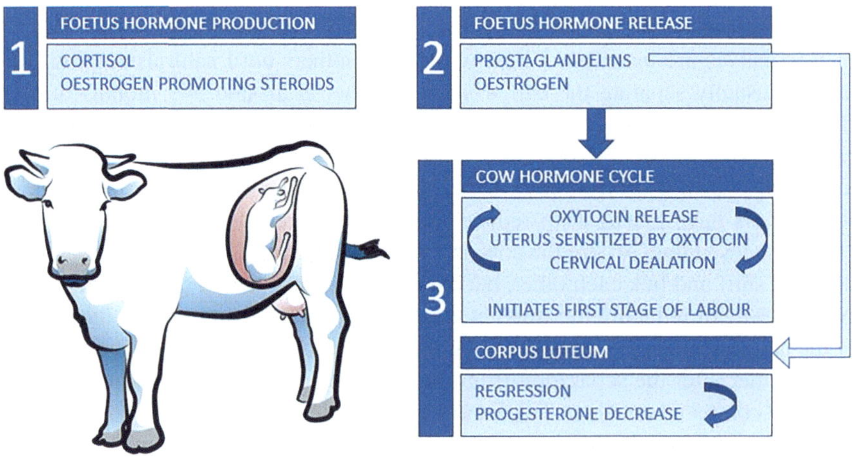

Fig. 6 An overview of the normal stress reaction of the calf that initiates the calving process in the cow. (Flowchart by Anna Lidfors Lindqvist)

indication that PGF2-alpha is involved in the prepartum luteolysis of the corpus luteum of pregnancy, which is in agreement with other reports (Edqvist et al. 1978; Kindahl et al. 1980). This increase in PGF concentration appears to be extremely important for normal uterine involution.

The PGF2-alpha levels parallel the rate of uterine involution with a peak on day 4 postpartum, and thereafter they remain elevated for up to 20 days (Kindahl et al. 1980; Lindell et al. 1982). Progesterone levels decline rapidly during the last 48 h before parturition and remain at very low levels (<0.5 ng/mL) throughout the early postpartum period (Hoffman et al. 1973; Kesler et al. 1977). This decline is related to regression of the corpus luteum of pregnancy. After parturition, the plasma estrogen levels decrease sharply to values below those found during the normal estrous cycle (Erb et al. 1981).

Follicle-stimulating hormone (FSH) is maintained at relatively constant levels throughout the life of the postpubertal dairy cow (Schams et al. 1978). Plasma FSH concentrations fluctuate between 30 and 70 ng/mL and are somewhat independent of other hormonal changes. FSH in response to exogenous GnRH injections is reduced during the period from 9 days before to 6 days after parturition (Schallenberger et al. 1978). FSH levels are slightly increased and are maintained at constant levels through parturition and into the postpartum period (Dobson 1978).

Several investigators have studied luteinizing hormone (LH) concentrations during various reproductive stages. LH follows a release pattern that is characterized by a rapid increase followed by a more gradual return to normal levels (Beck and Convey 1976; Rahe et al. 1980). These spasmodic episodes of rapid increase in LH concentration are described as pulsatile LH release (Foster et al. 1980). A tonic release maintains basal LH levels. The same basic LH concentrations and release patterns found during the normal estrous cycle have been observed in cows during the early postpartum period (Schallenberger et al. 1978; Carruthers and Hafs 1980; Foster et al. 1980; Webb et al. 1980). It has been suggested that the ovary regulates LH secretion in cattle (Rahe et al. 1980). However, from studies in primates, it has been proposed that LH secretion and ovulation are under neural control (Knobil 1980). In sexually mature female cattle, both estradiol 17-beta (Short et al. 1979) and GnRH (Schallenberger et al. 1978) can induce the release of LH from the anterior pituitary gland. The interaction of these two hormones to cause a LH surge depends primarily on the estradiol 17-beta concentration.

Gradually increasing estrogen concentrations to markedly elevated levels may have an inhibitory effect on the production and/or release of a LH surge by the pituitary. This effect has been demonstrated in late pregnancy and is different from the effect of a synchronous rapid elevation in estrogen during late proestrus of the normal cycle. In late pregnancy from 40 to 60 days prepartum until 6 days after calving, estrogen levels are high and there is a decrease in LH peak values. Several researchers have examined the pituitary response to exogenous GnRH during the prepartum and postpartum periods and have described LH release in response to GnRH injection to be substantially reduced from 5 weeks before until 8–10 days after parturition (Fernandes et al. 1978; Kesler et al. 1977; Schallenberger et al. 1978). Full responsiveness returned by day 10 postpartum. A full LH response has been artificially produced on day 2 postpartum by using a combination of estrogens, progesterone, and then GnRH (Azzazi et al. 1980).

High levels of progesterone appear to suppress the hypothalamus and not the pituitary. Progesterone levels are constantly elevated during gestation (Pope et al. 1968), except for a small decrease from day 60 to day 90 after conception (Robertson 1972). Yet, the pituitary can fully respond to exogenous GnRH throughout pregnancy (Schallenberger et al. 1978).

In beef cows, the return to cyclic ovarian function is markedly delayed, the cause of which is poorly understood. The presence of a sucking calf postpones the commencement of cyclic activity (Radford et al. 1978). Suckling inhibits the pulsatile release of LH and assures continued high prolactin (PRL) levels. In dairy cows, PRL increases at parturition (Convey 1974), but declines to lower levels within 2 or 3 days after calving (Tucker 1979). Elevated PRL levels have been recorded in a selected group of high-production dairy cows (Thatcher et al. 1980). The suppression of PRL does not alter the rate of return to cyclic function (Schallenberger et al. 1978). Thus, its role in controlling the onset of cyclic activity is suggested to be minimal.

Placenta Expulsion Postpartum

Placenta expulsion occurred at 224.5 ± 73.2 min postpartum in dairy cows that could keep their calf and at 277.1 ± 139.1 min in dairy cows separated from their calf immediately postpartum (Lidfors 1996). There was no significant difference between these treatments, and further, there were no correlations between cow parity and time to placenta expulsion (Lidfors 1996). Out of 30 observed cows, 77% of them ate the placenta, and the majority ate most of it, while a few only took small parts of it (Lidfors 1996). In a study on multiparous beef cows, 82% of the cows ate their placenta (George and Barger 1974). Galhotra and Gautam (1971) suggested that the reasons for bovines eating their placenta are cleanliness of the nest, provision of food, cover up of the young ones, and supply of minerals (placenta is high in calcium, phosphorus, and other trace elements). A clinical investigation of eight cows that had been eating placenta showed that they had developed indigestion, tympany, and other systemic disorders (Galhotra and Gautam 1971).

Some cows can show a retained placenta, which may have to be removed by a veterinarian. The reasons for retained placenta have been suggested to be uterus infections, nutritional, genetic, and physiological etiologies (Leslie 1983). Many cows get uterus infections with bacterial growth that are most common the first 45 days postpartum (Leslie 1983; Sheldon and Dobson 2004; Sheldon et al. 2008). The infection may cause subfertility, which, apart from direct effects of the uterus, involves perturbation of the hypothalamus, pituitary, and ovary, and these effects may persist even after treatment of the disease (Sheldon and Dobson 2004). The bacterial infection can have negative effects on reproductive hormones and ovulation, and it is therefore important to diagnose and treat uterine disease early (Sheldon and Dobson 2004). To find uterine infections, examination of the vagina for the presence of pus and the character and odor of the vaginal mucus should be scored. The bacteria involved in uterus infection are *Escherichia coli* and *Arcanobacterium pyogenes*, and the virus bovine herpesvirus 4 (BoHV-4) can cause a rapid cytopathic effect (Sheldon et al. 2008). The cows' immune system can eliminate the pathogens, but it is dependent on pattern recognition receptors that bind to pathogen-associated molecules (Sheldon et al. 2008). Uterine disease leads to damage of the endometrium and disruption of the cyclic activity of the ovaries, thus causing infertility (Sheldon et al. 2008).

7 The Behavior of Dairy Calves and Cows in Different Systems

Group Housing of Cows During Calving

It is possible to keep cows together in group housing during calving, and this is a common method in housed dairy cattle. However, some cows may interact with a cow during calving and even disturb her in the middle of giving birth (Lidfors et al. 1994a). Cows can also have more difficulties to maintain proximity to their calf when other cows intervene (Finger and Brummer 1969; Lidfors et al. 1994a). Most cows lick a newborn alien calf if the opportunity exists, but fewer lick an older alien calf (Edwards 1982). Most cows lost interest in an alien calf when their own calf was born, but a few multiparous cows abandoned their own newborn calf and remained with an alien one that they had adopted prior to calving (Edwards 1982).

Alien calves often distract the mother, and alien cows interfere with the mother-infant behavior (Finger and Brummer 1969). This frequently appears to hinder progress toward suckling by directing the attention of either mother or calf away from its partner (Finger and Brummer 1969). The group housed calves directed 34% of their teat-seeking toward alien cows (Edwards 1982). There were no differences in total suckling time of calves in the two housing types, but 6% of the suckling time observed under group housing involved a cow other than the mother (Edwards 1982). Michanek and Ventorp (1993) found that 11 out of 21 calves born in group housing suckled both their mother and alien cows, 4 suckled only alien cows, and 1 suckled only its mother. The remaining five calves did not manage to suckle any cows. Calves had lower plasma IgG level on the farm with group housing during calving compared to two other farms where cows were individually kept during calving (Michanek and Ventorp 1993). The calves suckling on cows not having colostrum caused this.

The amount of time cows spent licking their own calf after birth did not differ between individual and group housed heifers, whereas it differed for multiparous cows (Edwards 1982). Group housed cows approached their calf more than individually housed cows and less often remained lying when the calf was standing

(Edwards 1982). Butting of a calf by its mother was less common in group housing (Edwards 1982).

Based on own research (Lidfors et al. 1994a), it is recommended to move cows to individual calving pens before calving in order to give them isolation and undisturbed calving. However, allowing dairy cows to partly seek isolation and not be completely separated from the herd may reduce stress at calving (Rørvang et al. 2018). According to Mee (2008), good calving management is a critical determinant in reducing peri-parturient losses. Important parts of this are predicting accurately when calving is due, moving heifers and cows to the maternity unit on time, discrete calving supervision, and critically knowing when and how to intervene during calving (Mee 2008).

Effects of Deprivation of Maternal Care on the Newborn Calf

Only a few studies have compared the behavior of calves kept with their mothers and calves separated immediately after birth. Metz and Metz (1985) found that a long-lasting delivery is correlated with a delay in first standing in separated calves but not in calves kept with their mothers. This was attributed to the mother that stimulates her calf to early activity in spite of weakness. Moreover, mothers were licking their calves more often when they started to search for the udder later (Metz and Metz 1985). Calves kept with their mother had a shorter time until first attempts at standing and first time standing for more than 1 min compared to calves removed from their mother directly postpartum (Lidfors 1996). Because the mother continuously licked her calf, the calves kept with their dams urinated and defecated earlier and more frequently than the separated calves (Kovalčik et al. 1980; Metz and Metz 1986). In the calves kept with their mothers, there was more motor activity than in the separated calves, which were housed in small pens and had no chance of moving from place to place (Kovalčik et al. 1980).

Calves kept with their mothers were standing for a significantly longer time the first and second day of life than the separated calves, and they hardly emitted mooing sounds, while the separated calves showed a very high frequency of vocalization (Metz and Metz 1985). This was considered as a sign of unsatisfied behavioral needs in the separated calves. Because the cow spent a rather small amount of time licking her calf and the frequency of vocalization was strongly reduced during the second day, Metz and Metz (1985) concluded that the presence of the mother might be important for the young calf independently of the amount of maternal care that she gives.

During the first 10 days of life, calves kept with their mother suffered less from heavy diarrhea and reached a significantly higher daily weight gain than the separated calves (Metz and Metz 1985). Moreover, calves kept with their mother for the first 3 months of life distinguished themselves from group-reared calves separated from their mother by more threatening and more butting when being placed in a group with other calves of similar age. This was strongly correlated with their higher body weight (Metz and Metz 1985). Furthermore, calves kept with their mothers showed more prepuce contact behavior (licking, sniffing) than the group-reared herd mates, and this was correlated with their higher body weight (Metz and Metz 1985).

Effects of Cow-Calf Separation in Dairy Cattle at Different Times Postpartum

Removing the dairy calf after bonding may induce acute stress in both cows and calves. Hopster et al. (1995) found that when the calf was separated from the mother 3 days postpartum, cows vocalized initially and moved to the feeding rack but started feeding shortly after that. Heart rate was initially increased, but effects were restricted to the first minutes after separation, and there were no effects on the serum cortisol (Hopster et al. 1995). Calves separated at 4 days postpartum made significantly more movements in the pen and spent more time standing and more time with the head out of the pen than calves separated 6 h or 1 day after birth (Weary and Chua 2000). A similar pattern was observed for the cows, i.e., cows with younger calves moved more frequently in the pen and vocalized at much higher rates After separation, cows separated day 4 postpartum called at approximately four times the rate of those separated at 6 h or 1 day (Weary and Chua 2000). In a follow-up study, Flower and Weary (2001) showed that after separation at day 14 postpartum, cows had higher rates of vocalization, movement, and placing the head outside the pen than cows separated at day 1 postpartum. When introduced to an unfamiliar calf at 6 weeks of age, calves from the late-separation group showed more intense social behavior toward the unfamiliar calf than did those calves separated early (Flower and Weary 2001).

Marchant-Forde et al. (2002) separated dairy calves from their mothers at 24 h postpartum and 24 h later recorded their responses to playback vocalizations of the mother versus the calf compared with white noise. The cows had higher percentage of mean and peak heart rate, higher number of ear movements and head movements, higher proportion of walking and orientation toward the loudspeaker, and lower proportion of eating during the playback (Marchant-Forde et al. 2002). The calves had higher percentage heart rate change and tendencies for higher number of ear movements and lower number of head movements (Marchant-Forde et al. 2002). Sandem and Braastad (2005) found that an indicator of frustration in cows, i.e., the proportion of eye white, increased when cows were separated from their calf at day 4 postpartum, but decreased when they were reunited with the calf.

Fostering Calves onto Foster Cows

In different studies, researchers have tried to foster calves onto foster cows using various methods. Rosecrans and Hohenboken (1982) fostered two calves onto cows within 5 min after parturition. Wyatt et al. (1977) fostered a second newborn calf onto cows at the time of birth of their natural calf. Hudson (1977) removed the cow's own calf immediately after birth and introduced four calves to her within 2 min after parturition. Kilgour (1972) fostered three or four calves onto cows from the fourth day of lactation. In my own research, we have always removed the cows' own calf before fostering four alien calves to them (Lidfors 2000; Loberg and Lidfors 2001; Loberg et al. 2007, 2008).

Cows accepted foster calves and allowed them to suckle (Hudson 1977; Rosecrans and Hohenboken 1982), suggesting that successful maternal-offspring bonding had occurred (Rosecrans and Hohenboken 1982). However, adopted calves nursed less frequently, for a significantly shorter total interval daily (Wyatt et al. 1977), and were 38–43 kg lighter at weaning than their natural-born mates (Wyatt et al. 1977; Rosecrans and Hohenboken 1982). Wyatt et al. (1977) found that the cows were reluctant to readily accept adopted calves, and they thought that this might be a major factor contributing to their low level of performance.

A high percentage of cross suckling was observed in the foster herd (Kilgour 1972; Wyatt et al. 1977; Rosecrans and Hohenboken 1982). When that occurred, alien calves usually approached and suckled from the rear of the cow or alongside another calf already suckling (Rosecrans and Hohenboken 1982). Occasionally, as many as four calves suckled simultaneously (Wyatt et al. 1977; Rosecrans and Hohenboken 1982). If an alien suckled along with a cow's foster calf, it was not usually rejected (Rosecrans and Hohenboken 1982). The 5 min time lapse between birth and removal of a calf and its replacement with alien calves for fostering was possibly of sufficient duration to interfere with formation of an effective maternal young bond between a cow and her two foster calves (Rosecrans and Hohenboken 1982). Hudson (1977), on the other hand, found that a rapid specific maternal bond was developed between the cow and her foster calves, resulting in no cross suckling and more uniform growth rates. This was probably due to the method of removing the cow's own calf and fostering four calves onto the cow within a short time after birth, which did not subject the calves to the stresses associated with traditional fostering methods (Hudson 1977). It is important to make sure that the foster cow has enough milk to support four calves, around 32 l/day.

Instead of multiple fostering, Price et al. (1985) induced twinning in cows of beef breed through embryo transfer. They found that initially twins spent more time suckling their mothers than did single calves, presumably due to insufficient milk intake, but later they adapted by utilizing natural forage and obtaining milk from alien cows. By 15–20 weeks of age, twins were suckling their dams for less than one-quarter of the time spent at 3–6 weeks, whereas the suckling time of single calves did not decline. Price et al. (1985) made the conclusion that twins may experience a natural early weaning from the mother's milk. Mothers bearing twins groomed their offspring less than mothers with singles, and twins were normally found at greater distances from their mother than singles (Price et al. 1985). In a follow-up study, Price et al. (1986) studied the cow-calf relationships of Hereford beef cattle after calves were temporarily separated from their dams during the first week after parturition. They found that contact and contact-seeking behaviors between twins and their mothers began to decline by the fifth or sixth day postpartum, whereas contact behaviors shown by single calves and their dams remained stable or increased in frequency. Mothers of twins were less responsive to their calves than mothers of singles, both during and following separation, and twins interacted with alien females more frequently than single-born young did. In a study on naturally born twins, it was found that twin calves required three times as long as single-born calves to begin sucking (Owens et al. 1985). It was also found that twin-bearing cows had significantly shorter gestation lengths than those with single calves (Owens et al. 1985). Hafez and Lineweaver (1968) noted that twin calves were weaker than single calves.

8 Conclusions

This chapter has described the natural behavior of cows at calving, calves' behavior from birth until weaning, and the mother-young attachment both in low managed free-ranging cattle and in housed dairy cattle. Some information has also been given on the physiology of the cow around parturition. It shows that cattle are very flexible and adapt their behavior depending on the circumstances when possible. However, most dairy cows do not have much contact with their offspring, and their calf is raised artificially on milk replacer or cow milk. There is an increasing interest in allowing dairy cows and calves to be together during the first weeks of the calves' life, and research is ongoing to better understand how this may be managed.

References

Azzazi F, Krause GF, Garverick HA. Alteration of GnRH induced LH release early post partum in dairy cattle with estrogen and progesterone. Ann Meet ASAS Ithaca N Y. 1980;72:367.

Baskin LM, Stepanov NA. Behaviour of free-keeping cattle in conditions of boreal forest. In: Nichelmann M, Wirenga HK, Braun S, editors. Proceedings of the international congress of applied ethology Berlin: KTBL in Göttingen; 1993. p. 405–7.

Beck TW, Convey EM. Serum LH in heifers after ovariectomy. J Anim Sci. 1976;43:274.

Berteaux D, Micol T. Population studies and reproduction of the feral cattle (Bos taurus) of Amsterdam Island, Indian Ocean. J Zool. 1992;228(2):265–76.

Brownlee A. Studies in the behaviour of domestic cattle in Britain. Bull Anim Behav. 1950;8:11–20.

Carruthers TD, Hafs HD. Suckling and four times daily-milking: influences on ovulation, estrus and serum luteinizing hormone, glucocorticoids and prolactin in postpartum Holsteins. J Anim Sci. 1980;50:919–25.

Clutton-Brock J. A natural history of domesticated mammals. 2nd ed. Cambridge: Cambridge University Press; 1999. p. 1999.

Convey EM. Serum hormone concentrations in ruminants during mammary growth, lactogenesis and lactation: a review. J Dairy Sci. 1974;57:905–17.

Cook R. Ranking of countries with the most cattle. 2021. https://beef2live.com/story-world-cattle-inventory-ranking-countries-0-106905.

da Costa MJRP, Albuquerque LG, Eler JP, de Vasconcelos Silva JAI. Suckling behaviour of Nelore, Gir and Caracu calves and their crosses. Appl Anim Behav Sci. 2006;101(3–4):276–87.

Das SM, Redbo I, Wiktorsson H. Effect of age of calf on suckling behaviour and other behavioural activities of Zebu and crossbred calves during restricted suckling periods. Appl Anim Behav Sci. 2000;67(1–2):47–57.

Das SM, Redbo I, Wiktorsson H. Behaviour of zebu and crossbed cows in restricted suckling groups. Appl Anim Behav Sci. 2001;72(3):263–70.

De La Torre MP, Briefer EF, Ochocki BM, McElligott AG, Reader T. Mother–offspring recognition via contact calls in cattle, Bos taurus. Anim Behav. 2016;114:147–54.

Del Río NS, Stewart S, Rapnicki P, Chang YM, Fricke PM. An observational analysis of twin births, calf sex ratio, and calf mortality in Holstein dairy cattle. J Dairy Sci. 2007;90(3):1255–64.

Dobson H. Radioimmunoassay of FSH in the plasma of post-partum dairy cows. J Reprod Fertil. 1978;52:45–9.

Dufty JH. Clinical studies on bovine parturition – foetal aspects. Aust Vet J. 1973;49:177–82.

Edqvist LE, Kindahl H, Stabenfeldt G. Release of prostaglandin F_{2x} during the bovine peripartal period. Prostaglandins. 1978;16:111–9.

Edwards SA. Factors affecting the time to first suckling in dairy calves. Anim Sci. 1982;34(3):339–46.

Edwards SA, Broom DM. Behavioural interactions of dairy cows with their newborn calves and the effects of parity. Anim Behav. 1982;30(2):525–35.

Edwards SA, Broom DM, Collis SG. Factors affecting levels of passive immunity in dairy calves. Br Vet J. 1982;138(3):233–40.

Ekesbo I, Gunnarsson S. Farm animal behaviour – characteristics for assessment of health and welfare. 2nd ed. Glasgow: CAB International; 2018.

Erb RE, D'amico F, Chew BP, Malven PV, Zamet CN. Variables associated with peripartum traits in dairy cows. VII hormonal profiles associated with dystocia. J Anim Sci. 1981;52:346–58.

Espmark Y. Mother-young relationship and ontogeny of behaviour in reindeer (Rangifer tarandus L.). Z Tierpsychol. 1971;29(1):42–81.

Fernandes LC, Thatcher WW, Wilcox CJ, Call EF. LH release in response to GnRH during the postpartum period of dairy cows. J Anim Sci. 1978;46:443–8.

Finger KH, Brummer H. Beobachtungen uber das Saugverhalten mutterlos aufgezogener Kalber. Deut Tierarztl Wochensch. 1969;76:665–7.

Flower FC, Weary DM. Effects of early separation on the dairy cow and calf: 2. Separation at 1 day and 2 weeks after birth. Appl Anim Behav Sci. 2001;70(4):275–84.

Foster JP, Lamming GE, Peters AR. Short-term relationships between plasma LH, FSH and progesterone changes in postpartum dairy cows and the effect of GnRH injection. J Reprod Fertil. 1980;59:321–7.

Frasers AF. A monitored study of major physical activities in the perinatal calf. Vet Rec. 1989;125(2):38–40.

Galhotra AP, Gautam OP. Placentophagia in bovines. Haryana Veterinarian. 1971;10(1):33–7.

Gáspárdy A, Sheridan J, Ari M, Gulyás L. Twin calving and its connection to other economically important traits in dairy cattle. In: Abubakar M, editor. Ruminants-the husbandry, economic and health aspects. London: IntechOpen; 2018.

George JM, Barger IA. Observations of bovine parturition. Proc Aust Soc Anim Prod. 1974;10:314–7.

Green A, Clark C, Favaro L, Lomax S, Reby D. Vocal individuality of Holstein-Friesian cattle is maintained across putatively positive and negative farming contexts. Sci Rep. 2019;9(1):1–9.

Green AC, Clark CE, Lomax S, Favaro L, Reby D. Context-related variation in the peripartum vocalisations and phonatory behaviours of Holstein-Friesian dairy cows. Appl Anim Behav Sci. 2020;2020(231):105089.

Green AC, Lidfors LM, Lomax S, Favaro L, Clark CE. Vocal production in postpartum dairy cows: temporal organization and association with maternal and stress behaviors. J Dairy Sci. 2021;104(1):826–38.

Hafez ESE, Lineweaver JA. Suckling behaviour in natural and artificially fed neonate calves. Z Tierpsychol. 1968;25(2):187–98.

Haley DB, Bailey DW, Stookey JM. The effects of weaning beef calves in two stages on their behavior and growth rate. J Anim Sci. 2005;83:2205–14.
Hall SJG. Studying the Chillingham wild cattle [UK]. Ark. 1979;6:72–9.
Hall SJG. Chillingham cattle: social and maintenance behaviour in an ungulate that breeds all year round. Anim Behav. 1989;38(2):215–25.
Hall SJG. Behaviour of cattle. In: Jensen P, editor. The ethology of domestic animals: an introductory text. Wallingford: CAB International; 2002. p. 131–43.
Hall SJG, Moore GF. Feral cattle of Swona, Orkney Islands. Mammal Rev. 1986;16(2):89–96.
Hall SJG, Vince MA, Walser ES, Garson PJ. Vocalisations of the Chillingham cattle. Behaviour. 1988;104(1/2):78–104.
Hermann E, Stenum N. Mother-calf behaviour during the first six hours after parturition. In: Welfare and husbandry of calves. Curr Top Vet Med Anim Sci. 1982;19:3–23.
Hoffman B, Schams D, Bopp R, Ender ML, Giminez T, Karg H. Changes of PG, total estrogens and progesterone at parturition in the cow. Acta Endocrinol. 1973;73:385–95.
Hopster H, O'Connell JM, Blokhuis HJ. Acute effects of cow-calf separation on heart rate, plasma cortisol and behaviour in multiparous dairy cows. Appl Anim Behav Sci. 1995;44(1):1–8.
Hötzel MJ, Ungerfeld R, Quintans G. Behavioural responses of 6-month-old beef calves prevented from suckling: influence of dam's milk yield. Anim Prod Sci. 2010;50(10):909–15.
Hudson SJ. Multiple fostering of calves onto nurse cows at birth. Appl Anim Ethol. 1977;3(1):57–63.
Hudson SJ, Mullord MM. Investigations of maternal bonding in dairy cattle. Appl Anim Ethol. 1977;3(3):271–6.
Illmann G, Špinka M. Maternal behaviour of dairy heifers and sucking of their newborn calves in group housing. Appl Anim Behav Sci. 1993;36(2–3):91–8.
Kesler DJ, Garverick HA, Youngquist RS, Elmore RG, Bierschwal CJ. Effect of days post partum and endogenous reproductive hormones on GnRH induced LH release in dairy cows. J Anim Sci. 1977;46:797–806.
Kiley M. The vocalizations of ungulates, their causation and function. Z Tierpsychol. 1972;31:171–222.
Kiley-Worthington M, de la Plain S. The behaviour of beef suckler cattle (Bos Taurus). Tierhaltung 14. Basel: Birkhäuser Verlag; 1983.
Kilgour R. Some observations on the suckling activity of calves on nurse cows. Proc N Z Soc Anim Prod. 1972;32:132–6.
Kindahl H, Lindell JO, Edqvist LE. Luteolysis in domestic animals: control of $PGF_{2\alpha}$ release. Proceedings of the 9th international congress on animal reproduction and artificial insemination Madrid. 1980;11:17–26.
Klopfer PH, Adam DK, Klopfer MS. Maternal "imprinting" in goats. Proc Nat Acad Sci. 1964;52:911–4.
Knobil E. The neuroendocrine control of the menstrual cycle. Recent Prog Horm Res. 1980;36:53–39.
König I, Raasch ML, Hühn R. Behaviour pattern of lying out and animal welfare of suckler cows. In Saloniemi H. Proceedings of the 9th international congress of animal hygiene, 1997 17–21 August, Helsinki, Finland. p. 394–97.
Kovalčik K, Kovalčiková M, Brestenský V. Comparison of the behaviour of newborn calves housed with the dam and in the calf-house. Appl Anim Ethol. 1980;6(4):377–80.
Le Neindre P. Cow-calf relationships: the effect of management systems. Curr Top Vet Med Anim Sci. 1982;19:24–35.
Lent PC. Mother-infant relationships in ungulates. In: Geist V, Walther F, editors. The behaviour of ungulates and its relation to management, vol. 24. IUCN, New Series; 1974. p. 14–55.
Leslie KE. The events of normal and abnormal postpartum reproductive endocrinology and uterine involution in dairy cows: a review. Can Vet J. 1983;24:67–71.
Leuthold W. African ungulates: a comparative review of their ethology and behavioral ecology. In: Farner DS, coordinating, editors. Zoophysiology & ecology, vol. 8. Berlin/Heidelberg: Springer Verlag; 1977.

Lewandrowski NM, Hurnik JF. Nursing and cross-nursing behavior of beef cattle in confinement. Can J Anim Sci. 1983;63(4):849–53.
Lidfors L. Mother-young behaviour in cattle: Parturition, development of cow-calf attachment, suckling and effects of separation [dissertation]. Skara: Swedish University of Agricultural Sciences, Report 33; 1994.
Lidfors LM. Behavioural effects of separating the dairy calf immediately or 4 days post-partum. Appl Anim Behav Sci. 1996;49(3):269–83.
Lidfors L. Behavioural studies of keeping dairy calves on foster cows. Proceedings of the 34th international congress of international society of applied ethology; 2000 17–20 October; Florianópolis, Brazil. p. 95.
Lidfors L, Jensen P. Behaviour of free-ranging beef cows and calves. Appl Anim Behav Sci. 1988;20(3–4):237–47.
Lidfors LM, Moran D, Jung J, Jensen P, Castren H. Behaviour at calving and choice of calving place in cattle kept in different environments. Appl Anim Behav Sci. 1994a;42(1):11–28.
Lidfors LM, Jensen P, Algers B. Suckling in free-ranging beef cattle—temporal patterning of suckling bouts and effects of age and sex. Ethology. 1994b;98(3–4):321–32.
Lindell JO, Kindahl H, Jansson L, Edqvist LE. Post-partum release of prostaglandin $F_{2\chi}$ and uterine involution in the cow. Theriogenology. 1982;17:237–45.
Loberg J, Lidfors L. Effect of stage of lactation and breed on dairy cows' acceptance of foster calves. Appl Anim Behav Sci. 2001;74(2):97–108.
Loberg JM, Hernandez CE, Thierfelder T, Jensen MB, Berg C, Lidfors L. Reaction of foster cows to prevention of suckling from and separation from four calves simultaneously or in two steps. J Anim Sci. 2007;85(6):1522–9.
Loberg JM, Hernandez CE, Thierfelder T, Jensen MB, Berg C, Lidfors L. Weaning and separation in two steps—a way to decrease stress in dairy calves suckled by foster cows. Appl Anim Behav Sci. 2008;111(3–4):222–34.
Lott DF, Galland JC. Parturition in American bison: precocity and systematic variation in cow isolation. Z Tierpsychol. 1985;69:66–71.
Marchant-Forde JN, Marchant-Forde RM, Weary DM. Responses of dairy cows and calves to each other's vocalisations after early separation. Appl Anim Behav Sci. 2002;78(1):19–28.
Mayntz M. Ontogeny of suckling in cattle and its implications for milking. Proceedings of the Symposium on Milk Synthesis, Secretion and Removal in Ruminants. Bern, Switzerland, April 1996. p. 59–4.
Mayntz M. Stepwise development towards physiological milk harvest-a consequence of suckling technique. In: Tancin V, Mihina S, Uhrincat M, editors. Physiological and technical aspects of machine milking, ICAR Technical Series 10. Nitra Slovak Republic; 2005. 26–28 April.
Mayntz M, Costa R. Effect of pharmacologically induced changes in milk ejection on suckling in Bos taurus. Physiol Behav. 1998;65(1):151–6.
Mee JF. Managing the calf at calving time. In: Smith RA editor. Proceedings of the 41st Annual Conference of American Association of Bovine Practitioners; 2008 September 25–27; Charlotte, North Carolina. p. 46–3.
Metz J, Metz JHM. Die bedeutung der mutter ind der umwelt des neugeborenen kalbes. Vorträg 16 Int. Arbeitst. Angew. Ethologie Nutzt. Deutschen Veterinärmedizinischen Gesellschaft, Freiburgx/Breisgau 1984. KTBL, Darmstadt. 1985;307:188–98.
Metz J, Metz JHM. Maternal influence on defecation and urination in the newborn calf. Appl Anim Behav Sci. 1986;16(4):325–33.
Michanek P, Ventorp M. Passive immunization of new-born dairy calves on three farms with different housing systems. Swed J Agric Res. 1993;23(1):37–43.
Nicol AM, Sharafeldin MA. Observations on the behaviour of single-suckled calves from birth to 120 days. Proc N Z Soc Anim Prod. 1975;35:221–30.
Owens JL, Edey TN, Bindon BM, Piper LR. Parturient behaviour and calf survival in a herd selected for twinning. Appl Anim Behav Sci. 1985;13(4):321–33.

Papini M, Tenucci M, Vitale A, Lovari S. An outline of social behavior of free-ranging Maremma calves. Ital J Zool. 1983;17(2):203–4.

Poindron P, Le Neindre P. Comparaison des relations mère-jeune observées lors de la tétée chez la brebis (Ovis aries) et chez la vache (Bos taurus). Annales de Biologie Animale Biochimie Biophysique. 1975;15(3):495–501.

Pope GS, Gupta SK, Munro TB. Progesterone levels in the systemic plasma of pregnant, cycling and ovariectomized cows. J Reprod Fertl. 1968;20:369–81.

Price EO, Martinez CL, Coe BL. The effects of twinning on mother—offspring behavior in range beef cattle. Appl Anim Behav Sci. 1985;13(4):309–20.

Price EO, Smith VM, Thos J, Anderson GB. The effects of twinning and maternal experience on maternal-filial social relationships in confined beef cattle. Appl Anim Behav Sci. 1986;15(2):137–46.

Price EO, Harris JE, Sween ML, Connor JM. Fenceline contact of beef calves with their dams at weaning reduces the negative effects of separation on behavior and growth rate. J Anim Sci. 2003;81:116–21.

Radford MJ, Nancarrow CD, Mattner PE. Ovarian function in suckling and nonsuckling beef cows post partum. J Reprod Fertl. 1978;54:49–56.

Rahe CH, Owens RE, Fleeger JL, Newton HJ, Harms PG. Pattern of plasma luteinizing hormone in the cyclic cow: dependence upon the period of the cycle. Endocrinology. 1980;107:498–505.

Rasmussen MD, Mayntz M. Pressure in the teat cistern and the mouth of the calf during suckling. J Dairy Res. 1998;65(4):685–92.

Reinhardt V. Reproductive performance in a semi-wild cattle herd (Bos indicus). J Agric Sci. 1982;98(3):567–9.

Reinhardt V, Reinhardt A. Natural sucking performance and age of weaning in zebu cattle (Bos indicus). J Agric Sci. 1981a;96(2):309–12.

Reinhardt V, Reinhardt A. Cohesive relationships in a cattle herd (Bos indicus). Behaviour. 1981b;77:121–51.

Reinhardt V, Reinhardt A, Mutiso FM. Cow-calf relationship in Massai cattle. In Proceedings of the 28th annual meeting of the European association for animal production; 1977 22–25 August, Belgium. Paper M/I.04/1-7.

Reinhardt V, Mutiso FM, Reinhardt A. Social behaviour and social relationships between female and male prepubertal bovine calves (Bos indicus). Appl Anim Ethol. 1978;4(1):43–54.

Reinhardt C, Reinhardt A, Reinhardt V. Social behaviour and reproductive performance in semi-wild Scottish Highland cattle. Appl Anim Behav Sci. 1986;15(2):125–36.

Robertson HA. Sequential changes in plasma progesterone in the cow during the estrous cycle, pregnancy, at parturition and postpartum. Can J Anim Sci. 1972;52:645–58.

Rørvang MV, Herskin MS, Jensen MB. The motivation-based calving facility: social and cognitive factors influence isolation seeking behaviour of Holstein dairy cows at calving. PLoS One. 2018;13(1):e0191128.

Rosecrans JG, Hohenboken WD. Suckling activity and calf growth in a group of crossbred cows each rearing two foster calves. Appl Anim Ethol. 1982;9(2):131–40.

Sambraus HH. Zum mutter-kind-verhalten der wiederkäuer. Berliner und Münchener Tierärtzliche Wochenschrift. 1971;84:24–7.

Sambraus HH. Zum beitrag sauberlecken oder nicht. Praktische Tierartz. 1973;54:108.

Sandem AI, Braastad BO. Effects of cow–calf separation on visible eye white and behaviour in dairy cows—a brief report. Appl Anim Behav Sci. 2005;95(3–4):233–9.

Santos R, Cachapa A, Carvalho GP, Silva CB, Hernández L, Costa L, et al. Mortality and morbidity of beef calves in free-range farms in Alentejo, Portugal—a preliminary study. Vet Med Int. 2019;2019:3616284.

Schallenberger E, Schams D, Zottmeier K. Response of lutropin (LH) and follitropin (FSH) to administration of gonadoliberin (GnRH) in pregnant and post-partum cattle including experiments with prolactin suppression. Theriogenology. 1978;10:35–53.

Schams D, Schallenberger E, Menzer CH, Stangl J, Zottmeier K, Hoffman B, Karg H. Profiles on LH, FSH and progesterone in post-partum dairy cows and their relationship to the commencement of cyclic function. Theriogenology. 1978;10:453–68.

Scheurmann E. Untersuchungen über die Aktivität und Ruheverhalten bei neugeborenen Kälbern. Zuchthygiene. 1974;9:58–68.

Schloeth R. Cycle annuel et comportement social du taureau de Camargue. Mammalia. 1958;22(1–4):121–39.

Schloeth R. Das Sozialleben des Camargue-Rindes: Qualitative und quantitative Untersuchungen über die sozialen Beziehungen—insbesondere die soziale Rangordnung—des halbwilden französischen Kampfrindes 1. Z Tierpsychol. 1961;18(5):574–627.

Selman IE, McEwan AD, Fisher EW. Studies on natural suckling in cattle during the first eight hours post partum I. Behavioural studies (dams). Anim Behav. 1970a;18:276–83.

Selman IE, McEwan AD, Fisher EW. Studies on natural suckling in cattle during the first eight hours post partum II. Behavioural studies (calves). Anim Behav. 1970b;18:284–9.

Sheldon IM, Dobson H. Postpartum uterine health in cattle. Anim Reprod Sci. 2004;82:295–306.

Sheldon IM, Williams EJ, Miller AN, Nash DM, Herath S. Uterine diseases in cattle after parturition. Vet J. 2008;176(1):115–21.

Shenavai S, Preissing S, Hoffmann B, Dilly M, Pfarrer C, Zalp GR, et al. Investigations into the mechanisms controlling parturition in cattle. Reproduction. 2012;144(2):279.

Short RE, Randel RD, Staigmiller RB, Bellows RA. Factors affecting estrogen induced LH release in the cow. Biol Reprod. 1979;21:683–9.

Somerville SH, Lowman BG. Observations on the nursing behaviour of beef cows suckling Charolais cross calves. Appl Anim Ethol. 1979;5(4):369–73.

Swedish Dairy Association. Livestock characteristics. Swedish Dairy Association; 2007.

Thatcher WW, Lewis SG, Eley RM, Bazer FW, Fields MJ, Williams WF, Wilcox CJ. Contribution of the bovine conceptus to the endocrinological phenomenon existing at implantation, during gestation and around parturition. In: Proceedings of the 9th international congress on animal reproduction and artificial insemination Madrid, vol. 11; 1980. p. 9–22.

Tucker HA. Endocrinology of lactation. Semin Perinatol. 1979;3:199–208.

Ventorp M, Michanek P. Cow-calf behaviour in relation to first suckling. Res Vet Sci. 1991;51(1):6–10.

Ventorp M, Michanek P. The importance of udder and teat conformation for teat seeking by the newborn calf. J Dairy Sci. 1992;75(1):262–8.

Vitale AF, Tenucci M, Papini M, Lovari S. Social behaviour of the calves of semi-wild Maremma cattle, Bos primigenius taurus. Appl Anim Behav Sci. 1986;16(3):217–31.

Wagnon KA. Behaviour of beef cows on a California range. California Agricultural Experiment Station, USDA, FOLLETO 1338;799(04). 1963.

Walker DM. Observations on behaviour in young calves. Bull Anim Behav. 1950;8:5–10.

Walker DE. Suckling and grazing behaviour of beef heifers and valves. N Z J Agric Res. 1962;5:331–8.

Weary DM, Chua B. Effects of early separation on the dairy cow and calf: 1. Separation at 6 h, 1 day and 4 days after birth. Appl Anim Behav Sci. 2000;69(3):177–88.

Webb R, Lamming GE, Haynes NB, Foxcroft GR. Plasma progesterone and gonadotrophin concentrations and ovarian activity in post-partum dairy cows. J Reprod Fertl. 1980;59:133–43.

Witt M. Technique and economics of keeping nurse cows. Zuchtungskunde. 1963;35:145–57.

Wood-Gush DGM, Hunt KA, Carson K, Dennison SGC. The early behaviour of suckler calves in the field. Biol Behav. 1984;9:295–306.

Wyatt RD, Gould MB, Totusek R. Effects of single vs simulated twin rearing on cow and calf performance. J Anim Sci. 1977;45(6):1409–14.

Nonhuman Primate Paternal Care: Species and Individual Differences in Behavior and Mechanisms

Toni E. Ziegler, Stacey R. Tecot, Eduardo Fernandez-Duque, Anne Savage, and Charles T. Snowdon

Abstract Direct care of offspring by the father (sire) is relatively rare in primates. Besides humans, there are a number of species where the male is essential for the survival of offspring: marmosets, tamarins, titis and owl monkeys, some lemurs, and siamangs. All these species show reduced sexual dimorphism, territoriality, and biparental care. However, timing and levels of direct care may vary among these species. Here, relying on both lab and field data, we address the variability found in father's involvement with his infants, the behavioral, neuroendocrine and sensory systems that are a cause and consequence of paternal care, and social bonds between the breeding pair. We integrate studies of laboratory animals (where detailed observations and experimentation are possible) with field studies (which illuminate the ecological and evolutionary functions of paternal care) and discuss the future directions for examining the proximate and ultimate mechanisms of paternal care in nonhuman primates.

Keywords Paternal care · Nonhuman primates · Neuroendocrinology · Olfactory · Auditory · Behavior · Social bonding

T. E. Ziegler (✉)
Wisconsin National Primate Research Center, University of Wisconsin-Madison, Madison, WI, USA
e-mail: ziegler@primate.wisc.edu

S. R. Tecot
School of Anthropology, University of Arizona, Tucson, AZ, USA

E. Fernandez-Duque
Department of Anthropology and School of the Environment, Yale University, New Haven, CT, USA

A. Savage
Proyecto Tití, Inc, Orlando, FL, USA

C. T. Snowdon
Department of Psychology, University of Wisconsin-Madison, Madison, WI, USA

G. González-Mariscal (ed.), *Patterns of Parental Behavior*, Advances in Neurobiology 27, https://doi.org/10.1007/978-3-030-97762-7_7

1 Introduction

Primates are highly social animals that have a variety of social structures, including multiple systems of breeding. Unique to a few species are high investment by the father (sire) in his offspring – particularly in the direct care of infants. Social and ecological conditions vary in those species exhibiting paternal care which may explain the selective pressures resulting in father's involvement. Likewise, parenting skills and motivation vary between individual fathers within the same species. In captivity, where paternal care can be more closely observed, we find differences in success rate of rearing infants that may be caused in part from fathers' neuroendocrine differences.

While most nonhuman primate sires show indirect care of their offspring through protection and provisioning, only a few species show direct paternal care where fathers are carrying, protecting, and/or involved in food sharing. Direct care of offspring is mainly seen among siamangs, marmosets, tamarins, lion tamarins, titis, owl monkeys, and some lemur species such as fat-tailed dwarf lemurs and red-bellied lemurs (Fernandez-Duque et al. 2009; Tecot et al. 2013). All these species show some form of biparental care, reduced sexual dimorphism, and territoriality, but levels and timing of direct care may vary (Fernandez-Duque et al. 2009). Additionally, conditions of monogamy are also varied, for example, serial monogamy may occur due to takeovers by intruders of either sex, and some species also display polyandry. Polyandrous species are rare and may be an alternative reproductive strategy in a small proportion of groups in a primarily monogamous species, such as the mustached tamarins (*Saguinus mystax*). However, even in this species, the mating system is polyandry, but genetic analyses determined that paternity is often monopolized by a single male per group (Huck et al. 2005; Huck et al. 2014).

Several Central and South American monkeys include genera that display biparental care, are pair-living, and are sexually monogamous, and some show cooperative infant care where more than the parents care for infants. These species are excellent models for understanding the physical and neuroendocrine bases of positive parenting. Detailed studies on captive monkeys have examined the proximate mechanisms of paternal care through behavioral neuroendocrinology assessments and tests of males for their motivation to parent. For example, studies on the cotton-top tamarins (*Saguinus oedipus*) and common marmosets (*Callithrix jacchus*) have allowed the examination of how positive parenting is modulated by the sensory and neuroendocrine systems. Studies of other species in captivity and the wild, such as the owl monkeys (*Aotus*), titi monkeys (*Callicebus*), and red-bellied lemurs (*Eulemur*), have also contributed to our understanding of direct care, individual variation in paternal care, and the socioendocrinology of paternal care. In particular, species with facultative paternal care can help us understand what conditions select for direct paternal care. See Fig. 1 for pictures of our species.

While our understanding of the proximate mechanisms of paternal care is progressing, we believe there is still much we can learn from improved bridging of the gap between laboratory and field studies of nonhuman primates. Only through interpreting laboratory mechanistic findings within an ecological-evolutionary

Fig. 1 Our study subjects. These four species have all been studied in their native habitat for their paternal care behaviors, and two of these species have been extensively studied in the laboratory to understand the behavioral neuroendocrine mechanisms of paternal care. Left to right: The red-bellied lemur, *Eulemur rubriventer*, studied by Stacey Tecot in Ranomafana, Madagascar, photo by VELONTSARA Jean Baptiste; the cotton-top tamarin, *Saguinus oedipus*, studied by Anne Savage in Northeastern Colombia, photo by Suzi Eszterhas; the common marmoset, *Callithrix jacchus*, from Northeastern Brazil, photo by Jordana Lennon, Wisconsin National Primate Research Center; the owl monkey, *Aotus azare*, studied by Eduardo Fernandez-Duque in Argentina, photo by E. White/Owl Monkey Project, Formosa-Argentina

perspective will we understand the ultimate consequences of a father's care of his offspring. To this end, we have organized this chapter to present different aspects of paternal care where we address our knowledge of mechanisms from laboratory studies and present data on the natural behavior and physiology of wild animals from our field studies.

Using data from our own research, we will address the following questions: (1) What is paternal care, and why is it important? (2) How is paternal care activated and controlled? (3) What have we learned about the basic biology of reproductive and parenting hormones and the role of the sensory systems in paternal care? (4) How does paternal care vary within and between the five species that we study? (5) What are some directions for future research on paternal care that integrate behavioral neuroendocrinology and field studies? We hope this work will provide neuroscientists and evolutionary biologists with the framework to continue to advance research on paternal care.

The authors have broad experience studying paternal care in nonhuman primates. Collectively, we study species that represent the variety of direct paternal care seen in nonhuman primates. Three of the species show cooperative infant care (cotton-top tamarins, *Saguinus oedipus*; common marmosets, *Callithrix jacchus*; and red-bellied lemur, *Eulemur rubriventer*) where not only fathers, but other group members, actively care for infants. The other two species include the owl monkey (*Aotus azarae*) and the titi monkey (*Callicebus discolor*) and display biparental care. All these species have been studied in their natural field settings, and cotton-top tamarins, common marmosets, and another species of titi monkey (*Callicebus moloch*) have also been studied extensively under laboratory conditions. Thus, we are able to integrate knowledge from studies of captive individuals (where detailed observations and experimentation are possible) with field studies (which illuminate the ecological and evolutionary functions of paternal care). Tamarins, marmosets, owl monkeys, and titi monkeys are all native to Central and South America, whereas red-bellied lemurs are endemic to the island of Madagascar.

2 What Is Paternal Care, and Why Is It Important?

Paternal care has a special interest to biologists because it is relatively rare among mammals (Garcia de la Chica et al. 2017; Kvarnemo 2006; Stockley and Hobson 2016; West and Capellini 2016). We first describe the varieties of paternal care in our study species, address the costs and benefits of paternal care, and then discuss the role that pair-bonding may have on the expression of paternal care.

It has been traditional to distinguish between direct and indirect paternal care (Clutton-Brock 1991). In direct care, fathers carry infants, share food with them, retrieve them, and are involved in social play, whereas in indirect paternal care, males may guard females and infants, alert group members to predators, prevent intrusion of potentially infanticidal males, but are not directly involved in infant care. We can see direct paternal care in each of our species. In tamarins and marmosets, males assist mothers in carrying infants from the first days after birth and increase the proportion of time they carry the infants as the infants get older (Washabaugh et al. 2002; Zahed et al. 2008). Males will retrieve infants and, at weaning, distract infants from nursing by sharing food with them often using specific calls to attract them (Joyce and Snowdon 2007). In captivity, titi monkey males are the primary infant carriers and retrieve straying infants (Mendoza and Mason 1986). Studies of wild populations have found that in both Azara's owl monkeys (*Aotus azarae*) and red titis (*Plecturocebus discolor*), males carry infants most of the time (ca. 80–90%) from very soon after birth, and infants only transfer to the mother for brief periods usually around the time of active nursing bouts (Tecot and Baden 2018). This pattern has been documented for every group of owl monkey and titis that has been studied in the field and in captivity (Fernandez-Duque et al. 2020; Evans et al. 2022). Adult owl monkey males share food with infants more frequently than do mothers (Wolovich et al. 2008) and are also the first source of support when an infant faces a challenging situation (e.g., crossing a canopy gap) (Rotundo et al. 2005). In wild, red-bellied lemurs, biological fathers vary considerably in how much care they provide, ranging from negligible to as much, or more, than mothers (Tecot and Baden 2018). Fathers, which huddle, hold, carry, groom, and play with infants, do not carry them immediately after birth, but provide increasing amounts of care as infants develop (Overdorff 1991). Overdorff (1991) also found that as paternal care increases, maternal care decreases, until infants only receive care from fathers and siblings. Fathers are also a source of support when an infant faces a challenging situation (e.g., crossing a large gap or after a fall to the ground; Tecot, personal observation).

Paternal care in mammals is considered surprising since a male can never be 100% certain of paternity, and females have an extensive commitment to reproduction due to gestation and lactation. This means that there should be some benefits to males engaging in infant care along with reduction of costs to mothers and/or benefits to infants. A review of primate allomaternal care (Mitani and Watts 1997) found that the presence of helpers decreased interbirth intervals and increased fecundity of mothers. However, a later analysis found that this pattern in primates was driven by the callitrichines (Isler and van Schaik 2012). Marmosets and tamarins usually give birth to twins that weigh up to 20% of the mother at birth. In captivity, and in some wild populations, conception can occur within a few weeks after birth while the first set of twins are still nursing, thus increasing energetic stress for mothers that can be alleviated by the father and other helpers. This relationship between allomaternal care and reproductive output (interbirth interval, fetal and infant growth rates) was not present in a comparative study of lemurs (Tecot et al. 2012), suggesting that the ultimate benefits of paternal care could differ across taxa. In particular, taxa in which paternal care is facultative may not have relatively shorter interbirth intervals or faster growth rates compared with other species, but paternal care may still confer reproductive benefits on an individual level. For example, red-bellied lemurs often twin but usually bear singletons, and paternal care may be more or less critical depending on litter size (Tecot 2010). Studies of male reproductive success are difficult with long-lived species, but determining the effects of paternal care on infant outcomes at the individual level is a necessary step in determining the ultimate benefits, and the evolutionary pathways, of paternal care.

In callitrichids, there are costs and benefits to fathers engaging in infant care. Not only is there the lack of certainty of paternity for males, but there may also be a severe energetic cost. For cotton-top tamarins, captive males lose up to 10% of their body weight while carrying infants (Achenbach and Snowdon 2002; Sánchez et al. 1999). The presence of additional helpers mitigates the weight loss in proportion to the number of additional helpers (Achenbach and Snowdon 2002). However, a longitudinal field study has found no effect of group size on infant survival as long as at least two adult males are present (Savage et al. 2009, 2022).

A strong relationship within a pair serves to increase the likelihood that the male will be the biological father of the infants and that the mother will be assured of help at the birth of infants. Lukas and Clutton-Brock (2013) suggested that male care is likely a consequence of social monogomy in mammals. There are several ways in which pair-bonds have been evaluated in captivity and the field: sociosexual behavior and mating limited to one partner, territorial defense (aggressive response to intruders), short-term separations increasing stress responses and reunions reducing stress, and preference for a specific partner (see Bales et al. 2021 for a complete description).

In wild cotton-top tamarins, copulations have only been observed between the breeding female and one particular male in the group. When there is more than one adult male in the group and the suspected "breeding male" dies, then the "other" male may assume the role. Occasionally, groups of cotton-tops had more than one pregnant female; however, only one female gave birth to offspring that survived. These multiple pregnancies typically occurred during the formation of a new group or if a new male entered an established group (Savage et al. 1996, 2022). In newly formed groups, the adult male and female spend a lot of time grooming and huddling. Captive cotton-top tamarins are managed so that groups are formed with one adult male and one adult female, but grooming and huddling, along with nonconceptive sexual behavior, are common.

Among owl monkeys that give birth to singletons, there is only one reproductively active pair in each group (Garcia de la Chica et al. 2022). Males and females are regularly replaced by intruding adult individuals, with both males and females being replaced equally often (Fernandez-Duque and Huck 2013). These forced changes of partner reduced the reproductive success of the remaining pair-mate in the same group. An adult, male or female, that only had one partner throughout life produced approximately 25% more offspring per decade of tenure than one that was forced to switch partners. In both titis and owl monkeys, the presence of infants changes the dynamics between pair-mates. Pair-mates groom and huddle less after an infant is born, and sex differences in time budgets become accentuated (Spence-Aizenberg et al. 2018).

Red-bellied lemur groups consist of an adult male, an adult female, and their immature offspring (Tecot et al. 2016). Groups are stable and last for multiple years. They give birth to singletons and twins and genetic analyses indicate that they have strict monogamy (Merenlender 1993). Rarely have we seen a resident adult of either sex disappear and red-bellied lemurs are extremely cohesive throughout the year (Overdorff 1988; Tecot and Romine 2012). Adults are usually within 5 m of other group members (including offspring) over 90% of the time, and adult males and females are each other's nearest neighbors (within 5 m) 43% of the time (Overdorff and Tecot 2006). These pair-bonds are actively maintained by both sexes and all individuals in the group are equally likely to initiate interactions, and agonism within the group is extremely rare (Overdorff and Tecot 2006), with only two aggressive interactions between pair-mates observed in two >1-year studies (Overdorff 1991; Tecot 2008).

In all studied species, there is evidence that the reproducing adults defend the pair against intruders. In captive cotton-top tamarins, both pair-mates react to intruders of both sexes. Males displayed aggressive behavior and females typically scent-marked close to the intruding animal (French and Snowdon 1981). Groups in the wild are very territorial, and there are often vocal exchanges at territorial boundaries that sometimes lead to physical fighting. When a tamarin, or tamarins have been evicted, they may often try and enter another group or form a new group. Temporary associations of individuals that remain on the periphery of a group (if related to animals in the existing group), or attempting to enter new groups, are common. However, these individuals are actively repelled by the adult animals in the resident group. In general, established groups actively repelled neighbors or potential intruders, but if a breeding female/male dies or is evicted, novel individuals are allowed to enter groups and assume the breeding position (Savage et al. 2022) with little or no aggression observed.

In the laboratory, familiarity among groups of Wied's black tufted-ear marmoset (*Callithrix kuhlii*) was sufficient to modulate aggression during encounters (French et al. 1995). Aggression was higher in males than females. Familiarity among individuals in a neighboring group appears to modulate aggression during intergroup encounters. However, two studies on captive common marmosets have shown that males react differently to a novel female depending upon whether his mate is present or not. When a novel female intruder was presented to paired common marmosets, both the male and female engaged in aggressive behavior toward the intruder. However, when the male of the pair was tested alone with a novel female, he demonstrated affiliative and little aggressive behavior (Evans and Poole 1983). When males were tested separately from their mates, but with the mates visible behind a one-way window, males displayed much more aggression and less affiliation toward a novel female than when they were tested without seeing their mate (Anzenberger 1985). Similar results have been seen in the field where the male breeding adult and nonbreeders of both sexes frequently engaged in sexual behavior during intergroup encounters, but the breeding female was never observed to engage in extra-group copulations (Lazaro-Perea 2001). These studies suggest that male marmosets display pair-bonded behavior in the presence of their mate, but are responsive to novel females when the mate is gone.

We tested the response of Azara's owl monkey pairs toward calls of unfamiliar male and female unpaired individuals in the field. As evidenced by higher rates of sociosexual and vocal responses, movement toward the speaker, and intergroup encounters during and after the experiments, both sexes were more reactive to the calls of unfamiliar males than females. Paired males reacted more strongly than females. Females guarded their mates more often during the simulated presence of unfamiliar males in comparison to unfamiliar females. Coupled with the fact that replacement of the putative father leads to lower infant survival, this suggests that females may guard their mates to secure paternal investment (Garcia de la Chica et al. 2021a).

In contrast to the other species, red-bellied lemurs scent-mark their territories frequently, but mostly avoid other groups. Agonism is almost nonexistent in this species; however, between-group agonism rates during periods of food scarcity, when groups may be forced to come into contact, are ten times higher than during food abundance (Tecot et al. 2016). The vast majority of interactions with other groups occurred between, or along, territorial borders at feeding sources, and both males and females participated (Overdorff and Tecot 2006).

Bonded pairs may also show behavioral and physiological signs of distress. In long-term captive marmoset pairs, social separation causes increased distress behaviors and cortisol concentrations (Cavanaugh et al. 2016). However, reunion with social partners reduced cortisol levels and distress-related behavioral patterns. In captive pairs of tamarins, brief separations lead to an increase in long calls and increased huddling and sexual behavior on reunion (Porter and Snowdon 1997). Laboratory titi monkeys also showed increased cortisol levels when separated from the mate (Mendoza and Mason 1986). Controlled studies of separation and reunion are not possible in field studies, but researchers who are present when study subjects are captured could leverage short-term separations that occur between capturing each adult to record separation-induced behaviors. During such an event, red-bellied lemur pairs showed signs of distress, including prolonged long/contact calls (Tecot, personal observation).

3 How Is Paternal Care Activated and Controlled?

In order to understand how paternal care is activated and controlled, it is important to first acknowledge and discuss the variation in infant care that has been reported between taxa. Evidence of variability in the amount of time and the onset of post-birth time that the father, or resident male, carries and cares for infant(s) has been reported both under captive and field conditions. In the genus *Callithrix*, there are reports of variation between species in father's participation in the first few weeks after the birth of the infant at this most critical time for its survival. In both the black tufted-ear marmoset, *Callithrix kuhlii*, and the white-faced marmoset, *Callithrix geoffroyi*, mothers primarily carried the infants during the first 2 weeks (French et al. 2008). In the buffy-headed marmoset, *Callithrix flaviceps*, the breeding males carried as much as the mother, and all adult males carried more than adult females (Ferrari 1992). However, common marmosets show individual variation when and how often they are carrying infants. For the first week, they may be carrying up to 50% of the time (Zahed et al. 2008).

In the genus *Saguinus*, *S. labiatus*, fathers spend more time carrying than mothers, and in cotton-top tamarin, males can spend up to 80% of their time carrying infants during the first few weeks following birth (Washabaugh et al. 2002; Savage et al. 1996; Pryce 1988; Ziegler et al. 1990). The Goeldi's monkey, *Callimico goeldii*, has cooperative infant care and resembles the tamarins in morphology, but only one infant is born with each birth. The mother carries the infant for the first 3 weeks before the father starts carrying, and when they carry, they do not always carry more than the mothers (Schradin and Anzenberger 2001). In other biparental species that have only one infant, males may spend 80–90% of their time carrying infants shortly after birth, and infants are only transferred to the mother for brief periods for nursing. This has been reported for Azara's owl monkeys, red titis, *Callicebus cupreus*, and red-bellied lemurs (Fernandez-Duque et al. 2020; Tecot and Baden 2018). In red-bellied lemurs, fathers with one infant carry extensively (Tecot and Baden 2018) as do those with twins (S. Tecot, personal observation). However, there is considerable variation among fathers on the timing of their participation in infant care (Tecot and Baden 2018). Some fathers provide as much care as the average mother, and there are months in which they provide more care than mothers; other fathers barely provide direct care at all (Tecot and Baden 2018). Overdorff (Overdorff, 1991) observed that eventually maternal care ceases and fathers provide all infant care. Some of the variation between genera and species may have to do with the level of infant care experience of the male. Males gain experience with infants prior to becoming a father, first-time fathers with their own infants, and additional experience by multiple births of his own infants.

Taking care of primate infants is a learned skill, in part, and starts by observing care of young early in life and practicing caring for young infants. This is most evident in cooperative infant care species where alloparents, or older siblings, assist parents with infant care (Cleveland and Snowdon 1984), as do red-bellied lemurs (Tecot and Baden 2018). All of the species of *Callithrix*, *Saguinus*, *Leontopithecus*, *Callimico*, and *Eulemur rubriventer* show sibling care while they are living in their natal groups. Cotton-top tamarins are very interested in carrying younger siblings and carry more in the wild than those in captivity perhaps because their help is needed more in the wild to share the energetic burden of twins (Savage et al. 2022). In captivity, common marmoset juveniles at 6 months of age are very interested in the newborn infants (Zahed et al. 2008). They will attempt to carry but appear to need time to adjust to the feel and weight of infants. Red-bellied lemur siblings help with infants and have been documented playing, grooming, holding, and huddling with them. However, they rarely carry them, perhaps due to body size constraints, and not all siblings help (Tecot and Baden 2018).

Experience with infant care as siblings in natal groups is important for later reproductive success in marmosets and tamarins. However, field data on cotton-top tamarins found no difference in survival of first litters born to females with infant care experience compared to those with no infant care experience (Savage et al. 2022), but experienced males were present in all litters. In captivity, tamarins typically delay their own reproduction while helping the breeding parents with the young, dependent infants and gain parenting experience important for rearing their own offspring (Snowdon and Ziegler 2007). This deferral of reproduction can be associated with a suppression of fertility in at least one sex in captivity. Interestingly, in both marmosets and tamarins, the eldest male offspring spend more time caring for infants than the females (Zahed et al. 2008, 2010). In some captive groups of marmosets and tamarins, males may remain in their families past the age of 4 years, and these eldest males will continue to care for their younger siblings. Male tamarins living with infants and juveniles with their family, regardless of whether they are fathers or eldest sons, show higher prolactin levels than males living with non-pregnant females, but not as elevated as prolactin levels are with experienced fathers or first-time fathers (Ziegler et al. 1996). Since contact through carrying infants raises prolactin levels in fathers (Dixson and George 1982), it appears that prior to being a father, males are responding hormonally to caring for infants. It is unknown whether older female siblings have elevated prolactin, but they do not carry as much as the older male siblings.

While males gain experience in caring for infants when they are in their natal groups as older offspring, first-time fathers also require experience upon becoming parents. First-time parents have a lower success rate of surviving offspring in tamarins (Baker and Woods 1992; Epple 1978; Johnson et al. 1991; Tardif et al. 1984). Much depends upon the father in the first 2 weeks. For example, in cotton-top tamarins, mothers carry their infants approximately 50% of their waking time, and in established families, they spend approximately 20% of their day nursing them (Washabaugh et al. 2002; Ziegler et al. 1990). Fathers are the primary carriers and paternal carry time gradually decreases with increasing age of the infant (Zahed et al. 2010). Since the cotton-top tamarin has cooperative infant care, experienced fathers in large families decrease their time spent carrying as the infant grows and older sibling helpers increase their time carrying for the young.

In common marmosets, first-time fathers did not always influence the success rate in offspring survival if the mother had some infant experience (Tardif et al. 1984). However, other studies have indicated that there is considerable variation in father's carrying time for both first-time fathers and experienced fathers (Zahed et al. 2008; Yamamoto et al. 1996). When comparing captive and wild groups, mothers and fathers equally carry their infants in the first few weeks and helpers carry more in the field setting than in captivity (Yamamoto et al. 1996). Captive common marmoset fathers carried around 25% of the time that the infant was carried, while mothers carried around 50% in the first week, but mothers reduced their carrying by the second week and siblings carried more. Thus, fathers' carrying time stayed the same for the 2 weeks while mothers decreased. Additionally, when fathers were tested while separated from their family in response to an infant distress cry, inexperienced males responded significantly less to the stimulus than the experienced fathers. However, experienced fathers varied greatly in their response as well (Zahed et al. 2008).

4 What Have We Learned About the Basic Biology of Reproductive and Parenting Hormones and the Role of the Sensory Systems in Paternal Care?

Our studies of the neuroendocrine changes in the common marmoset and cotton-top tamarin have contributed much to understand the basic biological changes in males even prior to being a parent. We have documented changes in male hormones and weight while his pair-mate is pregnant, in the hormonal responses to the postpartum ovulation with odor detection, and in hormonal responses to infant sensory signals using odor and auditory cues. These studies have allowed us to present a dynamic illustration of the social system in these species providing evidence of social bonding between the father and his mate, as well as between the father and his infant. Our research has contributed to elucidating the plasticity of the system to meet the ever-changing social needs of the family (Fig. 2). Further, several of our species are validating the same hormonal changes in the field as we see in the lab.

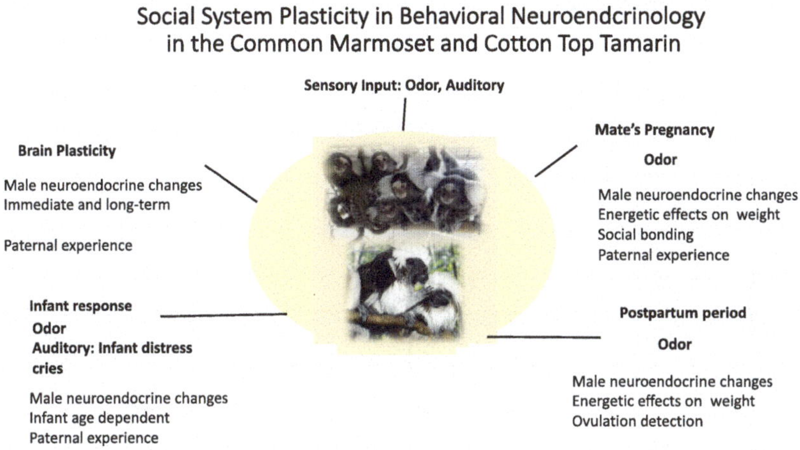

Fig. 2 External sensory signals drive changes in neuroendocrine and behavioral events

(a) Male's Hormonal Changes in Association with the Mate's Pregnancy in the Lab and Field.

Studies of both wild and laboratory primates demonstrate that males respond to their mate's pregnancy with changes in various hormone metabolites. Studies with laboratory primates further suggest that males gain weight during their partner's pregnancy, and these changes are facilitated by olfactory cues. Hormonal changes in male paternal primates during their mate's pregnancy may indicate that the communication between biparental mates may help the male prepare for becoming fathers. In the cotton-top tamarin at mid-gestation, male urinary hormones increase with highest change in concentration in the last month of pregnancy (Ziegler et al. 2004a). Of note is that estrogens, androgens, as well as prolactin and the glucocorticoid cortisol are elevated by midpregnancy (Fig. 3). Pregnant female tamarins also show a midpregnancy rise in glucocorticoids that are followed within 1–2 weeks by the male's peak in cortisol in all infant-experienced males (Ziegler et al. 2004a). In the field, red-bellied lemur males show elevated estradiol fecal metabolite levels that increased by fourfold by the end of their mate's pregnancy (Tecot, in preparation), elevated glucocorticoid metabolites in the mate's third trimester just after her midpregnancy rise (Tecot 2008), and elevated androgen metabolites during the peripartum period (Tecot and Baden 2018). Male owl monkeys also have elevated glucocorticoids during their mate's pregnancy (Corley et al. 2021). The midpregnancy rise of glucocorticoids in females may stimulate a glucocorticoid response in males and thereby activate other hormonal changes in males to prepare them for their parenting role. It is likely that this occurs through chemical/olfactory communication (Ziegler et al. 2004a). Additionally, male tamarins have high levels of circulating estrogens as do male owl monkeys, *Aotus trivirgatus* (Setchell and Bonney 1981),

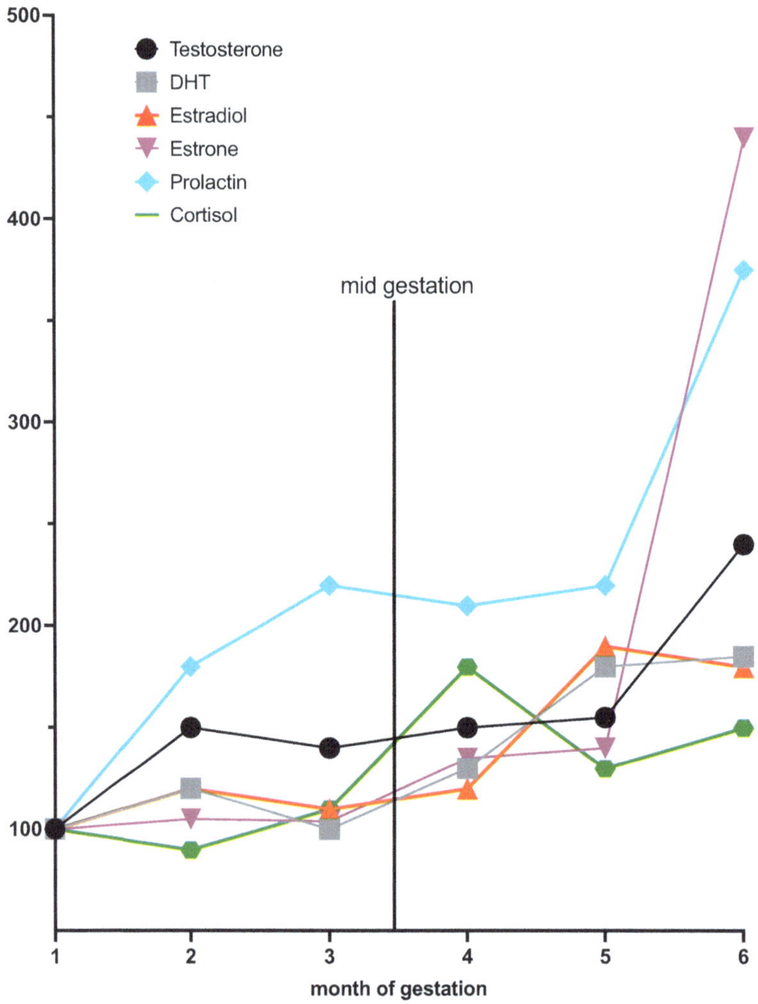

Fig. 3 Changes in steroids and prolactin in the urine of expectant fathers from the first month of the mate's gestation to the last month. All hormones increased in the second half of pregnancy. (Graph redrawn from data published on Cotton top tamarins, *Saguinus oedipus* (Ziegler et al. 2004a))

and common marmosets (Ziegler et al. 2009). In the common marmoset, serum levels of prolactin and testosterone are elevated at the end of the mate's pregnancy relative to the postpartum period. At the time of birth of infants in cotton-top tamarins, urinary glucocorticoids, estrogens, and androgens that were high in males at the end of pregnancy plummet and remain low until prior to the female's postpartum ovulation, indicating that these steroids were elevated at late pregnancy in response to the female's reproductive condition (Ziegler et al. 2004b). Prolactin levels are elevated at the end of gestation in both tamarin and marmoset fathers.

Along with changing hormones, male marmosets and tamarins gain weight during their mate's gestation similar to the female (Ziegler et al. 2006). When compared to control males (living with a nonpregnant female), male marmosets and tamarins show weight gains by midpregnancy, and their weights are at their highest during the last month. Physical changes, such as male weight gain, may be incidental to the hormonal changes occurring to prepare males for their role in infant care as well as the weight loss that accompanies infant care. This may be a part of an evolutionary process that occurs in biparental species (Daly 1979). Selection affects biparental species by diverting male reproductive effort into parental investment (Emlen and Oring 1977). Extensive energetic investment made by these fathers in caring for their young offspring may require larger energy reserves (Achenbach and Snowdon 2002; Sánchez et al. 1999).

We see here that male marmosets, tamarins, and red-bellied lemurs show a hormonal response to their mate's pregnancy and marmosets and tamarins have weight gain during the latter half of pregnancy likely through the chemical cues of the female's pregnancy and thus are primed to show paternal care for their infants.

(b) Postpartum Period Through Odor Communication.

In the laboratory, female marmosets and tamarins show a post-birth ovulation as soon as 2–3 weeks while there are still dependent infants in the natal group. Both tamarin and marmoset males can detect the periovulatory period through presentation of isolated odors of an ovulating female and through steroid changes during this time. When tamarin and marmoset males were presented with periovulatory odors from unknown females while they were living with a pregnant mate, they increased frequency of mounts to their mate and penile erections (Ziegler et al. 1993; Ziegler et al. 2005). Steroid levels also increase in male cotton-top tamarins responding to a female's periovulatory period. All tamarin males have low steroids following birth and then significant increases in urinary androgens, testosterone and dihydrotestosterone (DHT), prior to the female's postpartum LH peak, indicating ovulation. All steroids, including estradiol and glucocorticoids, had increased by 3–7 days prior to the female's LH peak that coincided with the female's follicular period. It is likely these steroids are involved in response to the estrogen changes during the follicular phase in the female. Sexual communication between paired male and female cotton-top tamarins ensures mating at the time of optimum fertility. The elevated glucocorticoids may indicate reception of the signal, similar to the peak levels found in males at their mate's midpregnancy. Based on the daily sampling of testosterone and DHT, we determined that the androgens were flexible during the period of infant care to ensure increased steroids around the time when the pair-mate was fertile, but otherwise levels were low and not associated with male care of infants. However, not all biparental species have a postpartum ovulation where this flexibility is essential for the dual role fathers provide at this time. In fact, father's care of infants in wild red-bellied lemurs was associated with increased androgens despite the lack of a postpartum ovulation (Tecot and Baden 2018). The general pattern that we have observed, but requires further testing, is that androgens decrease as males shift into pairs and fatherhood and away from acquiring mates, but within that lower range of androgen levels, acute changes can occur in association with specific behaviors: elevating with certain behaviors and decreasing with others.

Weight gain during pregnancy in both tamarin and marmoset males provides evidence that males may need to increase their energetic resources prior to taking care of infants following birth. Additionally, the drop of glucocorticoids following birth instead of an increase may indicate the role of prolactin, but not glucocorticoids, in the energetics of infant caretaking in fathers. Glucocorticoids, measured in owl monkeys in the field, were lowest during the periods when males were providing care (Corley et al. 2021). The elevated pattern of glucocorticoids prior to birth appears to be a common pattern across a range of species, but the timing of glucocorticoid decline might vary due to the timing of when males begin to parent.

Prolactin, as well, is influenced by parenting. To address the role of prolactin in promoting paternal care behaviors, physical changes, and reproductive hormones, we examined male marmosets over three different gestational and postpartum periods while experimentally elevating or reducing prolactin levels, compared with unmanipulated prolactin (Ziegler et al. 2009). Prolactin elevation or reduction had no effect on a male's behavior in parenting in his family. However, parenting motivation was affected when infant distress calls were played to males without manipulating their prolactin. However, rather than having a direct effect on the expression of paternal care, prolactin influenced a male's weight during the infant care period. Elevated prolactin postpartum, when males are actively caring for infants, may work to prevent excessive weight loss during their period of added energetic demands. This study also showed an inverse relationship between prolactin and testosterone. While prolactin is elevated during a male's intensive parenting during the first month, testosterone levels are low and glucocorticoids are low. We also found that when we lowered prolactin levels, the males lost the most weight while caring for infants. Without any manipulation of prolactin during a normal postpartum period, the fathers lost weight, but not as much as when prolactin levels were reduced. There is an energetic cost to caring for infants as fathers experience a significant weight loss while carrying infants during the first 3 weeks postpartum, and prolactin has a major role to play in preparing fathers for these energetic demands.

(c) Response to Infants Through Odor and Auditory Communication.

Parental recognition of offspring odors plays a role in determining the difference between offspring and non-offspring. Kin recognition is an assessment of genetic relatedness and can lead to differential treatment of conspecifics based on cues that correlate with relatedness (Gamboa et al. 1991). Recognition of odor signals from offspring requires the production of the specific label, and the recognition of the labels through the parent learning of the odor signature (Mateo 2004). Little is known about paternal-offspring recognition in biparental mammalian species. Males do not undergo pregnancy and parturition where hormonal facilitation of brain plasticity is produced, with neurogenesis occurring to promote olfactory involvement in offspring recognition and facilitation of infant care. However, as described above, neuroendocrine changes are recorded from males of biparental species during the gestational phase of their offspring or in response to infants in biparental rodents and cooperative breeding nonhuman primates.

Marmoset infant odor cues may facilitate recognition of offspring as primer odors by influencing paternal hormones. Fathers show reduced serum testosterone levels within 20 min of contact with an isolated infant scent (Prudom et al. 2008). In contrast, parentally inexperienced males have shown no changes in testosterone levels. Testosterone responsiveness to infant odor may indicate kin recognition and a role in the promotion of paternal behaviors. Common marmoset males showed a decrease in testosterone and an increase in estrogen levels in response to odors from their own infants but not to those of other infants (Ziegler et al. 2011). However, once infants are past 3 months of age and are no longer dependent on father's direct care, fathers were no longer hormonally responsive to their offspring odors. Chemical signals found in the scent secretions may only be relevant when the infant is totally dependent upon being carried. The ability to respond quickly and the flexibility of the androgens to olfactory cues indicate changes in the neurochemistry of the brain that occur to promote immediate responses toward infants.

In contrast, exposing males to auditory signals of distressed infant vocalizations elicited a higher response to infant distress calls with an increase in testosterone levels compared to control vocalizations (Zahed et al. 2008; Ziegler et al. 2017; Ziegler and Sosa 2016). The androgenic response to infant distress calls was not specific to a male's own infants as we used calls from unrelated 2-week-old infants. Experienced males that were the most responsive to infant distress cries also showed increased survival rates and better health outcomes for their offspring as indexed by increased infant weight gain (Ziegler et al. 2017). Further responsiveness of experienced males was increased with estrogen treatment showing that estrogens are important in controlling a male's motivation to respond to infant cries in experienced fathers (Ziegler and Sosa 2016). These studies show different roles for auditory and olfactory cues from infants in influencing fathers. Olfactory cues from one's own infants provide recognition and indicate the need to provide care for dependent infants. Infant distress cries may activate the motivation to respond to any infant's needs (Fig. 4).

Several studies have shown that oxytocin is also involved in regulating parenting and motivation in parenting behaviors. Finkenwirth and colleagues (Finkenwirth et al. 2015) reported that urinary oxytocin increased in all common marmoset family members following birth of infants. In early infancy, oxytocin levels were associated with increased infant-licking, and in late infancy, oxytocin was associated with increased proactive food sharing, suggesting that oxytocin reflects intrinsic care motivation in fathers. Based on infant retrieval response studies, fathers have higher motivation for parenting than mothers (Saito et al. 2011). Additionally, with infusion of oxytocin, adult male marmosets increased tolerance toward their offspring in transferring food (Saito and Nakamura 2011). Giving nasal oxytocin also enhances the responsiveness of males to infant auditory vocalizations (Taylor and French 2015). Males, but not females, showed an increase in responsiveness to infant vocalizations with oxytocin. However, since most of their subjects were not

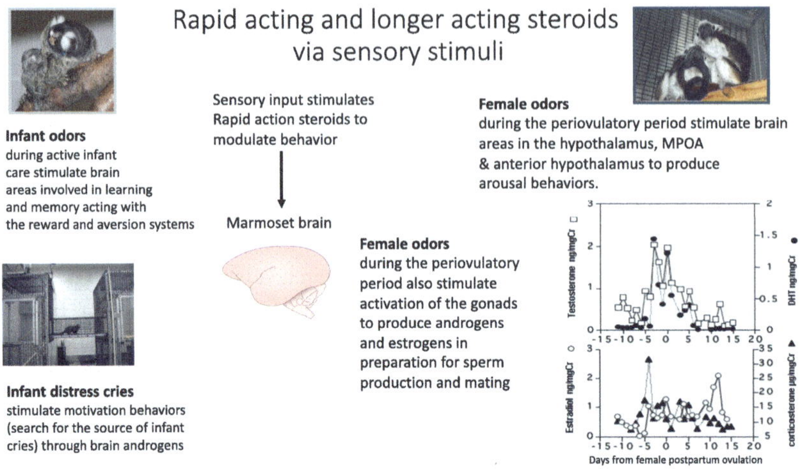

Fig. 4 The postpartum period where males are parenting and responding to the female's periovulatory period. (Photo of the common marmosets by Jordana Lennon, WNPRC, and of the cotton-top tamarin by Carla Boe, University of Wisconsin-Madison. Graph redrawn from data published (Ziegler et al. 2004b))

experienced fathers nor had recently been parents, this suggests that oxytocin is involved in social support and has a role in responsiveness to infants. Unfortunately, measuring oxytocin in urine under field conditions makes it difficult to associate specific behaviors with oxytocin levels without daily serial urine sampling to allow determining the timing of urinary oxytocin levels and behaviors, although this has been performed with chimpanzees (Crockford et al. 2013).

(d) Brain Plasticity in Neuroendocrine Changes Providing Immediate and Long-Term Alterations.

Most of what we know about parenting and the brain has been from studies on maternal brain changes, but a few studies have looked into short-term and long-term alterations in the brain of breeding male marmosets. This work is conducted with primates under laboratory conditions, as it is not possible to collect these data from wild primates. Neural and endocrine differences in male marmosets have been associated with paternal experience. Both first-time and infant-experienced marmoset fathers have an increased density of dendritic spines on pyramidal neurons in the prefrontal cortex of the brain, while nonfathers do not. This area has been implicated in goal-directed behaviors (Kozorovitskiy et al. 2006). Fathers of newborn marmoset infants have increased vasopressin V1a receptors, suggesting vasopressin in this brain area is increased during direct parenting, with the receptors declining as infant age increases, indicating a relationship with direct care of dependent offspring. This demonstrates the plasticity of the paternal brain to provide short-term changes in response to dependent infants.

We examined long-term changes to the brain in male marmosets with varying parenting experience (Ziegler et al. 2011). Using hypothalamic tissue from infant-experienced fathers with no current young and males that had never been fathers, we examined the role of prolactin on the secretion of dopamine, oxytocin, and vasopressin and with estradiol stimulation to these neurocrines. Prolactin is stored in the anterior pituitary and released into the periphery and the hypothalamus in the brain where it can inhibit dopamine release. Prolactin and oxytocin are often released simultaneously, and estradiol is a known releaser of both oxytocin and prolactin. Secretions released from the cultured neurons were measured and comparisons made between infant-experienced fathers and nonfathers. Dopamine levels were significantly lower in paternally experienced fathers compared to inexperienced males, while oxytocin and prolactin were significantly higher. Differences in hypothalamic secretion of dopamine and oxytocin may be related to paternal experience in male marmosets. Estradiol stimulation suppressed dopamine release. Additionally, prolactin entering the hypothalamus may also be altered by parenting experience. Changes in the neuroendocrine secretion in the experienced fathers are likely the result of long-term changes in the male paternal brain, as occurs in females (Kinsley and Meyer 2010). The neuroplasticity and re-patterning of the brain have provided many benefits to experienced rodent mothers. For fathers of biparental species, these benefits might also be available with parental experience. For instance, common marmoset parents have been shown to learn which trees contain food containers significantly better than nonparents (Kinsley and Lambert 2006). This suggests that, in biparental species, neuroplasticity and altered brain hormones may be occurring in both sexes with long-term benefits.

Brain imaging studies in the marmoset have shown male brain activation when exposed to periovulatory odors isolated from female marmosets providing further evidence that marmoset males are highly responsive to the cues from females during the ovulatory period. When exposed to the odors of ovulatory females compared to ovariectomized females during functional magnetic resonance imaging (fMRI), parentally inexperienced males showed activation in the medial preoptic area of the hypothalamus (MPOA) and the anterior hypothalamus which are two areas critical for male precopulatory and copulatory behavior (Ferris et al. 2001). Furthermore, several areas associated with arousal and sexual motivation, such as temporal cortex, MPOA, cingulate, insula, caudate, and cerebellum, showed significant deactivation in response to the ovariectomized odor relative to the vehicle control (Ferris et al. 2004). It is likely that olfactory cues of reproductive status are most important to a male evaluating his response to an unfamiliar female, whereas with a pair-bonded mate learned responses to other features override the olfactory cues.

In a follow-up study, we found that a male's social condition influenced his response to novel female ovulatory scents (Ziegler et al. 2005). We tested male common marmosets for behavioral and hormonal responses to an ovulatory scent collected from novel females and compared their response to a vehicle control scent. Males were either fathers living with their paired mates and offspring, males that were living with a paired mate but no offspring, or males that were living alone prior to being paired with a mate. All males were tested with scents in their home cages without other marmosets present, and then a blood sample was taken for testosterone measurement. Males showed arousal behaviors after smelling ovulatory odors, but the amount of testosterone response to the odor from ovulating females differed by social condition. Males with infants displayed no testosterone increase from control odor, while paired and single housed males showed a significant increase in testosterone. Under stable family conditions, there may be an inhibitory process that prevents males from exhibiting a full response to a novel ovulating female, and this could be part of their monogamous social system since the fathers in this study were directly involved in infant care at the time of testing.

5 How Does Paternal Care Vary Within and Between the Species That We Study?

Much of the variation we see between the marmosets, tamarins, owl monkeys, and red-bellied lemurs has to do with whether there are other helpers in the group. Since the marmosets, tamarins, and lemurs have more helpers, it is expected that a father does not have to invest as heavily in care of their offspring when others are there to help. Although there are older offspring that could help in owl monkey and titi monkey groups, they do not; one would expect these fathers to invest comparatively more in the energetics of infant care, yet this does not explain why there is within-species variation. In common marmosets, cotton-top tamarins, and red-bellied lemurs, fathers vary in the extent of their care of infants, as reported above. One factor that can explain this variation is individual experience. In common marmosets and cotton-top tamarins, prior experience with infants and specifically infant carrying can impact how much care fathers provide infants. In addition, the amount of grooming and sexual behavior vary across pairs and correlate with hormonal levels of oxytocin and prolactin levels in both partners (Snowdon et al. 2010; Snowdon and Ziegler 2015). It is difficult to determine the role of experience in paternal care in wild populations without long-term data on known individuals as they transition from subadult to adult, and from nulliparous to parous, and then over the course of successive births. So far, we lack this information for wild titi monkeys and red-bellied lemurs.

Across species, there is variability in whether a postpartum ovulation occurs. Owl monkeys and red-bellied lemurs do not ovulate or gestate while in the early stages of parenting. In the lab, both marmosets and tamarins have a postpartum ovulation within a few weeks of birth; therefore, there is a high energetic demand on the breeding female, necessitating helpers in caring for the present infants. Cotton-top tamarins can regulate their reproduction to maximize infant survival by giving birth so that their infants are weaned at the time of highest fruit availability. Cotton-top tamarins studied in the field have been observed to give birth twice a year, particularly if their infants die or there are social changes within their group (Savage et al. 2022). Other tamarin species in less seasonal rain forest within the

Amazon basin may give birth twice a year. Common marmosets do show postpartum ovulation in both the field and in captivity and have a high percentage of births occurring every 6 months with a 5-month gestation period.

In common marmosets, even with infant-experienced males, there are differences in the hormonal and behavioral responses that adult males have to infant distress calls. Males differ in motivation to respond to a distressed infant. The strength of the social bond between the mated pair could possibly influence the male's response to infants and could affect the male's ability to process the sensory cues of the female's reproductive state. Such variation between breeding males in the field has not been examined.

Field studies have demonstrated that relationships between pair-living adults may include serial monogamy and may be dependent upon group stability. For cotton-top tamarins, infant survival is dependent upon group stability (Savage et al. 2022), with stable social groups having increased infant survival. A common factor to all of our studies is that if one of the breeding pair is evicted or dies, a replacement is observed entering the group, and a new pair is formed.

6 Directions for Future Research

While we are beginning to understand the dynamics of maintaining a social bond with a mate while simultaneously responding to infant needs, more work is needed to advance our understanding of brain plasticity along with its interactions with the social environment. Our aforementioned studies in the cotton-top tamarin and the common marmoset in captivity, during the postpartum period when males are responding to their mates and infants, provide a unique opportunity to demonstrate the social interactions and their impact on neurological activity in the brain. As illustrated in Fig. 4, males have a flexible response during this critical time of simultaneous parenting and mating.

(a) Measuring Biological Variables in Field Settings.

There have been considerable advancements in determining reproductive hormones, stress responses, and genetic determinants of relatedness by the use of non-invasive collection methods. Applying these methods to examine paternal care in the field will allow us to validate assumptions made in laboratory studies, as well as to examine a variety of other factors that impact the responses of wild animal responses to caring for infants. One area of interest that field studies with hormonal data can provide is the role of the natural environment. In the field, we can associate individual variation in hormonal and behavioral relationships with environmental factors that may alter paternal care.

The development of methods for extracting steroids from feces has provided an important tool as these steroids act as indicators of social and environmental responses and as determinants of reproductive state. Fecal extracts can only measure the metabolites of the important steroids, and, therefore, laboratory work is needed to ensure observed changes can be related to their expected physiological responses. Measurement of reproductive steroids, such as estrogens and progestins, has allowed field primatologists to assess the timing of the periovulatory period and pregnancy of females (Corley et al. 2017; Fernandez-Duque et al. 2011). Androgens and glucocorticoids have been used to understand male seasonal reproduction and the energetics of mate competition (Tecot and Baden 2018; Tecot 2008; Corley et al. 2021). However, as has been reported in captive tamarins and marmosets, and recently in wild owl monkey males and red-bellied lemurs, fathers do not show sustained elevated glucocorticoids during the intensive time of caring for their infants after birth, demonstrating that the measured glucocorticoids may not be adequately reflecting the energetics of fathering as was previously assumed. Additionally, by associating short-term changes in androgen levels with paternal care, we also know that androgens can facilitate certain types of infant care and that they do not completely inhibit it (Tecot and Baden 2018; Ziegler and Sosa 2016).

Methods have been developed and validated for other essential physiological changes that will hopefully be addressed in the field. The protein and peptide neurohormones can be measured in primate urine and will, undoubtedly, provide an excellent source for understanding the physiology of pair-bonding and infant care. Urinary oxytocin, LH, FSH, chorionic gonadotropin, C-peptide, and prolactin all have been validated and used on nonhuman primate urine. However, collecting urine in arboreal primates may be difficult, and keeping samples frozen may not be possible at all field sites. Importantly, associating urinary hormonal changes with acute changes in behavior requires a very large dataset, as behaviors of interest may not occur within the window of time reflected in each urine sample.

Genetic data from the field have been a great resource for studies of paternal care for some time. This work is made even easier by the ability to extract DNA from preserved fecal samples, allowing for relatively easy, repeated data collection with little risk to the primates. These data have been used to determine relatedness between adult males and offspring (Huck et al. 2014) and can help determine differences in males' reproductive success relative to their investment in offspring care.

(b) Is Experience with Infants as Important as It Seems in Captivity?

Both captive and wild owl monkey pairs seldom have infants in their first season together (Garcia de la Chica et al. 2021b). With the ability to collect serial fecal samples from males and females, it would be possible to study, simultaneously, the hormonal changes occurring in both the male and female from the timing of their union to their first successful birth and provide this essential information. However, this requires long-term and extensive work in the field.

In cotton-top tamarins, infant care experience may vary between group members at the time of infant care. Given that wild cotton-tops typically are found in groups from 3 to 7 adults, there are some animals that have infant caretaking experience, thus increasing the likelihood of infant survival (Savage et al. 2022). Parity of the female impacts infant survival more than infant caretaking experience. Thus, the strength of the social bond between the breeding pair may be more important than infant caretaking experience when raising infants in the wild.

(c) Are Helpers Important?

Owl monkeys and titi monkeys rarely have helpers. There is only occasional participation of siblings in infant care in those circumstances (Fernandez-Duque et al. 2008; Jantschke et al. 1998). In wild cotton-top tamarins, the number of male or female helpers in a group was not found to be important for infant survival. However, groups with only one male had lower survival, 33%, compared with groups with more than one male, 57–70% survival (Savage et al. 2022). In the red-bellied lemur, helpers may be valuable, especially in the event of twin births (Tecot 2010), but we do not yet know the impact of helpers on the amount of care provided by each parent, nor infant outcomes.

(d) What Can Be Learned Through Periodic Captures and Weighing?

Efforts to habituate wild primates to periodic weighing in the wild would be beneficial to assess many factors. However, efforts to habituate animals to such an activity require new technology that will accurately and rapidly capture weights remotely and in an arboreal environment. In most field studies that use telemetry to locate animals in the field, captures are conducted annually. During these capture events, physical assessments are conducted and provide a snapshot of overall physical health and weight. This information can provide useful data in helping us to understand how primates adapt to environmental and social pressures. In owl monkeys, it is difficult to plan "periodic" captures and weighing (see Fernandez-Duque et al. 2022). Attracting them to platforms and capturing is still pending. Developing capture methods in captivity provides many advantages. Red-bellied lemurs are not captured in our research, though information on weight changes could be helpful in determining relationships between estradiol and paternal care, as well as infant growth. A noninvasive proxy for measuring body mass would be a great advance for this area of research.

(e) Can Sensory Cues Shown to Be Important in Captivity Be Studied in the Wild?

The subtleties of sensory cues observed in captivity may allow a field researcher to make inferences about sensory cues that are difficult to notice in the field. Although isolating certain modalities is difficult, multimodal signaling can be studied (see (Singletary and Tecot 2020)). One relevant area of study is how multiple signals may increase the likelihood of engaging fathers (on the part of the infant and/or mother) in affiliative or caregiving interactions.

Moving forward with research into nonhuman primate paternal care in both the laboratory setting and the field will require a broader span of researchers and time to expand the focus on wild paternal primates. It is hoped that the information we have discussed in this chapter will provide a basis for future studies. For example, examining the proximate mechanisms of the sensory systems in captive species, studies testing the sounds and odors from infants on adults using fMRI (functional magnetic resonance imaging) could be directly tied to activation of the key brain areas involved in sensory perception and processing and tied to the hormonal and behavioral measurements. Methods have already been developed to perform noninvasive imaging on awake marmosets. In the field, more funding for long-term continuous sampling and observations would allow for more precise data on both acute and chronic changes in the neuroendocrinology for nonhuman primate parenting.

References

Achenbach GG, Snowdon CT. Costs of caregiving: weight loss in captive adult male cotton-top tamarins (*Saguinus oedipus*) following the birth of infants. Int J Primatol. 2002;23(1):179–89.

Anzenberger G. How stranger encounters of common marmosets (*Callithrix jacchus jacchus)* are influenced by family members: the quality of behavior. Folia Primatol (Basel). 1985;45(3–4):204–24.

Baker AJ, Woods F. Reproduction of the emperor tamarin (*Saguinus imperator*) in captivity, with comparisons to cotton-top and golden lion tamarins. Am J Primatol. 1992;26(1):1–10.

Bales KL, Ardekani CS, Baxter A, Karaskiewicz CL, Kuske JX, Lau AR, et al. What is a pair bond? Horm Behav. 2021;136:105062.

Cavanaugh J, Carp SB, Rock CM, French JA. Oxytocin modulates behavioral and physiological responses to a stressor in marmoset monkeys. Psychoneuroendocrinology. 2016;66:22–30.

Cleveland J, Snowdon CT. Social development during the first twenty weeks in the cotton-top tamarin (*Saguinus o. oedipus*). Anim Behav. 1984;32(2):432–44.

Clutton-Brock TH. The evolution of parental care. Princeton University Press; 1991.

Corley M, Valeggia C, Fernandez-Duque E. Hormonal correlates of development and natal dispersal in wild female owl monkeys (*Aotus azarae*) of Argentina. Horm Behav. 2017 Nov;96:42–51.

Corley M, Perea-Rodriguez JP, Valeggia C, Fernandez-Duque E. Associations between fecal cortisol and biparental care in a pair-living primate. Am J Phys Anthropol. 2021;

Crockford C, Wittig RM, Langergraber K, Ziegler TE, Zuberbühler K, Deschner T. Urinary oxytocin and social bonding in related and unrelated wild chimpanzees. Proc Biol Sci. 2013 Mar 22;280(1755):20122765.

Daly M. Why don't male mammals lactate? J Theor Biol. 1979;78(3):325–45.

Dixson AF, George L. Prolactin and parental behaviour in a male New World primate. Nature. 1982;299(5883):551–3.

Emlen ST, Oring LW. Ecology, sexual selection, and the evolution of mating systems. Science. 1977;197(4300):215–23.

Epple G. Reproductive and social behavior of marmosets with special reference to captive breeding. Primates Med. 1978;10:50–62.

Evans S, Poole TB. Pair-bond formation and breeding success in the common marmoset *Callithrix jacchus jacchus*. Int J Primatol. 1983;4(1):83–97.

Evans S, Wolovich C. The owl monkeys of the DuMond Conservancy in Miami, USA – a scientific, educational and community asset. In: Fernandez-Duque E, editor. Owl monkeys: Evolution, behavioral ecology and conservation. Springer; 2022.

Fernandez-Duque E, Huck M. Till death (or an intruder) do us part: intrasexual-competition in a monogamous primate. PLoS One. 2013;8(1):e53724.

Fernandez-Duque E, Juárez CP, Di Fiore A. Adult male replacement and subsequent infant care by male and siblings in socially monogamous owl monkeys (*Aotus azarai*). Primates. 2008;49(1):81–4.

Fernandez-Duque E, Valeggia CR, Mendoza SP. The biology of paternal care in human and nonhuman primates. Annu Rev Anthropol. 2009;38:115–30.

Fernandez-Duque E, Burke K, Schoenrock K, Wolovich CK, Valeggia CR. Hormonal monitoring of reproductive status in monogamous wild female owl monkeys (*Aotus azarai*) of the Argentinean Chaco. Folia Primatol Int J Primatol. 2011;82(3):143–53.

Fernandez-Duque E, Huck M, Belle SV, Fiore AD. The evolution of pair-living, sexual monogamy, and cooperative infant care: insights from research on wild owl monkeys, titis, sakis, and tamarins. Am J Phys Anthropol. 2020;171(S70):118–73.

Ferrari SF. The care of infants in a wild marmoset (*Callithrix flaviceps*) group. Am J Primatol. 1992;26(2):109–18.

Ferris CF, Snowdon CT, King JA, Duong TQ, Ziegler TE, Ugurbil K, et al. Functional imaging of brain activity in conscious monkeys responding to sexually arousing cues. Neuroreport. 2001;12(10):2231–6.

Ferris CF, Snowdon CT, King JA, Sullivan JM Jr, Ziegler TE, Olson DP, et al. Activation of neural pathways associated with sexual arousal in non-human primates. J Magn Reson Imaging Off J Int Soc Magn Reson Med. 2004;19(2):168–75.

Finkenwirth C, van Schaik C, Ziegler TE, Burkart JM. Strongly bonded family members in common marmosets show synchronized fluctuations in oxytocin. Physiol Behav. 2015 Nov 1;151:246–51.

French JA, Snowdon CT. Sexual dimorphism in responses to unfamiliar intruders in the tamarin, *Saguinus oedipus*. Anim Behav. 1981;29(3):822–9.

French JA, Schaffner CM, Shepherd RE, Miller ME. Familiarity with intruders modulates agonism towards outgroup conspecifics in Wied's black-tufted-ear marmoset (Callithrix kuhli: primates, Callitrichidae). Ethology. 1995;99(1–2):24–38.

French J, Fite J, Ross C. Family life in marmosets: causes and consequences of variation in caregiving. Neurobiol Parent Brain. 2008:461–77.

Gamboa GJ, Reeve HK, Holmes WG. Conceptual issues and methodology in kin-recognition research: a critical discussion. Ethology. 1991;88(2):109–27.

Garcia de la Chica, Corley M, Fernandez-Deque E. Alloparental care. In Fuentes A. The international encyclopedia of primatology, 3 volume set. Wiley; 2017.

Garcia de la Chica A, Wood DB, Rotundo M, Fernandez-Duque E. Responses of a pair-living, sexually monogamous primate to the simulated presence of solitary individuals: A field playback experiment. Ethology. 2021a.

Garcia de la Chica A, Fernandez-Duque E, Williams LE. Behavioral biology of owl monkeys. In: Coleman K, Shapiro SJ, editors. Behavioral biology of laboratory animals. CRC Press; 2021b.

Garcia de la Chica A, Spence-Aizenberg A, Wolovich CK, Evans S, Fernandez-Duque E. The social life of owl monkeys. In: Fernandez-Duque E, editor. Owl monkeys: Evolution, behavioral ecology and conservation. Springer; 2022.

Huck M, Löttker P, Böhle U-R, Heymann EW. Paternity and kinship patterns in polyandrous moustached tamarins (*Saguinus mystax*). Am J Phys Anthropol Off Publ Am Assoc Phys Anthropol. 2005;127(4):449–64.

Huck M, Fernandez-Duque E, Babb P, Schurr T. Correlates of genetic monogamy in socially monogamous mammals: insights from Azara's owl monkeys. Proc R Soc Lond B Biol Sci. 2014;281(1782):20140195.

Isler K, van Schaik CP. Allomaternal care, life history and brain size evolution in mammals. J Hum Evol. 2012;63(1):52–63.

Jantschke B, Welker C, Klaiber-Schuh A. Rearing without paternal help in the Bolivian owl monkey *Aotus azarae boliviensis*: a case study. Folia Primatologica. 1998;69:115–20.

Johnson LD, Petto AJ, Sehgal PK. Factors in the rejection and survival of captive cotton top tamarins (*Saguinus oedipus*). Am J Primatol. 1991;25(2):91–102.

Joyce SM, Snowdon CT. Developmental changes in food transfers in cotton-top tamarins (*Saguinus oedipus*). Am J Primatol Off J Am Soc Primatol. 2007;69(9):955–65.

Kinsley CH, Lambert KG. The maternal brain. Sci Am. 2006;294(1):72–9.

Kinsley CH, Meyer EA. The construction of the maternal brain: Theoretical comment on Kim et al. Behavioral Neuroscience. 2010;124:710–4.

Kozorovitskiy Y, Hughes M, Lee K, Gould E. Fatherhood affects dendritic spines and vasopressin V1a receptors in the primate prefrontal cortex. Nat Neurosci. 2006;9(9):1094–5.

Kvarnemo C. Evolution and maintenance of male care: is increased paternity a neglected benefit of care? Behav Ecol. 2006;17(1):144–8.

Lazaro-Perea C. Intergroup interactions in wild common marmosets, *Callithrix jacchus*: territorial defence and assessment of neighbours. Anim Behav. 2001 Jul 1;62(1):11–21.

Lukas D, Clutton-Brock TH. The evolution of social monogamy in mammals. Science. 2013 Aug 2;341(6145):526–30.

Mateo JM. Recognition systems and biological organization: the perception component of social recognition. In: Annales Zoologici Fennici. JSTOR; 2004. p. 729–45.

Mendoza SP, Mason WA. Parental division of labour and differentiation of attachments in a monogamous primate (*Callicebus moloch*). Anim Behav. 1986;34(5):1336–47.

Merenlender AM. The effects of sociality on the demography and genetic structure of Lemur fulvus rufus (polygamous) and Lemur rubriventer (monogamous) and the conservation implications. University of Rochester; 1993.

Mitani JC, Watts D. The evolution of non-maternal caretaking among anthropoid primates: do helpers help? Behav Ecol Sociobiol. 1997;40(4):213–20.

Overdorff D. Preliminary report on the activity cycle and diet of the red-bellied lemur (*Lemur rubriventer*) in Madagascar. Am J Primatol. 1988;16(2):143–53.

Overdorff DJ. Ecological correlates to social structure in two prosimian primates: *Eulemur fulvus rufous* and *Eulemur rubriventer* in Madagascar [PhD Thesis]. Duke University; 1991.

Overdorff DJ, Tecot SR. Social pair-bonding and resource defense in wild red-bellied lemurs (*Eulemur rubriventer*). In: Lemurs. Springer; 2006. p. 235–54.

Porter TA, Snowdon CT. Female reproductive status and male pairmate behavior in cotton-top tamarins. Ann N Y Acad Sci. 1997;807(1):556–8.

Prudom SL, Broz CA, Schultz-Darken N, Ferris CT, Snowdon C, Ziegler TE. Exposure to infant scent lowers serum testosterone in father common marmosets (Callithrix jacchus). Biol Lett. 2008;4(6):603–5.

Pryce CR. Individual and group effects on early caregiver-infant relationships in red-bellied tamarin monkeys. Anim Behav. 1988 Sep 1;36(5):1455–64.

Rotundo M, Fernandez-Duque E, Dixson AF. Infant development and parental care in free-ranging *Aotus azarai azarai* in Argentina. Int J Primatol. 2005;26(6):1459–73.

Saito A, Nakamura K. Oxytocin changes primate paternal tolerance to offspring in food transfer. J Comp Physiol A. 2011;197(4):329–37.

Saito A, Izumi A, Nakamura K. Fathers have higher motivation for parenting than mothers in common marmoset (*Callithrix jacchus*). Behaviour. 2011;148(11–13):1199–214.

Sánchez S, Peláez F, Gil-Bürmann C, Kaumanns W. Costs of infant-carrying in the cotton-top tamarin (*Saguinus oedipus*). Am J Primatol Off J Am Soc Primatol. 1999;48(2):99–111.

Savage A, Giraldo LH, Soto LH, Snowdon CT. Demography, group composition, and dispersal in wild cotton-top tamarin (*Saguinus oedipus*) groups. Am J Primatol. 1996;38(1):85–100.

Savage A, Soto L, Medina F, Emeris G, Soltis J. Litter size and infant survivorship in wild groups of cotton-top tamarins (*Saguinus oedipus*) in Colombia. Am J Primatol Off J Am Soc Primatol. 2009;71(8):707–11.

Savage A, Snowdon CT, Soto L, Medina F, Emeris G, Guillen R. Factors influencing the survival of wild cotton-top tamarin (*Saguinus oedipus*) infants. Am J Primatol. 2022;83(7):e23262. https://doi.org/10.1002/ajp.23262

Schradin C, Anzenberger G. Infant carrying in family groups of Goeldi's monkeys (*Callimico goeldii*). Am J Primatol. 2001;53(2):57–67.

Setchell KDR, Bonney RC. The excretion of urinary steroids by the owl monkey (*Aotus trivirgatus*) studied using open tubular capillary column gas chromatography and mass spectrometry. J Steroid Biochem. 1981 Jan 1;14(1):37–43.

Singletary B, Tecot S. Multimodal pair-bond maintenance: a review of signaling across modalities in pair-bonded nonhuman primates. Am J Primatol. 2020;82(3):e23105.

Snowdon CT, Ziegler TE. Growing up cooperatively: family processes and infant care in marmosets and tamarins. J Dev Process. 2007;2(1):40–66.
Snowdon CT, Ziegler TE. Variation in prolactin is related to variation in sexual behavior and contact affiliation. PLoS One. 2015;10(3):e0120650.
Snowdon CT, Pieper BA, Boe CY, Cronin KA, Kurian AV, Ziegler TE. Variation in oxytocin is related to variation in affiliative behavior in monogamous, pairbonded tamarins. Horm Behav. 2010 Sep;58(4):614–8.
Spence-Aizenberg A, Williams LE, Fernandez-Duque E. Are olfactory traits in a pair-bonded primate under sexual selection? An evaluation of sexual dimorphism in *Aotus nancymaae*. Am J Phys Anthropol. 2018 Aug;166(4):884–94.
Stockley P, Hobson L. Paternal care and litter size coevolution in mammals. Proc R Soc B Biol Sci. 2016;283(1829):20160140.
Tardif SD, Richter CB, Carson RL. Effects of sibling-rearing experience on future reproductive success in two species of Callitrichidae. Am J Primatol. 1984;6(4):377–80.
Taylor JH, French JA. Oxytocin and vasopressin enhance responsiveness to infant stimuli in adult marmosets. Horm Behav. 2015;75:154–9.
Tecot S. Seasonality and predictability: the hormonal and behavioral responses of the red-bellied lemur, *Eulemur rubriventer*. Southeast Madag. Austin: University of Texas; 2008.
Tecot SR. It's all in the timing: birth seasonality and infant survival in *Eulemur rubriventer*. Int J Primatol. 2010;31(5):715–35.
Tecot SR, Baden AL. Profiling caregivers: hormonal variation underlying allomaternal care in wild red-bellied lemurs. Eulemur rubriventer Physiol Behav. 2018;193:135–48.
Tecot SR, Romine NK. Leading ladies: leadership of group movements in a pair-living, codominant, monomorphic primate across reproductive stages and fruit availability seasons. Am J Primatol. 2012;74(7):591–601.
Tecot SR, Baden AL, Romine NK, Kamilar JM. Infant parking and nesting, not allomaternal care, influence Malagasy primate life histories. Behav Ecol Sociobiol. 2012;66(10):1375–86.
Tecot SR, Baden AL, Romine N, Kamilar JM. Reproductive strategies and infant care in the Malagasy primates. In: Building babies. Springer; 2013. p. 321–59.
Tecot SR, Singletary B, Eadie E. Why "monogamy" isn't good enough. Am J Primatol. 2016;78(3):340–54.
Washabaugh KF, Snowdon CT, Ziegler TE. Variations in care for cottontop tamarin, *Saguinus oedipus*, infants as a function of parental experience and group size. Anim Behav. 2002;63(6):1163–74.
West HE, Capellini I. Male care and life history traits in mammals. Nat Commun. 2016;7(1):1–10.
Wolovich CK, Perea-Rodriguez JP, Fernandez-Duque E. Food transfers to young and mates in wild owl monkeys (*Aotus azarai*). Am J Primatol Off J Am Soc Primatol. 2008;70(3):211–21.
Yamamoto ME, Box HO, Albuquerque FS, de Fátima Arruda M. Carrying behaviour in captive and wild marmosets (*Callithrix jacchus*): a comparison between two colonies and a field site. Primates. 1996;37(3):297–304.
Zahed SR, Prudom SL, Snowdon CT, Ziegler TE. Male parenting and response to infant stimuli in the common marmoset (Callithrix jacchus). Am J Primatol Off J Am Soc Primatol. 2008;70(1):84–92.
Zahed SR, Kurian AV, Snowdon CT. Social dynamics and individual plasticity of infant care behavior in cooperatively breeding cotton-top tamarins. Am J Primatol Off J Am Soc Primatol. 2010;72(4):296–306.
Ziegler TE, Sosa ME. Hormonal stimulation and paternal experience influence responsiveness to infant distress vocalizations by adult male common marmosets, *Callithrix jacchus*. Horm Behav. 2016;78:13–9.
Ziegler TE, Widowski TM, Larson ML, Snowdon CT. Nursing does affect the duration of the post-partum to ovulation interval in cotton-top tamarins (*Saguinus oedipus*). J Reprod Fertil. 1990;90(2):563–70.

Ziegler TE, Epple G, Snowdon CT, Porter TA, Belcher AM, Küderling I. Detection of the chemical signals of ovulation in the cotton-top tamarin, *Saguinus oedipus*. Anim Behav. 1993;45(2):313–22.

Ziegler TE, Wegner FH, Snowdon CT. Hormonal responses to parental and nonparental conditions in male cotton-top tamarins, *Saguinus oedipus*, a New World primate. Horm Behav. 1996;30(3):287–97.

Ziegler TE, Washabaugh KF, Snowdon CT. Responsiveness of expectant male cotton-top tamarins, *Saguinus oedipus*, to mate's pregnancy. Horm Behav. 2004a;45(2):84–92.

Ziegler TE, Jacoris S, Snowdon CT. Sexual communication between breeding male and female cotton-top tamarins (*Saguinus oedipus*), and its relationship to infant care. Am J Primatol Off J Am Soc Primatol. 2004b;64(1):57–69.

Ziegler TE, Schultz-Darken NJ, Scott JJ, Snowdon CT, Ferris CF. Neuroendocrine response to female ovulatory odors depends upon social condition in male common marmosets, *Callithrix jacchus*. Horm Behav. 2005;47(1):56–64.

Ziegler TE, Prudom SL, Schultz-Darken NJ, Kurian AV, Snowdon CT. Pregnancy weight gain: marmoset and tamarin dads show it too. Biol Lett. 2006 Jun 22;2(2):181–3.

Ziegler TE, Prudom SL, Zahed SR, Parlow AF, Wegner F. Prolactin's mediative role in male parenting in parentally experienced marmosets (*Callithrix jacchus*). Horm Behav. 2009 Oct;56(4):436–43.

Ziegler TE, Peterson LJ, Sosa ME, Barnard AM. Differential endocrine responses to infant odors in common marmoset (*Callithrix jacchus*) fathers. Horm Behav. 2011;59(2):265–70.

Ziegler TE, Sosa ME, Colman RJ. Fathering style influences health outcome in common marmoset (Callithrix jacchus) offspring. PLoS One. 2017;12(9):e0185695.

Recent Neuroscience Advances in Human Parenting

Magdalena Martínez-García, Sofia I. Cardenas, Jodi Pawluski, Susanna Carmona, and Darby E. Saxbe

Abstract The transition to parenthood entails brain adaptations to the demands of caring for a newborn. This chapter reviews recent neuroscience findings on human parenting, focusing on neuroimaging studies. First, we describe the brain circuits underlying human maternal behavior, which comprise ancient subcortical circuits and more sophisticated cortical regions. Then, we present the short-term and long-term functional and structural brain adaptations that characterize the transition to motherhood, discuss the long-term effects of parenthood on the brain, and propose several underlying neural mechanisms. We also review neuroimaging findings in biological fathers and alloparents (such as other relatives or adoptive parents), who engage in parenting without directly experiencing pregnancy or childbirth. Finally, we describe perinatal mental illnesses and discuss the neural responses associated with such disorders. To date, studies indicate that parenthood is a period of enhanced brain plasticity within brain areas critical for cognitive and social processing and that both parenting experience and gestational-related factors can prime such plasticity.

Keywords Neuroimaging · Neuroplasticity · Parental brain · Pregnancy · Alloparents

---

M. Martínez-García (✉) · S. Carmona
Instituto de Investigación Sanitaria Gregorio Marañón, Madrid, Spain

Centro de Investigación Biomédica en Red de Salud Mental (CIBERSAM), Madrid, Spain
e-mail: mmartinez@hggm.es

S. I. Cardenas · D. E. Saxbe
Department of Psychology, University of Southern California, Los Angeles, CA, USA

J. Pawluski
Univ Rennes, Inserm, EHESP, Irset (Institut de Recherche en Santé, Environnement et Travail), Rennes, France

G. González-Mariscal (ed.), *Patterns of Parental Behavior*, Advances in Neurobiology 27, https://doi.org/10.1007/978-3-030-97762-7_8

1 Introduction

The transition to parenthood entails adaptation to the demands of caring for a newborn. Early psychodynamic models describe motherhood as a transitional period with profound impact on the mother's identity (Erikson 1959; von Mohr et al. 2017). According to these models (von Mohr et al. 2017), psychological adaptations enable mothers to identify and bond with their infant and, consequently, sensitively respond to the baby's needs (Ainsworth et al. 1978; Pines 1972; Winnicott 1956). More recently, neuroimaging studies of motherhood have focused on neural circuits linked with reward processing, empathy for another person's emotions, mentalization (i.e., attributing thoughts, desires, and intentions to others), memory, executive functions, and emotion regulation (Swain et al. 2014). In line with early models, these studies suggest that mothers' neuropsychological adaptations might facilitate their sensitivity not only to perceive and respond to the infant but also to promote bonding and motivation for caretaking behavior (Rutherford and Mayes 2011; Swain 2011).

The study of the human parental brain, which examines the neurobiological factors associated with human caregiving behavior, dates back to the late twentieth century. This body of research stems from nonhuman animal studies showing that gestational factors, mainly hormonal fluctuations during gestation, parturition, and lactation, influence the function and structure of the brain (Numan 2020a). As neuroimaging techniques developed, studies began examining human parents' neural responses to infant stimuli to understand the neural correlates of human parental behavior. These studies largely utilized task-based functional magnetic resonance imaging (fMRI) to infer brain function. Nowadays, parental brain researchers consider the transition to parenthood a critical period of heightened adult neural plasticity (Barba-Müller et al. 2019; Hillerer et al. 2014; Feldman et al. 2015). As such, scholars aim to characterize the functional and structural brain changes during the parental transition and how perinatal mental illnesses impact such parental brain adaptations.

This chapter reviews recent neuroscience findings on human parenting. It reflects the current state of the literature, which primarily focuses on mothers that undergo pregnancy. We begin by grounding this chapter on the brain circuits underlying maternal behavior (Sect. 1). We survey the literature on mothers' functional (Sect. 2) and structural brain adaptations (Sect. 3), propose several plausible neural mechanisms behind such brain adaptations (Sect. 4), and explain investigations on the long-term effect of parenthood on the brain (Sect. 5). We also review those neuroimaging studies of biological fathers whose female partners undergo pregnancy (Sect. 6). We then expand the discussion of research on the neuroscience of alloparents (Sect. 7), which is the provision of caregiving by individuals who are not the biological parents such as other family members and adoptive parents. Finally, we discuss research on the parental brain in the context of perinatal mental illness (Sect. 8).

When interpreting this chapter, we urge readers to remember that the human parental brain literature is composed of heterogeneous studies, thus challenging our ability to integrate findings. For instance, study samples vary on a number of domains, including size, parity status (multiparous vs. primiparous), age of the infants, and birthing and feeding methods. These factors might not only affect the maternal experience but also its underlying neural substrate. With regard to study design, some studies use a longitudinal design (following the same subjects at different points across the motherhood transition) and others a cross-sectional design (comparing parents versus nonparents). Also, studies capture different moments across the parental transition (i.e., pregnancy, peripartum, early postpartum, or late postpartum) characterized by unique physiological changes, endocrine shifts, and immune responses. Furthermore, few longitudinal studies use a comparison group of nonparents to minimize the noise induced by image acquisition and other confounding variables. Finally, the studies have used different processing steps, statistical models, and thresholds to determine statistical significance, which can also affect the results (Aurich et al. 2015). Therefore, we stress the necessity of multicenter studies, systematic reviews, and meta-analyses to accurately determine the location and trajectories of the brain changes associated with pregnancy and childbearing.

2 Maternal Brain Circuits

Converging evidence from nonhuman animal models, mostly murine models, demonstrates that certain parts of the brain undergo important dynamic adaptations during the transition to motherhood (Numan 2020b). These dynamic adaptations mainly involve the hypothalamus, the amygdala, and the motivational mesolimbic dopaminergic (DA) circuit (Numan 2020b). More specifically, the physiological events of late pregnancy and parturition activate the medial preoptic area (mPOA) of the hypothalamus, which projects to the mesolimbic DA system – comprised by the ventral tegmental area (VTA), the nucleus accumbens, and the ventral pallidum – the amygdala, and the medial prefrontal cortex (MPFC). As the activations of this circuit facilitate the onset of maternal behavior, researchers refer to this set of regions as the maternal brain circuit. Importantly, gestational hormones play a key role in the activation of this circuit and the immediate onset of maternal behavior (Numan 2020a). However, once the behavior is established, it emancipates from hormonal stimulation and is maintained solely by infant stimuli.

Converging with animal models, studies in humans indicate that motherhood is also associated with pronounced brain adaptations. These adaptations affect ancestral subcortical circuits that we share with other mammals, which include regions such as the hypothalamus, amygdala, VTA, nucleus accumbens, and ventral pallidum, as well as other brain cortical circuits that involve more sophisticated mental functions, including emotional empathy, mentalizing, emotion regulation, and altruistic behavior (Numan 2020c). So far, researchers point to the physiological events – mainly hormonal fluctuations during gestation, parturition, and lactation – as the principal mediating factors that trigger and coordinate these brain changes (Martínez-García et al. 2021b). However, brain changes are likely mediated by a combination of both intrinsic physiological factors and extrinsic environmental postpartum factors that translate into internal signals.

3 Maternal Functional Brain Adaptations

The main neuroimaging technique to study human brain function is noninvasive functional MRI (fMRI) based on blood-oxygen-level-dependent (BOLD) response (Attwell and Iadecola 2002). BOLD-fMRI uses regional changes in blood oxygenation to indirectly infer neuronal activity. Thanks to this technique, we can infer which regions are activated when a person performs a specific task or is exposed to certain stimuli. This modality is called task-based fMRI, and for many years, it was the most widely used fMRI method to infer brain function.

Task-Based fMRI

The first study using task-based fMRI to investigate the maternal brain dates back to 1999 (Lorberbaum et al. 1999). In this pioneering study, Lorberbaum et al. (1999) examined the brain of four primiparous mothers whose children were between 3 weeks and 3.5 years of age. The four primiparous mothers listened to the sounds of infants crying. As compared to white noise, infants' cries activate regions involved in processing salient and rewarding stimuli, such as the anterior cingulate cortex (ACC) and the right orbitofrontal cortex (OFC).

Although this pioneering study by Lorberbaum et al. has methodological limitations that must be considered when interpreting their findings, the study has established the basis for the study of the maternal brain. Since the publication of Lorberbaum et al. (1999), dozens of neuroimaging studies have focused on analyzing which brain regions are activated in the mother's brain when she is presented with baby-related stimuli (Paul et al. 2019; Bjertrup et al. 2019). The sensory modality of the presented stimuli varies, from auditory stimuli to visual stimuli or even olfactory stimuli. Among the different experimental paradigms, one of the most widely used focuses on investigating which brain regions are activated when a mother is presented with photos or videos of her own baby compared to those of unfamiliar babies. As expected, mothers display greater brain activation in response to their own infant cues as compared to other infant stimuli (Paul et al. 2019; Bjertrup et al. 2019).

Researchers have conducted several reviews (Swain et al. 2014; Kim 2016; Paul et al. 2019; Bjertrup et al. 2019; Sacher et al. 2020) and meta-analyses (Rocchetti et al. 2014; Paul et al. 2019) to help distinguish which brain regions are consistently

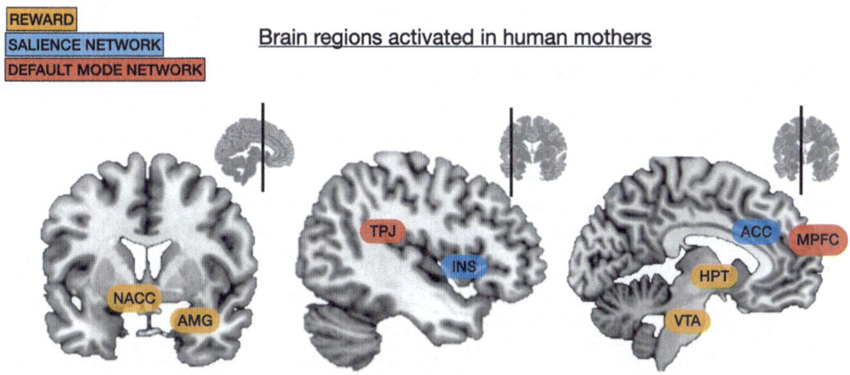

Fig. 1 Brain regions activated in human mothers when presented with baby-related stimuli. When mothers are presented with stimuli from their infant relative to stimuli from other unfamiliar infants, they show neural activation in the hypothalamus, ventral tegmental area (VTA), nucleus accumbens (NACC), and amygdala, key subcortical areas for reward and motivated behavior (marked in yellow); the anterior cingulate cortex (ACC) and insula (INS), which belong to the salience network (marked in blue); and the medial prefrontal cortex (MPFC) and temporoparietal junction (TPJ), which belong to the default mode network (marked in red). The location of the brain areas is approximative. For a comprehensive view of the neural circuits regulating maternal behavior in humans, see Bjertrup et al. (2019) and Numan (2020c)

activated when a mother is presented with stimuli from her infant (Fig. 1). In a recent systematic review, Bjertrup et al. (2019) observed that when mothers are presented with stimuli from their infant relative to stimuli from other unfamiliar infants, mothers show activation in regions involved in hormonal regulation, as well as in signaling stimuli as rewarding and relevant for guiding behavior. Specifically, they observed that own-baby faces elicit activation in the hypothalamus. This region is essential for maintaining homeostasis of the body by controlling the release and reception of hormones necessary for mating, pregnancy, birth, and suckling, such as oxytocin, progesterone, and estrogen, among others (Saper and Lowell 2014). Own-baby stimuli also activate the VTA and the nucleus accumbens of the mother, key areas of the mesolimbic dopaminergic circuits involved in reward processing and motivated behavior, as well as the amygdala, a region involved in marking and detecting stimuli as salient and relevant for behavior (Numan 2020c). As previously described, all these brain areas have been extensively linked to maternal behavior in murine models (Numan 2020b).

In addition to these ancient circuits, human mothers also show activation in more recently developed regions, namely, the ACC and the insula (Bjertrup et al. 2019). These regions belong to a system known as "salience network," also known as "ventral attentional network" or "homeostatic relevance network" (Seeley 2019). These cortical areas are connected with other subcortical regions, including the amygdala, hypothalamus, ventral striatum – including nucleus accumbens – thalamus, and brainstem (Seeley 2019). The "salience network" responds to stimuli that are homeostatically relevant at a visceral and autonomic level and reallocates cognitive and behavioral resources. When applied to maternal behavior, own-baby stimuli would elicit a visceral-autonomic response on the mother to direct all the cognitive and behavioral resources to them.

Besides the ACC and insula, when mothers are presented with stimuli or their infant, they also activate the MPFC and temporoparietal junction (TPJ; Bjertrup et al. 2019). Several studies suggest that these areas are important both for reflecting on oneself and for using the "self" as an anchor to simulate the mental states, intentions, and desires of others (Buckner and DiNicola 2019). When subjects are instructed to rest, their minds tend to reflect about the "self" and mentalize about other minds. These areas have been grouped under the acronym DMN, which stands for default mode network, alluding to a mental state activated by default at rest. In motherhood, these networks might help the mother to mentalize with her infant and promote bonding and altruistic behavior (von Mohr et al. 2017).

Resting-State fMRI

Resting-state fMRI (rsfMRI) allows us to extract different metrics about the spontaneous neural activity based on BOLD's low-frequency fluctuations (Murphy et al. 2013). This technique has the advantage of minimizing the effect that small variations on the paradigm selection, that is, the task, might have on the data as subjects do not perform any specific task – they are just instructed to lay on the scanner and stay awake during the acquisition. To date, there are still few studies that have used this type of acquisition to explore the maternal brain. Altogether, studies suggest that parenthood-induced brain changes are not restricted to the processing of infant-related stimuli. Indeed, they seem to affect the basal state of the brain in a global and widespread manner (Zheng et al. 2018; Dufford et al. 2019). Moreover, these changes seem to adapt over time to the changing demands of childbearing (Zhang et al. 2019b; Dufford et al. 2019) and to depend on the number of children raised (Orchard et al. 2021). At the moment, it appears that brain adaptations predominantly, but not exclusively, affect limbic and DMN regions involved in reward and mentalization processes, respectively (Sacher et al. 2020).

4 Maternal Structural Brain Adaptations

The first structural investigations of the human maternal brain focused on the pituitary gland, a small neuroendocrine organ that undergoes a striking transformation in preparation for motherhood. During pregnancy, lactotroph cells in the pituitary gland steadily produce prolactin, with high prolactin levels stimulating milk production for the postpartum breastfeeding (Duthie and Reynolds 2013). Such in-pregnancy hyperprolactinemia leads to a prolactin cell hypertrophy and a pituitary enlargement (Fischer 1931; Gonzalez et al. 1988; Elster et al. 1991; Dinç et al. 1998).

In the early twenty-first century, an MRI study assessed for the first time the global brain changes occurring during the transition to motherhood (Oatridge et al. 2002). To do so, the authors contoured with a semiautomatic method the brain and ventricles of two small groups of mothers (i.e., nine healthy pregnant women and five pregnant women with preeclampsia) at different time points across pregnancy and postpartum. In both the healthy and preeclamptic groups, the brain size decreased, and the ventricular space increased during pregnancy, peaked at parturition, and reversed by 6 months after delivery. This was the very first evidence that human pregnancy impacts the structure of the mother's brain. With the development of new neuroimaging techniques, parental brain researchers started analyzing finer aspects of the brain's morphology, such as changes in cortical volume, cortical thickness, surface area, and cortical folding. Structural MRI generates high-resolution images with well-defined contrast between gray and white matter interfaces, thus being the standard imaging technique to analyze morphometric aspects of the brain (Backhausen et al. 2021). Some MRI studies have explored whole-brain metrics (Kim et al. 2010, 2018; Hoekzema et al. 2017; Lisofsky et al. 2019; Zhang et al. 2019a; Carmona et al. 2019), while others have focused on specific brain regions, mainly subcortical areas (Hoekzema et al. 2020; Luders et al. 2021a, b). The captured timeframes varied among the studies. Most studies focused on changes during the postpartum period (Kim et al. 2010, 2018; Lisofsky et al. 2019; Zhang et al. 2019a; Luders et al. 2020, 2021a, b), whereas a group of studies investigated brain changes across the gestational period (comparing preconception vs. early postpartum) (Hoekzema et al. 2017; Carmona et al. 2019; Hoekzema et al. 2020). Some of the cortical regions that consistently change in mothers overlap with the DMN, including the MPFC and precuneus; the salience network, including the ACC and insula; and with subcortical regions, such as the amygdala, thalamus, caudate, hippocampus, and nucleus accumbens (Martínez-García et al. 2021b). The unified trajectories resulting from these longitudinal studies indicate a pre- to post-pregnancy gray matter (GM) volume decrease, a GM volume increase right after birth, and a GM decline at around 1 year postpartum (Martínez-García et al. 2021b). Follow-up studies indicate that the pre- to post-pregnancy GM reductions do not reverse after the first 2 years of maximal maternal investment, except for a partial hippocampal volume recovery (Hoekzema et al. 2017; Martínez-García et al. 2021a).

Studies further suggest that GM volume changes predict the quality of mother-to-infant attachment (Hoekzema et al. 2017) and correlate with the mother's positive perception of the baby (Kim et al. 2010) and empathic abilities (Zhang et al. 2019a). Based on these results, it could be argued that motherhood-related neuroadaptations facilitate the mother's ability to recognize the infant's needs and promote bonding through a rewarding process. However, additional experiences often associated with parenting, such as sleep deprivation and stress, can also influence how a mother's brain adapts to motherhood and thus might contribute to the observed maternal brain changes (Kim 2021). Thus, research suggests that both physiological and environmental factors impact maternal brain adaptations.

5 Proposed Neural Mechanisms Mediating the Genesis and Operation of the Human Parental Brain

We know very little about the biological and neural mechanisms that operate in the human maternal brain. In humans, inferring neural mechanisms from structural MR imaging is challenging. The standard 1 mm^3 resolution of MR anatomical images, the modality on which human maternal brain studies have mainly relied, is insufficient to capture cellular-level processes (Ribeiro et al. 2013). Recent interpretations point to neuroplasticity, myelination processes, and modifications in microglial density as the mechanisms behind brain changes with parenting in humans.

Neuroplasticity

The dynamic brain changes found in human mothers have been raised to reflect a hallmark for neuroplasticity. Ex vivo animal findings inform about several neuroplasticity mechanisms through which such macroscopic GM changes could arise. Such mechanisms are not necessarily mutually exclusive and are likely to vary among species in the timing and brain regions affected.

One neuroplasticity phenomenon that has been argued to contribute to volume GM reductions is a fine-tuning of synapses concomitant with a reduction in the surrounding neuropil and microglial cells. This neural mechanism, also regarded as synaptic pruning, refines brain circuits during another transitional period of behavioral adaptations and sex-steroid increases: the adolescence period (Spear 2013; Riccomagno and Kolodkin 2015; Mills and Tamnes 2014). A recent MRI longitudinal study compared the profile of brain morphometric changes (surface area, thickness, and gyrification index, among other measures) between female adolescents and first-time adult mothers and observed the same pattern of cortical flattening in both groups (Carmona et al. 2019). Based on these shared characteristics, similar neurobiological mechanisms might operate during both transitional periods.

Besides this idea of synaptic pruning, rodent studies point to other neuroplastic events as potential mechanisms behind volume changes in cortical and subcortical brain areas. For instance, mPOA cell bodies and dendritic branching increase in size in late-pregnant rats, when pregnancy-related hormones prime this region for the display of maternal behavior (Keyser-Marcus et al. 2001), hippocampal CA1 spine density increases during late pregnancy and early postpartum in rats (Kinsley et al. 2006), but hippocampal neurogenesis is decreased during this same period, a time when dendritic branching is also reduced (Pawluski et al. 2016 for review, 2020) and increased dendritic spine density on pyramidal neurons of the MPFC occur during the postpartum period (Leuner et al. 2010). Thus, there are likely brain region-specific increases and decreases in aspects of neuroplasticity that vary during pregnancy and the postpartum period (for a recent review, see Pawluski et al. 2021a). Beyond these neuronal modifications, the macroscopic GM changes observed during the motherhood transition likely involve additional complex neural processes that translate into visible, pronounced brain changes in the maternal brain circuit.

Myelination

Another hallmark of neural refinement that might contribute to structural maternal brain changes is white matter (WM) myelination, a process that accelerates the electrical transmission through the axons (Bercury and Macklin 2015). Studies in nonhuman animals suggest that pregnancy-related hormones promote myelination and myelin repair capacity (Gregg et al. 2007). Enhanced myelination might also occur in human pregnancy since pregnant women with demyelinating diseases such as multiple sclerosis have a better prognosis during the gestational period (Alhomoud et al. 2021). In humans, increased myelination can lead to voxels at the WM and GM interface being misclassified as GM, leading to an apparent decrease in cortical volume (Paus 2010). Pregnancy-related GM reductions could also be concomitant to WM increases. Human research examining changes in WM has typically used T1-weighted MR images, which provide accurate WM volume estimates. Zhang et al. (2019a) detected increases in WM volume within the insula, postcentral gyrus, inferior parietal, and superior and middle temporal gyri during late postpartum. In contrast, other studies have not found changes either in WM volume (Hoekzema et al. 2017) or thickness (Carmona et al. 2019) when comparing mothers' brains before and after pregnancy. To improve the comprehension of the WM modifications of the maternal brain, we need to examine metrics extracted from diffusion-weighted imaging (DWI), a form of MR imaging that allows for more precise estimates of WM fiber integrity and orientation (Alexander et al. 2019). Notably, a recent rodent DWI study showed that pregnancy is associated with higher water diffusion, which could indicate enhanced WM integrity (Chan et al. 2015). This promising tool has been used in middle-aged mothers who gave birth decades before being scanned (Voldsbekk et al. 2021) but has not been applied to puerperal mothers.

Microglia, the Brain's Immune Cells

The immunological environment of the perinatal period might also impact the human maternal brain. The hypothesis, based on recent work in rodents, is that microglia, the brain's immune cells, proliferate less during gestation to minimize their immunological activity on the brain in a period when tolerance toward the fetus is crucial (Barth and de Lange 2020). Following this reasoning, the drastic perinatal shift in the immunological profile (from inflammatory-resistant to pro-inflammatory) (Barth and de Lange 2020) might elicit a restoration from the in-pregnancy microglial reduction, which could translate into a volume recovery later in the postpartum period. In rat models, research shows decreased microglial density and cell number in regions such as the MPFC, amygdala, nucleus accumbens, and hippocampus during gestation and the early postpartum compared to virgin rats, and most of these reductions reversed by the time of weaning (Haim et al. 2017). Thus, it seems plausible that similar changes occur in human mothers. Notably, the location and trajectories of these changes in microglial density are resemblant to what has been found with regard to structural brain changes in human mothers (Martínez-García et al. 2021b), suggesting that neuroimmune factors might also operate in the human maternal brain adaptations.

6 The Late-Life Parental Brain

Beyond the immediate effects of motherhood on a woman's brain, reproductive experience also impacts the brain aging trajectories and the long-term mental health vulnerability of a mother. Studies with large-scale databases of middle-aged women (including data from thousands of participants) found that a higher parity (i.e., number of previous children) is associated with less apparent brain aging, as measured by a machine-learning algorithm (de Lange et al. 2019). Specifically, compared to nulliparous women, middle-aged women with more children display higher WM integrity (Voldsbekk et al. 2021), higher water diffusion within brain tissue (Voldsbekk et al. 2021), and less brain aging signs in cortical thickness late in life (Orchard et al. 2020) and key maternal subcortical structures such as the nucleus accumbens, putamen, thalamus, hippocampus, and amygdala (de Lange et al. 2019). Additionally, the greater the number of children parented, the less similar the patterns of rsfMRI connectivity are to those characterizing age-related cognitive decline (Orchard et al. 2021). However, these "protective" effects of parity on brain aging are less pronounced in grand-parous mothers (de Lange et al. 2019; Voldsbekk et al. 2021), in agreement with the higher risk of Alzheimer disease (Jang et al. 2018) and mortality (Zeng et al. 2016) in mothers with more than four pregnancies. Notably, some of the positive associations between cognition and functional connectivity and the number of children parented are also present in elderly fathers (Ning et al. 2020; Orchard et al. 2021), suggesting that factors other than pregnancy-related hormonal fluctuations might mediate the long-term benefits of reproductive experience on the brain aging trajectories. Candidate mediators are perinatal-related immunological (Eid et al. 2019) or other factors associated with parental lifestyle (Ning et al. 2020). It is also plausible that some of the effects detected in late life are driven by the cumulative parenting experience, greater social integration in general (which is known to affect brain health), or by active participation in grandchild care. Longitudinal studies are needed to draw solid causal conclusions on how parenting impacts brain health for life.

7 Non-birthing Parents

Another approach to understanding how the human brain adapts to parenthood is to investigate caregivers beyond the biological, gestational mother or birthing parent. Studies in rodents have shown that virgin female rats can behave maternally and even exhibit brain plasticity after a period of sensitization, indicating that offspring caregiving can be triggered by factors other than the pregnancy and birth, such as the interaction with the pup (Rosenblatt et al. 1967; Leuner et al. 2010; Stolzenberg and Champagne 2016; Pawluski and Galea 2007). Rodent literature has also documented significant brain changes in fathers in biparental species and alloparents (for a review, see Glasper et al. 2019).

In humans, studying biological fathers and partners may help to disentangle pregnancy-induced from experience-induced influences on the brain, since fathers experience the parental transition without directly experiencing pregnancy or childbirth. Another approach is to consider neurobiological adaptations in alloparents, which is loosely defined as caregivers besides biological parents, as a response to their degree of caregiving exposure. This section reviews a broad range of research on both the paternal and alloparental brain in human models.

The Paternal Brain

Human societies have witnessed an increase in fathers' investment in childcare (Henz 2019; Li 2020). But sociocultural and individual differences still produce a wide variety of paternal behaviors ranging from an absent father to being the primary caregiver (Abraham and Feldman 2018). This variability might have discouraged researchers from examining the neural mechanisms that support paternal caregiving behavior. However, studying fathers is an excellent opportunity to track how experiential factors shape the parental brain. Animal research has shown that, for those mammals that naturally display paternal behavior, factors outside the boundaries of pregnancy prime subcortical neural circuits comparable to those that underlie maternal behavior (Numan et al. 2020d). Based on this evidence, it is plausible that brain changes also take place in human fathers.

Functional Brain Adaptations

In humans, most of the neuroimaging studies of the paternal brain have examined fathers' neural responses to infant stimuli, such as photographs, infant cry sounds (Rilling et al. 2021), or video clips of the infant (Abraham and Feldman 2018; Atzil et al. 2012; Feldman 2015). For instance, Abraham et al. (2014) scanned first-time mothers and fathers while watching videos of themselves playing with their infants and a stranger playing with a strange infant (Abraham et al. 2014). They compared three groups of first-time parents with varying degrees of involvement in childrearing: primary caregiver mothers in a heterosexual relationship, secondary caregiver fathers in a heterosexual relationship, and primary caregiver fathers in a homosexual relationship raising infants without maternal involvement. Compared to secondary caregiver fathers, primary caregiver mothers and primary caregiver fathers showed greater activation in the amygdala, a central hub of the ancient parental reward circuits. Compared to primary caregiver mothers, both groups of fathers exhibited greater activation in the superior temporal sulcus, an area widely associated with the DM network. Notably, the more time that fathers reported spending with the child, the higher the functional connectivity between these two DM and motivational hubs. These findings suggest that the neurobiology of fatherhood has some overlap and some distinction from that of mothers and that the degree of caregiving behavior may play a role in how caregivers process infant-related stimuli.

More recent research on the paternal brain has studied the neural underpinnings of specific and directed paternal behaviors. Rilling et al. (2021) scanned first-time fathers whose infants were on average 4 months old. They either passively listened to infant cries or actively listened to infant cries and selected different soothing strategies to calm the infant. Furthermore, fathers who successfully soothed an infant showed greater neural activation in areas involved in action-outcome learning, mentalization, and salience processing (e.g., ACC and PCC). This study provides evidence of neural mechanisms underlying active caregiving behavior, allowing researchers to gain further insight into real-life caregiving behavior.

Although literature leans toward the premise that the paternal brain begins to form after childbirth, some evidence shows that the expectant fathering brain may forecast the adjustment to parenthood. fMRI studies suggest that fathers' prenatal neural responses to infant stimuli starting before birth may be associated with parenting-relevant hormones and predictive of postpartum parenting outcomes (Cardenas et al. 2022). For example, Cardenas et al. (2021) found that expectant fathers who showed higher neural activation on a mentalizing task displayed higher prenatal oxytocin levels and endorsed a more attunement-oriented parenting philosophy after their child's birth. Likewise, Khoddam et al. (2020) found that first-time expectant fathers with higher prenatal testosterone showed stronger neural activation to cry sounds in both auditory processing and mentalizing network regions. Another prospective, longitudinal study found that first-time fathers displayed similar neural responses to infant threat from prenatal to postpartum and that fathers' postpartum neural responses to infant threat were associated with their parental protectiveness in daily life (van 't Veer et al. 2019).

Structural Brain Adaptations

After the substantial evidence for structural brain changes in human mothers, interest in characterizing paternal neural plasticity during the fatherhood transition grew. To date, only two longitudinal MRI studies have examined the structural GM changes of fathers (Kim et al. 2014; Paternina-Die et al. 2020). Kim et al. (2014)

looked at structural GM changes in fathers' brains during the postpartum period from 2–4 weeks postpartum to 12–16 weeks postpartum (Kim et al. 2014). The study found that fathers had increased GM volume in regions associated with reward and attachment such as the hypothalamus, the amygdala, and the striatum (which includes the VP, caudate, and putamen) and mentalization (e.g., PFC, subgenual cingulate, and superior temporal gyrus), similar to what Kim et al. (2010) found in mothers scanned at a comparable postpartum interval. They also found that the greater increase in GM volume within a cluster containing the striatum, amygdala, and subgenual cingulate cortex, the lower levels of depressive symptoms, particularly those related to physical symptoms such as sleep and fatigue. Additionally, they found reductions in GM volume in regions involved in processing threat and parenting stress (e.g., orbitofrontal cortex), salience (e.g., insula), and mentalization (e.g., posterior cingulate cortex, precuneus, and MPFC). They also found that decreases in the orbitofrontal cortex GM volume were associated with higher intrusiveness levels in fathers when manipulating their infants' bodies. The authors concluded that there are specific paternal brain areas affected by depressive symptoms and related to early father-infant attachment. More recently, the study of Paternina-Die et al. (2020) scanned first-time fathers before their partners' pregnancy and at 2 months postpartum and compared the longitudinal changes to those of a control group of non-fathers. They found "preconception-to-postpartum" cortical volume and thickness reductions within the precuneus – a DM node – and the more the reductions, the higher the father's neural response to pictures of his baby. These reductions were less pronounced and affected fewer regions compared to those observed in their pregnant female partners (Hoekzema et al. 2017). Plausible interpretations of these findings are that the parental brain becomes sensitive to experience factors in the absence of pregnancy-related priming signals or that both gestational factors and postpartum experience factors exert cumulative effects over the neuroadaptations to parenting. Either way, such signals seem to trigger neuroanatomical adaptations in the father's brain that are associated with the father-infant relationship, suggesting an adaptive role of the brain changes for the challenges posed by fatherhood.

The Alloparental Brain

Humans are cooperative breeders, which means that the care of children is a shared enterprise, and care is often performed by alloparents, or caregivers who are not the biological parents of the child. Alloparents can include kin, such as grandparents or siblings; adoptive or foster parents; and nonbiologically related caregivers such as babysitters and teachers. Alloparenting has been theorized as the reason that humans develop large, social brains that are capable of cooperation, and in fact, brain size is correlated with the degree of alloparenting care across a variety of mammalian species (Kenkel et al. 2017).

A considerable amount of childcare is performed by alloparents. The vast majority of young children spend time in non-parental childcare on a regular basis (NHES 2018). In the United States alone, about three million children are being raised by grandparents as their primary caregivers (Rapoport et al. 2020), and many more children cohabit with grandparents or are regularly cared for by them. Similarly, sibling care – especially care by older sisters – is common in many cultures (Weisner 2017).

Despite high involvement from alloparents in the care of children, there is minimal research on the neurobiology of the alloparenting brain. This gap in the literature may be due in part to a paucity of model organisms, given that <3% of mammalian species engage in alloparenting (Kenkel et al. 2017). The complexity and variety of alloparenting in humans make it important to develop a human neuroscience of alloparenting. This represents an important frontier for studying the parenting brain, because alloparents do not undergo the hormonal and physiological events of pregnancy, yet still participate in care. Studying alloparents therefore offers the potential to better disambiguate the influences of pregnancy and childbirth on the brain from the impact of caregiving experience itself. In the following section, we review recent neuroscientific work on grandparents, other kin, and on adoptive parents.

Grandparents

The "grandmother effect" hypothesis posits that grandmother allomaternal care is so beneficial to infant survival and that women gain a reproductive advantage by assisting their daughters with care (Herndon 2010). Given this, it would be reasonable to expect the grandmother's brain to show adaptations that benefit engagement in care. Indeed, there is cross-cultural evidence from Asian, European, and North American samples that cognitive functioning, including memory and language, is improved in older adults who spend time caring for grandchildren (Arpino and Bordone 2014; Amano et al. 2018; Ahn and Choi 2019). However, there has been limited research on the neural correlates of grandparenting. One functional study (Rilling, Gonzalez, & Lee, 2021) examined grandmothers' neural responses to images of their grandchildren (aged 3–12), contrasted with an unfamiliar child. Grandmothers activated the insula and sensorimotor areas when viewing their own grandchild. Moreover, grandmothers who desired more involvement with their grandchildren showed more activation in the TPJ and MPFC, both DM regions associated with social cognition/mentalization. Interestingly, the study compared the grandmother data with previously collected data on fathers viewing their own children, and found some distinct patterns, such that grandmothers activated subcortical regions associated with motivation (ventral striatum, caudate nucleus, medial orbitofrontal cortex) and cortical regions associated with salience and emotional empathy (dorsal ACC and insula) more strongly than did fathers, whereas fathers showed stronger activation in the visual cortex, cerebellum, and left dorsolateral PFC. This study suggests that the "grandmothering brain" may look closer to the maternal brain than the paternal brain, but much more research is needed to probe these interesting results.

As small the literature on the grandparenting brain may be, the human neuroscience literature on other forms of kin care is even smaller. For example, there is some evidence that sibling closeness may be linked with neural activation during risky decision-making in adolescence (Rogers et al. 2018), but we are not aware of any studies specifically examining the neural correlates of sibling or other kin care.

Adoptive and Foster Parents

As with the literature on kin alloparents, the literature on adoptive and foster parents is very small. In the neuroimaging fMRI study in which Abraham et al. (2014) compared primary caregiver and secondary caregiver fathers, the primary caregiver fathers were drawn from a sample of homosexual male couples who had children through surrogacy. As such, some of the fathers were not biologically related to their children; however, secondary analyses found no significant difference in oxytocin or brain activation between adoptive and biological fathers.

Besides the fMRI study of Abraham et al. (2014), no other research has applied MRI techniques to study the brain adaptations of adoptive parents. Instead, researchers have focused on recording event-related potentials (ERPs), which are postsynaptic voltages generated in response to specific stimuli, with electroencephalography (EEG). One study focusing on foster mothers found that oxytocin levels after cuddling their infants were associated with their ERPs in response to infant stimuli (Bick et al. 2013). Interestingly, in the first 2 months of the foster parent placement, oxytocin was linked with ERP responses to infants in general, but after infants had been placed with their foster mothers for a longer period, their oxytocin levels tracked with their ERP responses to their own infants. Moreover, both oxytocin and ERP responses were associated with greater behaviorally coded maternal delight when interacting with the foster infant. Another EEG study found that biological mothers and adoptive mothers showed generally similar responses to infant stimuli, which could be differentiated from non-mothers (Hernández-González et al. 2016). However, adoptive mothers exhibited higher reactivity to crying infants, suggesting that they may have needed to deploy more cognitive and affective resources to processing the cry stimuli.

8 Perinatal Mental Illness

The changes that occur in the parental brain are remarkable and necessary, but the transition to parenting is also a time when parents are vulnerable to several mental illnesses (Fisher 2017; Meltzer-Brody et al. 2018). These perinatal mental illnesses can have enduring effects on parental health, parent-infant interactions, and child development. Research is also showing that perinatal mental illnesses, such as perinatal depression (PND), anxiety, and postpartum psychosis, significantly impact the parental brain (Moses-Kolko et al. 2014; Pawluski et al. 2017, 2021a); however, much more research is needed to improve our understanding of these and other perinatal mental illnesses.

Perinatal Depression (PND)

The most talked about perinatal mental illness is PND with studies indicating that 10% of parents are faced with depression and it most significantly affects the mother (O'Hara and McCabe 2013; Shrestha et al. 2014; Goodman et al. 2016; Pawluski et al. 2017). The symptom profile of PND is similar to major depressive disorder (MDD) and includes sad mood, restlessness, and impaired concentration. PND, according to the DSM-V, is diagnosed as MDD identified during pregnancy or within 4 weeks postpartum (American Psychiatric Association 2013). Similar to MDD outside the perinatal period, PND is often comorbid with high levels of anxiety (Falah-Hassani et al. 2016). However, PND (as well as perinatal anxiety) is unique in its timing, some physiological and psychosocial risk factors, and consequences for the mother-infant dyad (Fleming et al. 1988; Field et al. 1990).

Recent studies of the neurobiology of PND have focused on postpartum changes in a mother's brain in relation to depressive symptoms (Pawluski et al. 2017; Duan et al. 2017; Stickel et al. 2019; Barba-Muller et al. 2019; Bjertrup et al. 2019; Pawluski et al. 2021b). Functional MRI, either using task-based or resting-state modalities, is the current preferred choice of study of PND. These functional studies have shown several brain changes in brain regions important for parenting and emotional and cognitive regulation.

rsfMRI investigations of PND report core brain regions in which brain activation at rest differs between depressed and nondepressed postpartum women. Throughout the cortex (frontal, parietal, temporal lobes, and posterior cingulate cortex), there is increased local resting-state homogeneity (ReHo) in the right frontal and parietal lobes and less ReHo in the left frontal and right temporal lobes in postpartum women with depression compared to postpartum women without depression (Wang et al. 2011). When looking at specific brain areas, women with depression symptoms (who also had significantly elevated levels of anxiety) show weaker functional connectivity between the amygdala, ACC, dorsolateral prefrontal cortex (dlPFC), and the hippocampus compared to nondepressed postpartum women (Deligiannidis et al. 2013; Chase et al. 2014). Together, this work indicates that women with depressive symptoms postpartum have significant changes in neural activity at rest in brain regions important for reward, salience, mentalization, and self-regulation.

When looking at brain responses to an emotional cue that is either infant-related or non-infant-related (i.e., emotional word) through fMRI tasks, research shows that neural activation changes in postpartum women with depression from postpartum women without depression and in relation to cue type. For example, when women with postpartum depression are exposed to an emotional cue, such as a negative word or adult face, there are a *decrease* in the activation of the right amygdala and striatum and an increase in the activation of the insula (Silverman et al. 2007, 2011). When exposed to their own infant versus other infant positive pictures, depressed mothers compared to healthy control mothers show increased activation of the right amygdala and decreased amygdala-insular cortex functional connectivity, the latter being associated with increasing symptoms of depression and anxiety (Wonch et al. 2016). Others have shown that when mothers are exposed to their own infant cry versus another infant, mothers with depressive symptoms have *increased* activation in the lenticular nucleus, left MPFC (Swain et al. 2008), and left amygdala and a decrease in left amygdala-nucleus accumbens functional connectivity (Ho and Swain 2017). This research underlines the importance of cue type and neural circuit analysis beyond basic activation maps to characterize the effects of PND on maternal brain and behavior.

The question of whether these neural activation patterns differ in women with postpartum depression and women with MDD outside of the postpartum period has been the topic of discussion as one considers the unique characteristics of PND. To date, one study has compared neural activation in mothers with PND and non-mothers with MDD to determine if brain activity differs in PND and MDD (Dudin et al. 2019). This work shows that women with PND had an *increased* right amygdala activation to infant picture compared to non-mothers with MDD and healthy postpartum or non-postpartum women. There were also minimal changes in right amygdala activation in women with MDD when exposed to a picture of an infant or scenery (Dudin et al. 2019). This work shows that depressive symptoms in mothers are linked to unique neural activity and further research must consider reproductive state when investigating the impact of depression on the brain.

Evidence of the brain structural aspects of PND is limited to one DWI study in mothers with PND. This study showed that, compared with healthy postpartum women, women with postpartum depression had significantly lower WM integrity in the left anterior limb of the internal capsule at 2–8 weeks postpartum (Silver et al. 2018). This suggests a disruption of fronto-subcortical circuits which are important for the adaptive human parental behavior. This study also found that the higher the depressive symptoms postpartum, the lower the WM integrity within the corpus callosum, suggesting a disruption of interhemispheric structural connectivity in women with elevated depression scores. This aberrant interhemispheric functional connectivity correlated with severity of depressive symptoms has been replicated in women diagnosed with postpartum depression 4 months after birth (Zhang et al. 2020).

On a molecular level, using PET scans or magnetic resonance spectroscopy (MRS), research is beginning to demonstrate a neurotransmitter dysregulation that occurs with PND in women. With a focus on monoamines, key findings show that PND is associated with an increase in monoamine oxidase A (MAO-A) density in the PFC and ACC (Sacher et al. 2015); an increase of glutamate in the MPFC (for a review, see Duan et al. 2017, 2019); a decrease in serotonin receptor binding (5-HT1A) in the ACC, orbitofrontal cortex, and mesotemporal cortex (Moses-Kolko et al. 2008); D2/3 receptor binding in the striatum (Moses-Kolko et al. 2012); and GABA reductions in the occipital cortex (Epperson et al. 2006).

Although there has been a rather large focus on the amygdala when investigating PND and brain activation in women, it is clear that many brain areas are involved in PND. In addition, we know that no one brain area is responsible for one behavioral outcome. Alterations in the activation of different parental brain areas with PND likely alter key neural networks associated with caregiving behavior involving empathy, stress regulation, motivation, and emotional reactivity (Robinson and Berridge 2015) and executive functioning (for reviews, see Barrett and Fleming 2011; Moses-Kolko et al. 2014).

Perinatal Anxiety Disorders

Recent research reports that clinical levels of anxiety exist in 13–21% of pregnant women and 11–17% of postpartum women (Thorsness et al. 2018). Perinatal anxiety does not currently have unique diagnostic criteria, and its symptom profile is most often characterized by the same symptoms of anxiety disorders that occur outside of the perinatal period. Recent meta-analyses show that high levels of anxious symptoms in mothers are associated with increased risk of birth complications, reduced breastfeeding, poorer caregiving behavior, as well as number of poor neurodevelopmental outcomes in children (Ho et al. 2014; Elmadih et al. 2016; Guo et al. 2018; Grigoriadis et al. 2019).

Despite the prevalence of anxiety disorders during the perinatal period, our understanding of the neurobiology of these disorders is limited. Research using EEG shows that maternal state anxiety during pregnancy is positively associated with a sustained processing to neutral infant faces, but not to sad faces (Malak et al. 2015; Rutherford et al. 2017). This research suggests that in mothers with high levels of anxiety, there is a bias toward interpreting threat in a potentially ambiguous infant's face.

Recent research using fMRI shows that mothers with high trait anxiety have similar levels of elevated amygdala activation when viewing pictures of their own infant and an unfamiliar infant expressing positive affect. Mothers with low trait anxiety show higher amygdala response to their own infant's face only (Barrett et al. 2012). Consistent with this result, women with high trait anxiety express lower positive feelings about their own infant. Further research by this same group shows that mothers with high trait anxiety (but not necessarily depression) have lower amygdala response to baby faces in general, as well as lower functional connectivity between the amygdala and the insular cortex (Wonch et al. 2016). The amygdala and the insula are involved in many functions including salience processing and empathy (Nieuwenhuys 2012; Seeley 2019), suggesting the amygdala-insula pathway may be the basis of anxious mothers' altered caregiving sensitivity (Hipwell et al. 2015). Importantly, higher maternal trait anxiety symptoms have been correlated with an infant's lower left amygdala-right ACC functional connectivity and lower cognitive development, suggesting that this functional network is sensitive to environmental cues during early life (Dufford et al. 2021). Together, these findings point to an important role of regions of the parental brain in perinatal anxiety, but much more research is needed to determine the distinct neurobiological features of anxiety disorders in parents.

Postpartum Psychosis (PP)

Postpartum psychosis (PP) is a severe psychiatric disorder associated with childbirth that affects up to 1–2 in 1000 women (Meltzer-Brody et al. 2018). This debilitating disorder requires immediate hospitalization and care. Recently, a growing body of research has been investigating how PP affects a mother's brain. One of the first neuroimaging investigations of PP found that, when compared to non-postpartum women with psychosis or bipolar disorders, PP women showed a general ventricular enlargement (Lanczik et al. 1998). Recent work by Dazzan and colleagues has extended these findings and shown that mothers with a recent episode of PP have smaller ACC, superior temporal gyrus, and parahippocampal gyrus compared to mothers at risk of developing PP (Fusté et al. 2017). Interestingly, mothers who were at risk of developing PP but did not have a psychotic episode postpartum had a larger volume of the inferior frontal gyrus than controls, which may be a protective factor that requires more investigation (Fusté et al. 2017). The brain areas structurally affected by PP are involved in empathy, emotional regulation, and decision-making. Further research by this group has shown that the dlPFC, a brain area typically associated with executive functions such as working memory and selective attention, is more functionally connected with other brain areas in mothers who had PP (Kowalczyk et al. 2021). This work is in contrast with other work showing that people with psychosis at other times in life have a dlPFC less connected with other brain areas (discussed in Kowalczyk et al. 2021). These findings suggest that understanding brain morphology may be a key in characterizing women at risk for PP and other perinatal mental illnesses.

Factors Influencing Perinatal Mental Illness

Hormones, neuropeptides, and neurotransmitters have been a focus of research, particularly in rodent models, on the physiological factors responsible for PND and other perinatal mental illnesses (Slattery and Hillerer 2016; Dickens and Pawluski 2018; Pawluski et al. 2021b). Unfortunately, little consistency exists in the literature as to the neurophysiological correlates of mental illnesses occurring during the perinatal period (for a review, see Dickens and Pawluski 2018).

There are several physiological, social, behavioral, and experiential (i.e., treatment) factors that have the potential to alter the neurobiology of perinatal mental illnesses. For example, imaging data has shown that complications associated with birth, such as the degree of intervention and medicalization, and the breastfeeding experience, both linked to a host of physiological changes in the mother, can significantly alter the neural response toward the infant in the maternal brain (Swain et al. 2008; Berman et al. 2020) and affect the mental well-being of the mother (Dekel et al. 2017, 2019). The extent to which these and other factors play a role in the neurobiology of perinatal mental illness remains to be determined.

Nonpregnant Parents

It should be mentioned that although the research has focused on perinatal mental illness in pregnant mothers, fathers also struggle with perinatal mental illness (Fisher 2017). Kim et al. (2014) found that, in biological fathers, postpartum structural changes were associated with levels of depressive symptoms and parenting intrusiveness, pointing to potential effects of perinatal mental illness on the brains of fathers. No human study to date has analyzed the alterations of functional or structural brain adaptations in nonpregnant parents diagnosed with a perinatal mental illness. Furthermore, research on the distinct changes in the parental brain of biological and nonbiological parents at risk for developing perinatal mental illness is needed.

9 Conclusions

This chapter has reviewed the current evidence on how the human parental brain adapts to parenthood. We have discussed how parents activate the brain when viewing their infants' stimuli, the parental behaviors related to these activations, and brain structural and functional adaptations detected in early and late parenthood and even prior to birth. We have also reviewed studies starting to explore the unique neurobiological profiles underlying perinatal mental illnesses. We centered our discussion on pregnant mothers, given that neuroimaging studies have found more extensive and pronounced neurobiological adaptations in the maternal brain compared to other non-gestational parents. Findings from structural, task-based, and resting-state MRI studies have jointly found that the brain remodeling in mothers affects cortical networks involved in salience, mentalization, and emotion regulation and subcortical regions involved in processing rewarding stimuli. We also explained the fewer studies on the paternal and alloparental brains to conclude that caregiving experience and other experiential factors not directly related to parenthood may also shape the parental brain. To date, studies indicate that parental transition is a period of increased brain plasticity within brain areas critical for parenthood's cognitive and social challenges ahead.

The above-cited published evidence is scant, and many gaps remain in the literature on the human parental brain. The precise neural mechanisms and function behind the observed macroscopic changes are still under debate. Also, more work is needed to disentangle the reproductive and experience influence on the parental brain. One strategy would be collecting specific hormonal and immunological information in parents and levels of parental investment, stress, and sleep quality, and to directly compare the brain adaptations of pregnant mothers and other non-gestational parents. Another important question is to explore the unique brain profiles of kin care, adoptive and foster parenting, and different types of alloparenting arrangements. Do care experiences such as sibling care, or as a caregiver employer – for example, as a nanny or teacher – help to prime the brain for future parenting experiences? Additionally, further work on adoptive or foster caregivers could inform targeted interventions to boost attachment and ease parenting transitions (Bick et al. 2013). In sum, the emerging literature on the human parental brain is intriguing and suggests many fruitful directions for further study.

References

Abraham E, Feldman R. The neurobiology of human allomaternal care; implications for fathering, coparenting, and children's social development. Physiol Behav. 2018;193:25–34. https://doi.org/10.1016/j.physbeh.2017.12.034.

Abraham E, Hendler T, Shapira-Lichter I, Kanat-Maymon Y, Zagoory-Sharon O, Feldman R. Father's brain is sensitive to childcare experiences. Cognit Sci. 2014;111(27):9792–7.

Ahn T, Choi KD. Grandparent caregiving and cognitive functioning among older people: evidence from Korea. Rev Econ Househ. 2019;17(2):553–86. https://doi.org/10.1007/s11150-018-9413-5.

Ainsworth MD, Blehar M, Waters E, Wall S. Patterns of attachment. Hillsdale: Erlbaum; 1978.

Alexander DC, Dyrby TB, Nilsson M, Zhang H. Imaging brain microstructure with diffusion MRI: practicality and applications. NMR Biomed. 2019;32(4):e3841. https://doi.org/10.1002/nbm.3841.

Alhomoud MA, Khan AS, Alhomoud I. The potential preventive effect of pregnancy and breastfeeding on multiple sclerosis. Eur Neurol. 2021;84(2):71–84. https://doi.org/10.1159/000514432.

Amano T, Shen H, Sun F. The influence of grandparenting on cognitive function and the mediating role of health factors. Innov Aging. 2018;2(suppl_1):116–7. https://doi.org/10.1093/geroni/igy023.430.

American Psychiatric Association. Diagnostic and statistical manual of mental disorders. 5th ed. Washington, DC: American Psychiatric Publishing; 2013.

Arpino B, Bordone V. Does grandparenting pay off? The effect of child care on grandparents' cognitive functioning: effect of child care on grandparents' cognition. Fam Relat. 2014;76(2):337–51. https://doi.org/10.1111/jomf.12096.

Attwell D, Iadecola C. The neural basis of functional brain imaging signals. Trends Neurosci. 2002;25(12):621–5. https://doi.org/10.1016/S0166-2236(02)02264-6.

Atzil S, Hendler T, Zagoory-Sharon O, Winetraub Y, Feldman R. Synchrony and specificity in the maternal and the paternal brain: relations to oxytocin and vasopressin. J Am Acad Child Adolesc Psychiatry. 2012;51(8):798–811. https://doi.org/10.1016/j.jaac.2012.06.008.

Aurich NK, Alves Filho JO, Marques da Silva AM, Franco AR. Evaluating the reliability of different preprocessing steps to estimate graph theoretical measures in resting state fMRI data. Front Neurosci. 2015;9:48. https://doi.org/10.3389/fnins.2015.00048.

Backhausen LL, Herting MM, Tamnes CK, Vetter NC. Best practices in structural neuroimaging of neurodevelopmental disorders. Neuropsychol Rev. 2021:1–19. https://doi.org/10.1007/s11065-021-09496-2.

Barba-Muller E, Craddock S, Carmona S, Hoekzema E. Brain plasticity in pregnancy and the postpartum period: links to maternal caregiving and mental health. Arch Womens Ment Health. 2019;22(2):289–99. https://doi.org/10.1007/s00737-018-0889-z.

Barrett J, Fleming AS. Annual research review: all mothers are not created equal: neural and psychobiological perspectives on mothering and the importance of individual differences. J Child Psychol Psychiatry. 2011;52(4):368–97. https://doi.org/10.1111/j.1469-7610.2010.02306.x.

Barrett J, Wonch KE, Gonzalez A, Ali N, Steiner M, Hall GB, Fleming AS. Maternal affect and quality of parenting experiences are related to amygdala response to infant faces. Soc Neurosci. 2012;7(3):252–68. https://doi.org/10.1080/17470919.2011.609907.

Barth C, de Lange A-MG. Towards an understanding of women's brain aging: the immunology of pregnancy and menopause. Front Neuroendocrinol. 2020;58:100850. https://doi.org/10.1016/j.yfrne.2020.100850.

Bercury KK, Macklin WB. Dynamics and mechanisms of CNS myelination. Dev Cell. 2015;32(4):447–58. https://doi.org/10.1016/j.devcel.2015.01.016.

Berman Z, Kaim A, Reed T, Felicione J, Hinojosa C, Oliver K, Rutherford H, Mayes L, Shin L, Dekel S. Amygdala response to traumatic childbirth recall is associated with impaired bonding behaviors in mothers with childbirth-related PTSD symptoms: preliminary findings. Biol Psychiatry. 2020;87(9):S142. https://doi.org/10.1016/j.biopsych.2020.02.378.

Bick J, Dozier M, Bernard K, Grasso D, Simons R. Foster mother-infant bonding: associations between Foster mothers' oxytocin production, electrophysiological brain activity, feelings of commitment, and caregiving quality. Child Dev. 2013;84(3):826–40. https://doi.org/10.1111/cdev.12008.

Bjertrup AJ, Friis NK, Miskowiak KW. The maternal brain: neural responses to infants in mothers with and without mood disorder. Neurosci Biobehav Rev. 2019;107:196–207. https://doi.org/10.1016/j.neubiorev.2019.09.011.

Buckner RL, DiNicola LM. The brain's default network: updated anatomy, physiology and evolving insights. Nat Rev Neurosci. 2019;20(10):593–608. https://doi.org/10.1038/s41583-019-0212-7.

Cardenas SI, Stoycos SA, Sellery P, Marshall N, Khoddam H, Kaplan J, Goldenberg D, Saxbe DE. Theory of mind processing in expectant fathers: associations with prenatal oxytocin and parental attunement. Dev Psychobiol. 2021;63(5):1549–67. https://doi.org/10.1002/dev.22115.

Cardenas SI, Morris AR, Marshall NA, Aviv L, Martínez-García M, Sellery P, Saxbe DE. Fathers matter from the start: role of expectant fathers on child development. Child Dev Perspect. 2022;16(1):54–59.

Carmona S, Martínez-García M, Paternina-Die M, Barba-Müller E, Wierenga LM, Alemán-Gómez Y, Pretus C, Marcos-Vidal L, Beumala L, Cortizo R, Pozzobon C, Picado M, Lucco F, García-García D, Soliva JC, Tobeña A, Peper JS, Crone EA, Ballesteros A, Vilarroya O, Desco M, Hoekzema E. Pregnancy and adolescence entail similar neuroanatomical adaptations: a comparative analysis of cerebral morphometric changes. Hum Brain Mapp. 2019;40(7):2143–52. https://doi.org/10.1002/hbm.24513.

Chan RW, Ho LC, Zhou IY, Gao PP, Chan KC, Wu EX. Structural and functional brain remodeling during pregnancy with diffusion tensor MRI and resting-state functional MRI. PLoS One. 2015;10(12):e0144328. https://doi.org/10.1371/journal.pone.0144328.

Chase HW, Moses-Kolko EL, Zevallos C, Wisner KL, Phillips ML. Disrupted posterior cingulate-amygdala connectivity in postpartum depressed women as measured with resting BOLD fMRI. Soc Cogn Affect Neurosci. 2014;9(8):1069–75. https://doi.org/10.1093/scan/nst083.

de Lange A-MG, Kaufmann T, van der Meer D, Maglanoc LA, Alnæs D, Moberget T, Douaud G, Andreassen OA, Westlye LT. Population-based neuroimaging reveals traces of childbirth in the maternal brain. Proc Natl Acad Sci USA. 2019;116(44):22341–6. https://doi.org/10.1073/pnas.1910666116.

Dekel S, Stuebe C, Dishy G. Childbirth induced posttraumatic stress syndrome: a systematic review of prevalence and risk factors. Front Psychol. 2017;8 https://doi.org/10.3389/fpsyg.2017.00560.

Dekel S, Ein-Dor T, Berman Z, Barsoumian IS, Agarwal S, Pitman RK. Delivery mode is associated with maternal mental health following childbirth. Arch Womens Ment Health. 2019;22(6):817–24. https://doi.org/10.1007/s00737-019-00968-2.

Deligiannidis KM, Sikoglu EM, Shaffer SA, Frederick B, Svenson AE, Kopoyan A, Kosma CA, Rothschild AJ, Moore CM. GABAergic neuroactive steroids and resting-state functional connectivity in postpartum depression: a preliminary study. J Psychiatr Res. 2013;47(6):816–28. https://doi.org/10.1016/j.jpsychires.2013.02.010.

Dickens MJ, Pawluski JL. The HPA axis during the perinatal period: implications for perinatal depression. Endocrinology. 2018;159(11):3737–46. https://doi.org/10.1210/en.2018-00677.

Dinç H, Esen F, Demirci A, Sari A, Resit GH. Pituitary dimensions and volume measurements in pregnancy and post partum MR assessment. Acta Radiol. 1998;39(1):64–9. https://doi.org/10.1080/02841859809172152.

Duan C, Cosgrove J, Deligiannidis KM. Understanding peripartum depression through neuroimaging: a review of structural and functional connectivity and molecular imaging research. Curr Psychiatry Rep. 2017;19(10):70. https://doi.org/10.1007/s11920-017-0824-4.

Duan C, Hare MM, Staring M, Deligiannidis KM. Examining the relationship between perinatal depression and neurodevelopment in infants and children through structural and functional neuroimaging research. Int Rev Psychiatry. 2019;31(3):264–79. https://doi.org/10.1080/09540261.2018.1527759.

Dudin A, Wonch KE, Davis AD, Steiner M, Fleming AS, Hall GB. Amygdala and affective responses to infant pictures: comparing depressed and non-depressed mothers and non-mothers. J Neuroendocrinol. 2019;31(9):e12790. https://doi.org/10.1111/jne.12790.

Dufford AJ, Erhart A, Kim P. Maternal brain resting-state connectivity in the postpartum period. J Neuroendocrinol. 2019;31(9) https://doi.org/10.1111/jne.12737.

Dufford AJ, Salzwedel AP, Gilmore JH, Gao W, Kim P. Maternal trait anxiety symptoms, frontolimbic resting-state functional connectivity, and cognitive development in infancy. Dev Psychobiol. 2021;63(6) https://doi.org/10.1002/dev.22166.

Duthie L, Reynolds RM. Changes in the maternal hypothalamic-pituitary-adrenal axis in pregnancy and postpartum: influences on maternal and fetal outcomes. Neuroendocrinology. 2013;98(2):106–15. https://doi.org/10.1159/000354702.

Eid RS, Chaiton JA, Lieblich SE, Bodnar TS, Weinberg J, Galea LAM. Early and late effects of maternal experience on hippocampal neurogenesis, microglia, and the circulating cytokine milieu. Neurobiol Aging. 2019;78:1–17. https://doi.org/10.1016/j.neurobiolaging.2019.01.021.

Elmadih A, Wan MW, Downey D, Elliott R, Swain JE, Abel KM. Natural variation in maternal sensitivity is reflected in maternal brain responses to infant stimuli. Behav Neurosci. 2016;130(5):500–10. https://doi.org/10.1037/bne0000161.

Elster AD, Sanders TG, Vines FS, Chen MY. Size and shape of the pituitary gland during pregnancy and post partum: measurement with MR imaging. Radiology. 1991;181(2):531–5. https://doi.org/10.1148/radiology.181.2.1924800.

Epperson CN, Gueorguieva R, Czarkowski KA, Stiklus S, Sellers E, Krystal JH, Rothman DL, Mason GF. Preliminary evidence of reduced occipital GABA concentrations in puerperal women: a 1H-MRS study. Psychopharmacology (Berl). 2006;186(3):425–33. https://doi.org/10.1007/s00213-006-0313-7.

Erikson EH. Identity and the life cycle: collected papers. New York: WW Norton; 1959.

Falah-Hassani K, Shiri R, Dennis CL. Prevalence and risk factors for comorbid postpartum depressive symptomatology and anxiety. J Affect Disord. 2016;198:142–7. https://doi.org/10.1016/j.jad.2016.03.010.

Feldman R. The adaptive human parental brain: implications for children's social development. Trends Neurosci. 2015;38(6):387–99. https://doi.org/10.1016/j.tins.2015.04.004.

Field T, Healy B, Goldstein M, Guthertz M. Behavior-state matching and synchrony in mother-infant interactions of nondepressed versus depressed dyads. Dev Psychobiol. 1990;26(1):7–14.

Fischer F. V. Schwangerschaftsveränderung der Hypophyse als Ursache bitemporaler Hemianopsie. Ophthalmologica. 1931;75(5–6):343–54. https://doi.org/10.1159/000297234.

Fisher SD. Paternal mental health: why is it relevant? Am J Lifestyle Med. 2017;11(3):200–11. https://doi.org/10.1177/1559827616629895.

Fleming AS, Ruble DN, Flett GL, Shaul D. Postpartum adjustment in first-time mothers: relations between mood, maternal attitudes and mother-infant interactions. Dev Psychobiol. 1988;24:71–81.

Fusté M, Pauls A, Worker A, Reinders AATS, Simmons A, Williams SCR, Haro JM, Hazelgrove K, Pawlby S, Conroy S, Vecchio C, Seneviratne G, Pariante CM, Mehta MA, Dazzan P. Brain structure in women at risk of postpartum psychosis: an MRI study. Transl Psychiatry. 2017;7(12):1286. https://doi.org/10.1038/s41398-017-0003-8.

Glasper E, Kenkel R, Bick WM, J. & Rilling, J. K. More than just mothers: the neurobiological and neuroendocrine underpinnings of allomaternal caregiving. Front Neuroendocrinol. 2019;53:100741. https://doi.org/10.1016/j.yfrne.2019.02.005.

Gonzalez JG, Elizondo G, Saldivar D, Nanez H, Todd LE, Villarreal JZ. Pituitary gland growth during normal pregnancy: an in vivo study using magnetic resonance imaging. Am J Med. 1988;85(2):217–20. https://doi.org/10.1016/s0002-9343(88)80346-2.

Goodman JH, Watson GR, Stubbs B. Anxiety disorders in postpartum women: a systematic review and meta-analysis. J Affect Disord. 2016;203:292–331. https://doi.org/10.1016/j.jad.2016.05.033.

Gregg C, Shikar V, Larsen P, Mak G, Chojnacki A, Yong VW, Weiss S. White matter plasticity and enhanced remyelination in the maternal CNS. J Neurosci. 2007;27(8):1812–23. https://doi.org/10.1523/JNEUROSCI.4441-06.2007.

Grigoriadis S, Graves L, Peer M, Mamisashvili L, Tomlinson G, Vigod SN, Dennis CL, Steiner M, Brown C, Cheung A, Dawson H, Rector NA, Guenette M, Richter M. A systematic review and meta-analysis of the effects of antenatal anxiety on postpartum outcomes. Arch Womens Ment Health. 2019;22(5):543–56. https://doi.org/10.1007/s00737-018-0930-2.

Guo C, Moses-Kolko E, Phillips M, Swain JE, Hipwell AE. Severity of anxiety moderates the association between neural circuits and maternal behaviors in the postpartum period. Cogn Affect Behav Neurosci. 2018;18(3):426–36. https://doi.org/10.3758/s13415-017-0516-x.

Haim A, Julian D, Albin-Brooks C, Brothers HM, Lenz KM, Leuner B. A survey of neuroimmune changes in pregnant and postpartum female rats. Brain Behav Immun. 2017;59:67–78. https://doi.org/10.1016/j.bbi.2016.09.026.

Henz U. Fathers' involvement with their children in the United Kingdom: recent trends and class differences. Demogr Res. 2019;40:865–96.

Hernández-González M, Hidalgo-Aguirre RM, Guevara MA, Pérez-Hernández M, Amezcua-Gutiérrez C. Observing videos of a baby crying or smiling induces similar, but not identical, electroencephalographic responses in biological and adoptive mothers. Infant Behav Dev. 2016;42:1–10. https://doi.org/10.1016/j.infbeh.2015.10.006.

Herndon JG. The grandmother effect: implications for studies on aging and cognition. Gerontology. 2010;56(1):73–9. https://doi.org/10.1159/000236045.

Hillerer KM, Jacobs VR, Fischer T, Aigner L. The maternal brain: an organ with peripartal plasticity. Neural Plast. 2014;2014:1–20. https://doi.org/10.1155/2014/574159.

Hipwell AE, Guo C, Phillips ML, Swain JE, Moses-Kolko EL. Right frontoinsular cortex and subcortical activity to infant cry is associated with maternal mental state talk. J Neurosci. 2015;35(37):12725–32. https://doi.org/10.1523/JNEUROSCI.1286-15.2015.

Ho SS, Swain JE. Depression alters maternal extended amygdala response and functional connectivity during distress signals in attachment relationship. Behav Brain Res. 2017;325(Pt B):290–6. https://doi.org/10.1016/j.bbr.2017.02.045.

Ho SS, Konrath S, Brown S, Swain JE. Empathy and stress related neural responses in maternal decision making. Front Neurosci. 2014;8:152. https://doi.org/10.3389/fnins.2014.00152.

Hoekzema E, Barba-Müller E, Pozzobon C, Picado M, Lucco F, García-García D, Soliva JC, Tobeña A, Desco M, Crone EA, Ballesteros A, Carmona S, Vilarroya O. Pregnancy leads to long-lasting changes in human brain structure. Nat Neurosci. 2017;20(2):287–96. https://doi.org/10.1038/nn.4458.

Hoekzema E, Tamnes CK, Berns P, Barba-Müller E, Pozzobon C, Picado M, Lucco F, Martínez-García M, Desco M, Ballesteros A, Crone EA, Vilarroya O, Carmona S. Becoming a mother entails anatomical changes in the ventral striatum of the human brain that facilitate its responsiveness to offspring cues. Psychoneuroendocrinology. 2020;112:104507. https://doi.org/10.1016/j.psyneuen.2019.104507.

Jang H, Bae JB, Dardiotis E, Scarmeas N, Sachdev PS, Lipnicki DM, Han JW, Kim TH, Kwak KP, Kim BJ, Kim SG, Kim JL, Moon SW, Park JH, Ryu S-H, Youn JC, Lee DY, Lee DW, Lee SB, Lee JJ, Jhoo JH, Yannakoulia M, Kosmidis MH, Hadjigeorgiou GM, Sakka P, Kim KW. Differential effects of completed and incomplete pregnancies on the risk of Alzheimer disease. Neurology. 2018;91(7):e643–51. https://doi.org/10.1212/WNL.0000000000006000.

Kenkel WM, Perkeybile AM, Carter CS. The neurobiological causes and effects of alloparenting: neurobiological causes and effects of neurobiology. Dev Neurobiol. 2017;77(2):214–32. https://doi.org/10.1002/dneu.22465.

Khoddam H, Goldenberg D, Stoycos SA, Horton KT, Marshall N, Cárdenas SI, Kaplan J, Saxbe D. How do expectant fathers respond to infant cry? Examining brain and behavioral responses and the moderating role of testosterone. Soc Cogn Affect Neurosci. 2020;15(4):437–46. https://doi.org/10.1093/scan/nsaa051.

Kim P. Human maternal brain plasticity: adaptation to parenting. New Dir Child Adolesc Dev. 2016;2016(153):47–58. https://doi.org/10.1002/cad.20168.

Kim P. How stress can influence brain adaptations to motherhood. Front Neuroendocrinol. 2021;60:100875. https://doi.org/10.1016/j.yfrne.2020.100875.

Kim P, Leckman JF, Mayes LC, Feldman R, Wang X, Swain JE. The plasticity of human maternal brain: longitudinal changes in brain anatomy during the early postpartum period. Behav Neurosci. 2010;124(5):695–700. https://doi.org/10.1037/a0020884.

Kim P, Rigo P, Mayes LC, Feldman R, Leckman JF, Swain JE. Neural plasticity in fathers of human infants. Soc Neurosci. 2014;9(5):522–35. https://doi.org/10.1080/17470919.2014.933713.

Kim P, Dufford AJ, Tribble RC. Cortical thickness variation of the maternal brain in the first 6 months postpartum: associations with parental self-efficacy. Brain Struct Funct. 2018;223(7):3267–77. https://doi.org/10.1007/s00429-018-1688-z.

Kinsley C, Trainer H, Stafisso-Sandoz R, Quadros G, Marcus P, Hearon LK, C. & Lambert, K. G. Motherhood and the hormones of pregnancy modify concentrations of hippocampal neuronal dendritic spines. Horm Behav. 2006;49(2):131–42. https://doi.org/10.1016/j.yhbeh.2005.05.017.

Kowalczyk OS, Pauls AM, Fuste M, Williams SCR, Hazelgrove K, Vecchio C, Seneviratne G, Pariante CM, Dazzan P, Mehta MA. Neurocognitive correlates of working memory and emotional processing in postpartum psychosis: an fMRI study. Psychol Med. 2021;51(10):1724–32. https://doi.org/10.1017/S0033291720000471.

Keyser-Marcus L, Stafisso-Sandoz G, Gerecke K, Jasnow A, Nightingale L, Lambert KG, Kinsley CH. Alterations ofmedial preoptic area neurons following pregnancy and pregnancy-like steroidal treatment in the rat. Brain researchbulletin, 2001;55(6):737–745.

Lanczik M, Fritze E, Hofmann E, Schulz C, Knoche M, Becker T. Ventricular abnormality in patients with postpartum psychoses. Arch Womens Ment Health. 1998;1(1):45–7.

Leuner B, Glasper ER, Gould E. Parenting and plasticity. Trends Neurosci. 2010;33(10):465–73. https://doi.org/10.1016/j.tins.2010.07.003.

Li X. Fathers' involvement in Chinese societies: increasing presence, uneven progress. Child Dev Perspect. 2020;14(3):150–6. https://doi.org/10.1111/cdep.12375.

Lisofsky N, Gallinat J, Lindenberger U, Kühn S. Postpartal neural plasticity of the maternal brain: early renormalization of pregnancy-related decreases? Neurosignals. 2019;27(1):12–24. https://doi.org/10.33594/000000105.

Lorberbaum JP, Newman JD, Dubno JR, Horwitz AR, Nahas Z, Teneback CC, Bloomer CW, Bohning DE, Vincent D, Johnson MR, Emmanuel N, Brawman-Mintzer O, Book SW, Lydiard RB, Ballenger JC, George MS. Feasibility of using fMRI to study mothers responding to infant cries. Depress Anxiety. 1999;10(3):99–104. https://doi.org/10.1002/(sici)1520-6394(1999)10:3<99::aid-da2>3.0.co;2-#.

Luders E, Kurth F, Gingnell M, Engman J, Yong E-L, Poromaa IS, Gaser C. From baby brain to mommy brain: widespread gray matter gain after giving birth. Cortex. 2020;126:334–42. https://doi.org/10.1016/j.cortex.2019.12.029.

Luders E, Gaser C, Gingnell M, Engman J, Sundström Poromaa I, Kurth F. Gray matter increases within subregions of the hippocampal complex after pregnancy. Brain Imaging Behav. 2021a; https://doi.org/10.1007/s11682-021-00463-2.

Luders E, Gaser C, Gingnell M, Engman J, Sundström Poromaa I, Kurth F. Significant increases of the amygdala between immediate and late postpartum: pronounced effects within the superficial subregion. J Neurosci Res. 2021b:24855. https://doi.org/10.1002/jnr.24855.

Malak SM, Crowley MJ, Mayes LC, Rutherford HJ. Maternal anxiety and neural responses to infant faces. J Affect Disord. 2015;172:324–30. https://doi.org/10.1016/j.jad.2014.10.013.

Martínez-García M, Paternina-Die M, Barba-Müller E, Martín de Blas D, Beumala L, Cortizo R, Pozzobon C, Marcos-Vidal L, Fernández-Pena A, Picado M. Do pregnancy-induced brain changes reverse? The brain of a mother six years after parturition. Brain Sci. 2021a;11(2):168.

Martínez-García M, Paternina-Die M, Desco M, Vilarroya O, Carmona S. Characterizing the brain structural adaptations across the motherhood transition. Frontiers in Global Women's Health. 2021b;2(742775) https://doi.org/10.3389/fgwh.2021.742775.

Meltzer-Brody S, Howard LM, Bergink V, Vigod S, Jones I, Munk-Olsen T, Honikman S, Milgrom J. Postpartum psychiatric disorders. Nat Rev Dis Primers. 2018;4:18022. https://doi.org/10.1038/nrdp.2018.22.

Moses-Kolko EL, Meltzer CC, Berga SL, Wisner KL. Postpartum depression: the clinical disorder and application of PET imaging research methods. In: Bridges RS, editor. Neurobiology of the parental brain. Burlington, MA: Elsevier; 2008. p. 175–99.

Moses-Kolko EL, Price JC, Wisner KL, Hanusa BH, Meltzer CC, Berga SL, Grace AA, di Scalea TL, Kaye WH, Becker C, Drevets WC. Postpartum and depression status are associated with lower [(1)(1)C]raclopride BP(ND) in reproductive-age women. Neuropsychopharmacol: Off Public Am Coll Neuropsychopharmacol. 2012;37(6):1422–32. https://doi.org/10.1038/npp.2011.328.

Moses-Kolko EL, Horner MS, Phillips ML, Hipwell AE, Swain JE. In search of neural endophenotypes of postpartum psychopathology and disrupted maternal caregiving. J Neuroendocrinol. 2014;26(10):665–84. https://doi.org/10.1111/jne.12183.

Murphy K, Birn RM, Bandettini PA. Resting-state fMRI confounds and cleanup. Neuroimage. 2013;80:349–59. https://doi.org/10.1016/j.neuroimage.2013.04.001.

Mills KL, Tamnes CK. Methods and considerations for longitudinal structural brain imaging analysisacross development. Developmental cognitive neuroscience, 2014;9:172–190.

NHES. National Household Education Surveys Program of 2016 Data File User's Manual. 2018. https://nces.ed.gov/pubsearch/pubsinfo.asp?pubid=2018100. Accessed 26 Sep 2021.

Nieuwenhuys R. The insular cortex: a review. Prog Brain Res. 2012;195:123–63. https://doi.org/10.1016/B978-0-444-53860-4.00007-6.

Ning K, Zhao L, Franklin M, Matloff W, Batta I, Arzouni N, Sun F, Toga AW. Parity is associated with cognitive function and brain age in both females and males. Sci Rep. 2020;10(1):6100. https://doi.org/10.1038/s41598-020-63014-7.

Numan M. Hormonal control of maternal behavior in nonhuman mammals. In: The Parental Brain. Oxford University Press; 2020a. https://doi.org/10.1093/oso/9780190848675.003.0003.

Numan M. Central neural circuits regulating maternal behavior in nonhuman mammals. In: The Parental Brain. Oxford University Press; 2020b. https://doi.org/10.1093/oso/9780190848675.003.0005.

Numan M. The parental brain in humans. In: The Parental Brain. Oxford University Press; 2020c. https://doi.org/10.1093/oso/9780190848675.003.0008.

Numan M. Alloparental behavior and paternal behavior in nonhuman mammals. In: The Parental Brain. Oxford University Press; 2020d. https://doi.org/10.1093/oso/9780190848675.003.0007.

O'Hara MW, McCabe JE. Postpartum depression: current status and future directions. Annu Rev Clin Psychol. 2013;9:379–407. https://doi.org/10.1146/annurev-clinpsy-050212-185612.

Oatridge A, Holdcroft A, Saeed N, Hajnal JV, Puri BK, Fusi L, Bydder GM. Change in brain size during and after pregnancy: study in healthy women and women with preeclampsia. Am J Neuroradiol. 2002;23(1):19–26.

Orchard ER, Ward PGD, Sforazzini F, Storey E, Egan GF, Jamadar SD. Relationship between parenthood and cortical thickness in late adulthood. PLoS One. 2020;15(7):e0236031. https://doi.org/10.1371/journal.pone.0236031.

Orchard ER, Ward PGD, Chopra S, Storey E, Egan GF, Jamadar SD. Neuroprotective effects of motherhood on brain function in late life: a resting-state fMRI study. Cereb Cortex. 2021;31(2):1270–83. https://doi.org/10.1093/cercor/bhaa293.

Paternina-Die M, Martínez-García M, Pretus C, Hoekzema E, Barba-Müller E, Martín de Blas D, Pozzobon C, Ballesteros A, Vilarroya Ó, Desco M, Carmona S. The paternal transition entails neuroanatomic adaptations that are associated with the father's brain response to his infant cues. Cereb Cortex Comm. 2020;1:tgaa082. https://doi.org/10.1093/texcom/tgaa082.

Paul S, Austin J, Elliott R, Ellison-Wright I, Wan MW, Drake R, Downey D, Elmadih A, Mukherjee I, Heaney L, Williams S, Abel KM. Neural pathways of maternal responding: systematic review and meta-analysis. Arch Womens Ment Health. 2019;22(2):179–87. https://doi.org/10.1007/s00737-018-0878-2.

Paus T. Growth of white matter in the adolescent brain: myelin or axon? Brain Cogn. 2010;72(1):26–35. https://doi.org/10.1016/j.bandc.2009.06.002.

Pawluski JL, Galea LAM. Reproductive experience alters hippocampal neurogenesis during the postpartum period in the dam. Neuroscience. 2007;149(1):53–67. https://doi.org/10.1016/j.neuroscience.2007.07.031.

Pawluski JL, Lambert KG, Kinsley CH. Neuroplasticity in the maternal hippocampus: relation to cognition and effects of repeated stress. Horm Behav. 2016;77:86–97. https://doi.org/10.1016/j.yhbeh.2015.06.004.

Pawluski JL, Lonstein JS, Fleming AS. The neurobiology of postpartum anxiety and depression. Trends Neurosci. 2017;40(2):106–20. https://doi.org/10.1016/j.tins.2016.11.009.

Pawluski JL, Paravatou R, Even A, Cobraiville G, Fillet M, Kokras N, Charlier TD. Effect of sertraline on central serotonin and hippocampal plasticity in pregnant and non-pregnant rats. Neuropharmacology. 2020;166:107950. https://doi.org/10.1016/j.neuropharm.2020.107950.

Pawluski JL, Swain JE, Lonstein JS. Neurobiology of peripartum mental illness. Handb Clin Neurol. 2021a;182:63–82. https://doi.org/10.1016/B978-0-12-819973-2.00005-8.

Pawluski JL, Hoekzema E, Leuner B, Lonstein JS. Less can be more: fine tuning the maternal brain. Neuroscience & Biobehavioral Reviews. 2021b. https://doi.org/10.1016/j.neubiorev.2021.11.045.

Pines D. Pregnancy and motherhood: interaction between fantasy and reality. Br J Med Psychol. 1972;45(4):333–43. https://doi.org/10.1111/j.2044-8341.1972.tb02216.x.

Rapoport E, Muthiah N, Keim SA, Adesman A. Family Well-being in grandparent- versus parent-headed households. Pediatrics. 2020;146(3):e20200115. https://doi.org/10.1542/peds.2020-0115.

Ribeiro M, Rumor M, Oliveira L, O'Neill M, J. G. & Maurício, J. C. STIR, SPIR and SPAIR techniques in magnetic resonance of the breast: A comparative study. J Biomed Sci Eng. 2013;06(03):395–402. https://doi.org/10.4236/jbise.2013.63A050.

Riccomagno MM, Kolodkin AL. Sculpting neural circuits by axon and dendrite pruning. Annu Rev Cell Dev Biol. 2015;31:779–805. https://doi.org/10.1146/annurev-cellbio-100913-013038.

Rilling JK, Richey L, Andari E, Hamann S. The neural correlates of paternal consoling behavior and frustration in response to infant crying. Dev Psychobiol. 2021;2021(00):1–14. https://doi.org/10.1002/dev.22092.

Rilling JK, Gonzalez A, Lee M. The neural correlates of grandmaternal caregiving. Proceedings of the Royal Society B, 2021;288(1963):20211997.

Robinson MJF, Berridge KC. Wanting versus needing. In: Wright JD, editor. International encyclopedia of the social & behavioral sciences. 2nd ed. Oxford: Elsevier; 2015. p. 351–6.

Rocchetti M, Radua J, Paloyelis Y, Xenaki L, Frascarelli M, Caverzasi E, Politi P, Fusar-Poli P. Neurofunctional maps of the 'maternal brain' and the effects of oxytocin: A multimodal

voxel-based meta-analysis. Psychiatry Clin Neurosci. 2014;68(10):733–51. https://doi.org/10.1111/pcn.12185.
Rogers CR, McCormick EM, van Hoorn J, Ivory SL, Telzer EH. Neural correlates of sibling closeness and association with externalizing behavior in adolescence. Soc Cogn Affect Neurosci. 2018;13(9):977–88. https://doi.org/10.1093/scan/nsy063.
Rutherford HJ, Mayes LC. Primary maternal preoccupation: using neuroimaging techniques to explore the parental brain. Psyche-Zeitschrift Fur Psychoanalyse Und Ihre Anwendungen. 2011;65(9–10):973–88.
Rutherford HJV, Byrne SP, Austin GM, Lee JD, Crowley MJ, Mayes LC. Anxiety and neural responses to infant and adult faces during pregnancy. Biol Psychol. 2017;125:115–20. https://doi.org/10.1016/j.biopsycho.2017.03.002.
Rosenblatt JS. Nonhormonal basis of maternal behavior in the rat. Science, 1967;156(3781):1512–1514.
Sacher J, Rekkas PV, Wilson AA, Houle S, Romano L, Hamidi J, Rusjan P, Fan I, Stewart DE, Meyer JH. Relationship of monoamine oxidase-A distribution volume to postpartum depression and postpartum crying. Neuropsychopharmacology. 2015;40(2):429–35. https://doi.org/10.1038/npp.2014.190.
Sacher J, Chechko N, Dannlowski U, Walter M, Derntl B. The peripartum human brain: current understanding and future perspectives. Front Neuroendocrinol. 2020;59:100859. https://doi.org/10.1016/j.yfrne.2020.100859.
Saper CB, Lowell BB. The hypothalamus. Curr Biol. 2014;24(23):R1111–6. https://doi.org/10.1016/j.cub.2014.10.023.
Seeley WW. The salience network: A neural system for perceiving and responding to homeostatic demands. J Neurosci. 2019;39(50):9878–82. https://doi.org/10.1523/JNEUROSCI.1138-17.2019.
Shrestha S, Adachi K, Petrini MA, Shrestha S. Factors associated with post-natal anxiety among primiparous mothers in Nepal. Int Nurs Rev. 2014;61(3):427–34. https://doi.org/10.1111/inr.12118.
Silver M, Moore CM, Villamarin V, Jaitly N, Hall JE, Rothschild AJ, Deligiannidis KM. White matter integrity in medication-free women with peripartum depression: a tract-based spatial statistics study. Neuropsychopharmacology. 2018;43(7):1573–80. https://doi.org/10.1038/s41386-018-0023-y.
Silverman ME, Loudon H, Safier M, Protopopescu X, Leiter G, Liu X, Goldstein M. Neural dysfunction in postpartum depression: an fMRI pilot study. CNS Spectr. 2007;12(11):853–62.
Silverman ME, Loudon H, Liu X, Mauro C, Leiter G, Goldstein MA. The neural processing of negative emotion postpartum: a preliminary study of amygdala function in postpartum depression. Arch Womens Ment Health. 2011;14(4):355–9. https://doi.org/10.1007/s00737-011-0226-2.
Slattery DA, Hillerer KM. The maternal brain under stress: consequences for adaptive peripartum plasticity and its potential functional implications. Front Neuroendocrinol. 2016;41:114–28. https://doi.org/10.1016/j.yfrne.2016.01.004.
Spear LP. Adolescent neurodevelopment. J Adolesc Health. 2013;52(2):S7–S13. https://doi.org/10.1016/j.jadohealth.2012.05.006.
Stickel S, Wagels L, Wudarczyk O, Jaffee S, Habel U, Schneider F, Chechko N. Neural correlates of depression in women across the reproductive lifespan – an fMRI review. J Affect Disord. 2019;246:556–70. https://doi.org/10.1016/j.jad.2018.12.133.
Stolzenberg DS, Champagne FA. Hormonal and non-hormonal bases of maternal behavior: the role of experience and epigenetic mechanisms. Horm Behav. 2016;77:204–10. https://doi.org/10.1016/j.yhbeh.2015.07.005.
Swain JE. The human parental brain: in vivo neuroimaging. Prog Neuropsychopharmacol Biol Psychiatry. 2011; https://doi.org/10.1016/j.pnpbp.2010.10.017.
Swain JE, Tasgin E, Mayes LC, Feldman R, Constable RT, Leckman JF. Maternal brain response to own baby-cry is affected by cesarean section delivery. J Child Psychol Psychiatry. 2008;49(10):1042–52.

Swain JE, Kim P, Spicer J, Ho SS, Dayton CJ, Elmadih A, Abel KM. Approaching the biology of human parental attachment: brain imaging, oxytocin and coordinated assessments of mothers and fathers. Brain Res. 2014; https://doi.org/10.1016/j.brainres.2014.03.007.

Thorsness KR, Watson C, LaRusso EM. Perinatal anxiety: approach to diagnosis and management in the obstetric setting. Am J Obstet Gynecol. 2018;219(4):326–45. https://doi.org/10.1016/j.ajog.2018.05.017.

van't Veer AE, Thijssen S, Witteman J, van IJzendoorn MH, Bakermans-Kranenburg MJ. Exploring the neural basis for paternal protection: an investigation of the neural response to infants in danger. Soc Cogn Affect Neurosci. 2019;14(4):447–57. https://doi.org/10.1093/scan/nsz018.

Voldsbekk I, Barth C, Maximov II, Kaufmann T, Beck D, Richard G, Moberget T, Westlye LT, de Lange A-MG. A history of previous childbirths is linked to women's white matter brain age in midlife and older age. Hum Brain Mapp. 2021; https://doi.org/10.1002/hbm.25553.

Von Mohr M, Mayes LC, Rutherford HJ. The transition to motherhood: psychoanalysis and neuroscience perspectives. Psychoanal Study Child. 2017;70(1):154–73. https://doi.org/10.1080/00797308.2016.1277905.

Wang X, Wang J, Liu Z, Ming Y, Zhang S. Increased posterior cingulate, medial frontal and decreased temporal regional homogeneity in depressed mothers. A resting-state functional magnetic resonance study. Procedia Environ Sci. 2011;8:737–43.

Weisner TS. Socialization for parenthood in sibling caretaking societies. In: Lancaster JB, Altmann J, Rossi AS, Sherrod LR, Lancaster JB, editors. Parenting across the life span. 1st ed. Routledge; 2017. p. 237–70.

Winnicott DW. Primary maternal preoccupation. The Maternal Lineage: Identification, Desire, and Transgenerational Issues; 1956. p. 59–66.

Wonch KE, de Medeiros CB, Barrett JA, Dudin A, Cunningham WA, Hall GB, Steiner M, Fleming AS. Postpartum depression and brain response to infants: differential amygdala response and connectivity. Soc Neurosci. 2016:1–18. https://doi.org/10.1080/17470919.2015.1131193.

Zeng Y, Ni Z, Liu S, Gu X, Huang Q, Liu J, Wang Q. Parity and all-cause mortality in women and men: a dose-response meta-analysis of cohort studies. Sci Rep. 2016;6(1):19351. https://doi.org/10.1038/srep19351.

Zhang K, Wang M, Zhang J, Du X, Chen Z. Brain structural plasticity associated with maternal caregiving in mothers: a voxel- and surface-based morphometry study. Neurodegener Dis. 2019a;19(5–6):192–203. https://doi.org/10.1159/000506258.

Zhang K, Wang M, Zhang J, Xie J, Su X, Du X, Chen Z. Dynamic alterations in spontaneous brain activity in mothers: a resting-state functional magnetic resonance imaging study. Neurosci Bull. 2019b;35(4):766–70. https://doi.org/10.1007/s12264-019-00392-7.

Zhang S, Wang W, Wang G, Li B, Chai L, Guo J, Gao X. Aberrant resting-state interhemispheric functional connectivity in patients with postpartum depression. Behav Brain Res. 2020;382:112483. https://doi.org/10.1016/j.bbr.2020.112483.

Zheng J-X, Chen Y-C, Chen H, Jiang L, Bo F, Feng Y, Tang W-W, Yin X, Gu J-P. Disrupted spontaneous neural activity related to cognitive impairment in postpartum women. Front Psychol. 2018;9:624. https://doi.org/10.3389/fpsyg.2018.00624.

Integrative Studies of the Effects of Mothers on Offspring: An Example from Wild North American Red Squirrels

Ben Dantzer, Stan Boutin, Jeffrey E. Lane, and Andrew G. McAdam

Abstract Animal species vary in whether they provide parental care or the type of care provided, and this variation in parental care among species has been a common focus of comparative studies. However, the proximate causes and ultimate consequences of within-species variation in parental care have been less studied. Most studies about the impacts of within-species variation in parental care on parental fitness have been in primates, whereas studies in laboratory rodents have been invaluable for understanding what causes inter-individual variation in parental care and its influence on offspring characteristics. We integrated both of these perspectives in our long-term study of North American red squirrels (*Tamiasciurus hudsonicus*) in the Yukon, Canada, where we have focused on understanding the impacts of mothers on offspring. This includes documenting the impacts that mothers or the maternal environment itself has on their offspring, identifying how changes in maternal physiology impact offspring characteristics, the presence of individual variation in maternal attentiveness toward offspring before weaning and its fitness consequences, and postweaning maternal care and its fitness consequences. We provide an overview of these contributions to understanding the impacts mothers have on their offspring in red squirrels using an integrative framework and contrast them with studies in the laboratory.

Keywords Glucocorticoids · HPA axis · Lifetime reproductive success · Maternal behavior · Parental care · Parental effects

---

B. Dantzer (✉)
Department of Psychology, University of Michigan, Ann Arbor, MI, USA

Department of Ecology and Evolutionary Biology, University of Michigan, Ann Arbor, MI, USA
e-mail: dantzer@umich.edu

S. Boutin
Department of Biological Sciences, University of Alberta, Edmonton, AB, Canada

J. E. Lane
Department of Biology, University of Saskatchewan, Saskatoon, SK, Canada

A. G. McAdam
Department of Ecology and Evolutionary Biology, University of Colorado, Boulder, CO, USA

G. González-Mariscal (ed.), *Patterns of Parental Behavior*, Advances in Neurobiology 27, https://doi.org/10.1007/978-3-030-97762-7_9

1 Introduction

Parental care describes all the ways in which parents influence offspring development in such a way that benefits the offspring's survival. Parental care manifests itself across species in diverse ways (Balshine 2012). In mammals, it can be described as the investment parents make toward their offspring prior to and after parturition (e.g., intrauterine growth and development, nest site construction and defense, nursing, post-parturition feeding) and attendance and attention to dependent offspring by parents (e.g., behavioral interactions, social learning, inheritance of material resources or social rank, etc.). The evolution of parental care has been a major focus of behavioral ecology for more than 50 years. Popular books and comparative studies have described the presence or absence of parental care across diverse taxonomic groups, its potential fitness costs and benefits, and the possible environmental circumstances favoring its evolution (Clutton-Brock 1991; Maestripieri and Mateo 2009; Royle et al. 2012; Clutton-Brock 2016). In many cases, this is done through theoretical models or phylogenetic comparative studies that focus on understanding the factors (environmental, life history, etc.) that may favor the evolution of different parental care patterns among closely related species (e.g., Badyaev and Ghalambor 2001; Mank et al. 2005; Gonzalez-Voyer et al. 2008; Brown et al. 2010; Klug and Bonsal 2010; Klug et al. 2012; Furness and Capellini 2019) or testing hypotheses and duration and type of specfific forms of parental care provided within specific taxonomic groups (Gross 2005; Schulz and Bowen 2005; Skibiel et al. 2013).

These studies have provided foundational knowledge on the causes of among-species variation in parental care, yet there are relatively few studies of within-species parental care and its fitness consequences in wild mammal species. Similar to other behavioral traits (Réale et al. 2007), females can exhibit consistent differences in their behavior toward offspring, which have been referred to as "maternal styles" (Fairbanks 1996). For example, in captive primates, laboratory rodents, and domesticated animals, mothers exhibit consistent differences in their level of rejection, attentiveness, contact, or protectiveness toward offspring (Fairbanks 1989; Schino et al. 1995; Maestripieri 1998; Albers et al. 1999; Dwyer and Lawrence 2000; Spinka et al. 2000; Champagne et al. 2003). However, it has been less common to formally examine if maternal styles exist in the wild other than in several primate species (e.g., Tanaka 1989; Berman 1990) and a few non-primate species (e.g., dolphins; Hill et al. 2007; Stewart and McAdam 2014) and relatively little is known about their fitness consequences. This is largely a consequence of the difficulties of observing interactions between parents and offspring in non-primate mammalian species in the wild. Much of these interactions occur in secret while parents and offspring are inside of a nest or otherwise obscured from view.

Even more surprising is the lack of attention paid to understanding the causes of individual differences in parental care within species in mammals. The latter is particularly noticeable given the number of studies investigating proximate causes of individual variation in parental care in laboratory rodents (Numan and Insel 2003;

Numan 2007). For example, studies in laboratory rodents (rats, mice, voles, gerbils) have highlighted a number of potential proximate mechanisms that contribute to individual variation in the expression of parental care, such as estrogen receptor alpha expression (Champagne et al. 2003; Li et al. 2015), androgen receptors or titers (Clark and Galef 1999; Trainor and Marler 2002), oxytocin receptor expression (Pedersen et al. 1994; Pedersen and Boccia 2003; Olazábal and Young 2006; Keebaugh et al. 2015), and hypothalamic-pituitary-adrenal (HPA) axis activity due to perinatal stress (Bales et al. 2006; Brummelte and Galea 2010; Harris et al. 2013), among other potential mechanisms. Similar work in a broader diversity of mammalian species and in free-living individuals has greatly lagged behind. This of course lessens our understanding of how this variation in maternal behavior is maintained, but also if the results from studies in laboratory rodents are idiosyncratic or reflect a general pattern found in other species.

Here, we argue that there is a gap in our understanding of the causes and consequences of within-species variation in parental care in mammalian species. Although studies in laboratory rodents, primates, and domestic animals have been invaluable for documenting the presence and proximate causes of individual variation in parental care, its fitness consequences are not well documented given the challenges of quantifying them in natural populations. This is problematic because it is intraspecific variation that ultimately can produce interspecific variation, yet we have little understanding of whether the same features (environmental or life history) operate across these two levels. For example, if certain types of ecological conditions give rise to interspecific differences in the presence or type of parental care, do they also give rise to the same intraspecific patterns in parental care? Addressing this issue can help connect microevolutionary to macroevolutionary processes and address if explanations for the evolution of parental care generated at one level (among species) provide predictive power for other levels (within species; *sensu* Levin 1992). Moreover, a lack of understanding of the proximate factors that modulate parental care in mammals limits the generality of findings from laboratory rodents where there is paucity of genetic variation compared to wild species.

We discuss research into documenting the causes and consequences of individual variation in maternal care in wild North American red squirrels (*Tamiasciurus hudsonicus*) in the Yukon, Canada (Fig. 1). Over the past 32 years, we have documented the individual life histories of both female and male red squirrels in this study area (McAdam et al. 2007; Dantzer et al. 2020a). This has included documenting the impacts that mothers or the maternal environment has on their offspring (referred to here as maternal effects), identifying how changes in maternal physiology impact offspring, the presence of individual variation in maternal attentiveness toward offspring before weaning and its fitness consequences, and postweaning maternal care (territory bequeathal) and its fitness consequences. Below we summarize these contributions to understanding of the impacts of mothers on offspring in wild red squirrels to provide a comprehensive assessment of what we currently know and do not know.

Fig. 1 As a part of the Kluane Red Squirrel Project, individually marked North American red squirrels (*Tamiasciurus hudsonicus*) in Kluane, Yukon, Canada, are followed from birth to death using a combination of live trapping and behavioral observations. A lactating female is shown in this image with a VHF radio collar around her neck. Her nest was located using radiotelemetry, and she watched us while we temporarily removed her offspring from her nest before placing them after processing. After returning pups to the nest, females will return and move them to a new nest. We quantify this latency to retrieve pups and use it as an estimate of maternal attentiveness (Westrick et al. 2020). (Photo by Ben Dantzer)

2 Kluane Red Squirrel Project

We study a natural population of red squirrels in Kluane, Yukon, Canada (61° N, 138° W), within the traditional territory of the Champagne and Aishihik First Nations. Further details of the study area can be found elsewhere (Boonstra et al. 2001;

Dantzer et al. 2020a). Since the start of the project, we have followed standardized and consistent protocols to document the life histories of individually marked red squirrels over their lifetime (McAdam et al. 2007; Dantzer et al. 2020a). All squirrels in our study areas (two to six different study areas depending upon the year) are marked permanently with metal ear tags with unique codes on them as well as small pieces of colored materials (such as electrical wire) threaded through these ear tags. The latter allows us to assess squirrel identities without capture as squirrels are diurnal and visually conspicuous (Williams et al. 2014a). Most squirrels receive these ear tags as juveniles when they are still in their natal nest with their mother. Throughout most of the year (March to October), squirrels are monitored through live capture and behavioral observations. Squirrels are captured and temporarily handled to assess identity, sex, weight, and reproductive status. Females are captured in the late stages of pregnancy or early lactation, and a VHF radio collar is applied. We then locate their nests (containing pups) when they are lactating and retrieve the pups two separate times. This provides us with information on birthdates, litter size, and litter sex ratio. During the second nest entry, pups receive their uniquely labeled metal ear tags. The change in body mass from the first nest entry (typically when pups are only 1–2 days of age) from the second nest entry (when pups are ~25 days of age) provides us with a measure of offspring growth that occurs during a linear period of growth when offspring are only consuming milk from their mother (McAdam et al. 2002). Offspring survival is assessed through the same procedures where behavioral observations and live trapping are used to confirm if offspring from previous years survived their first winter (McAdam and Boutin 2003). We then can estimate both annual and lifetime measures of reproductive success by quantifying either the number of offspring produced (either per year or over their lifetime; Haines et al. 2020; Martinig et al. 2020) or the number of offspring produced that acquire a territory and survive their first winter (number of recruits either annually or over lifetime; Dantzer et al. 2013; Hämäläinen et al. 2019; Fisher et al. 2019; McAdam et al. 2019; Siracusa et al. 2021).

In addition to these individual life history data, each year we also collect information on the population ecology of red squirrels at our study areas. In the spring and autumn of each year, we conduct an annual census of all squirrels on the study areas to measure squirrel density (Descamps et al. 2009; Dantzer et al. 2013, 2020a). In the autumn of each year, we estimate the availability of the major food source of red squirrels, seeds from white spruce (*Picea glauca*) trees (Fletcher et al. 2013; Ren et al. 2017), by counting the number of spruce cones on ~450 individual spruce trees distributed across our study areas (Boutin et al. 2006; LaMontagne and Boutin 2007; McAdam et al. 2019; Dantzer et al. 2020a). When the seeds inside the spruce cones are mature (usually in late August), squirrels collect and cache these cones underground (Fletcher et al. 2010). More recently, we have estimated the number of spruce cones that each squirrel collects and hoards (Larivée et al. 2010; Donald and Boutin 2011; Fisher et al. 2019). These data on squirrel density and food abundance have provided us with an opportunity to examine how these ecological factors induce selection on life history traits in squirrels, such as documenting plasticity in parturition dates, litter size, number of litters, or offspring growth rates and their fitness consequences (McAdam and Boutin 2003; Réale et al. 2003; Boutin et al. 2006; Dantzer et al. 2013; Fisher et al. 2017; McAdam et al. 2019).

3 North American Red Squirrels

Red squirrels at our study sites in the Yukon are a relatively small bodied species (adults are ~250 g) and exhibit a short lifespan (median lifespan of squirrels that acquire a territory as a juvenile = 3.5 years, maximum lifespan = 8 years; McAdam et al. 2007). They are mostly nonsocial and both females and males at our study site exhibit year-round territoriality (Smith 1968; Dantzer et al. 2012; Siracusa et al. 2019). Each individual squirrel defends a nonoverlapping territory containing a larder hoard of white spruce cones that is cached underground and at the center of the territory (known as a midden). Squirrels collect and cache these cones in the autumn of each year, though the number of spruce cones produced in any year is highly variable because white spruce is a masting tree species (Boutin et al. 2006; LaMontagne and Boutin 2007; Fletcher et al. 2010; Dantzer et al. 2020a). Squirrels defend their territories using specific vocalizations called "rattles" (Smith 1978; Siracusa et al. 2017a), and territorial intrusions are usually infrequent when the owner is at the territory (Dantzer et al. 2012; Siracusa et al. 2017b). Juveniles usually need to acquire a territory (containing a midden) to survive their first winter (Larsen and Boutin 1994; Hendrix et al. 2020), and once squirrels acquire a territory, they rarely move (Larsen and Boutin 1995; Berteaux and Boutin 2000).

The timing of mating at our study areas in Yukon occurs anywhere from the winter to summer months and is dependent upon squirrel density and spruce cone production in the previous year (Boutin et al. 2006; McAdam et al. 2019; Dantzer et al. 2020a). Red squirrels are polygamous with a scramble competition mating system (Lane et al. 2009). Females are thought to be spontaneous ovulators, and when they are in estrus, several to many males arrive at a female's territory and attempt to mate with her (Smith 1968; Lane et al. 2008). Litters often contain offspring from more than one sire (Lane et al. 2008, 2009).

Red squirrels in Yukon usually only produce one successful litter in each year (Boutin et al. 2006). Females may attempt a second litter if their first litter fails, but only in anticipation of mast cone crops do they produce two (or more) litters successfully (Boutin et al. 2006; Williams et al. 2014b; McAdam et al. 2019). Litter size is variable but is usually 2–3 pups except in anticipation of a mast cone crops where litter sizes increase (Boutin et al. 2006; McAdam et al. 2019; Dantzer et al. 2020a). Gestation is around 35 days, and offspring are altricial and reared by their mother inside of a nest constructed mostly of grass that may be attached to a tree limb, underground, or inside of a hollow tree. Only females provide parental care to offspring, and they do so through nest construction and selection (offspring are reared inside of the nest until they are mobile and independent), nursing, and, in some cases, bequeathing their territory (described below). Offspring first emerge from their nest on their own around ~42 days after birth, but they can continue to be nursed by their mother until they are ~70 days of age (Stuart-Smith and Boutin 1995; McAdam et al. 2007). Offspring often disperse from their natal territory after birth, unless the mother bequeaths the territory to them (see below). Offspring dispersal distances are usually quite low, often around 100 m away from their natal territory (Berteaux and Boutin 2000; Cooper et al. 2017). Mortality of these offspring is quite high, though it depends upon the year (Boutin and Larsen 1993; McAdam and Boutin 2003; McAdam et al. 2007; Hendrix et al. 2020), and, in most years, few offspring acquire a territory, which is again usually necessary for them to survive their first winter (Larsen and Boutin 1994; Hendrix et al. 2020).

4 Methods for Quantifying Effects of Mothers on Offspring and Maternal Behavior

Documenting the impacts of parents on offspring is anything but straightforward in most small mammal species given the difficulty of observing parental interactions. Thus, while a diversity of measures of parental care have been developed for laboratory rodents where these interactions can be readily observed and recorded, many of these are not easily applicable to wild mammals. In some situations, researchers have brought pregnant or lactating females into captivity to observe maternal-offspring interactions (Stewart and McAdam 2014), but this is not possible in red squirrels. Instead, we have focused on three major ways of identifying the impacts of mothers on their offspring, which we describe below.

Quantitative Genetics

When the degree of relatedness among individuals within a natural population is known, quantitative genetics approaches can be used to quantify the importance of additive genetic effects to total phenotypic effects as a heritability (Falconer and Mackay 1996; Charmantier et al. 2014). In populations for which offspring can be assigned to mothers and fathers either through behavioral observations or genetic similarity, the resulting pedigree can be used to infer relatedness and quantify sources of variation using "animal model" analyses (Kruuk 2004). This flexible statistical approach for quantifying sources of variation can also estimate the unique effect of the identity of the mother for the phenotypes of her offspring (i.e., maternal effects). Depending on the structure of the data, however, it can sometimes be challenging to disentangle genetic effects from maternal effects in unmanipulated populations (Kruuk and Hadfield 2007). For some species, it is possible to move eggs or offspring between nests to experimentally break apart associations between the parental environment and inheritance using cross-fostering experiments.

In our long-term study of red squirrels, we retrieve offspring from their mother's nest when they are still being nursed by their mother. Because adoption is very rare (Gorrell et al. 2010), pups that are inside of the nest are extremely likely to be the offspring of the mother that owns the nest where the pups were located and we assign paternity using microsatellite markers (Gunn et al. 2005). We have used this information on parentage to generate a 10-generation pedigree of over 6000 squirrels born into this population since 1989. We have used this multigenerational pedigree (and cross-fostering experiments in some circumstances) to quantify the sources of variation in offspring traits, such as offspring postnatal growth rates (McAdam et al. 2002; McAdam and Boutin 2003), behavioral characteristics (Taylor et al. 2012), or other traits (Réale et al. 2003; McFarlane et al. 2011, 2014; Fisher et al. 2019). We discuss the relevant results for quantifying sources of variation in offspring traits below and summarize them in Fig. 2.

In most of these analyses, we have then quantified the heritability of these traits (i.e., the proportion of total phenotypic variation in the trait that is caused by additive genetic effects) and the overall effect of the mother on offspring (i.e., the proportion of total variation that is caused by the identity of the mother). This variance-partitioning approach reflects the overall impact of the mother on the phenotypes of her offspring (Wolf and Wade 2009) but is agnostic to the particular traits of the mother or environments experienced by the mother that are causally

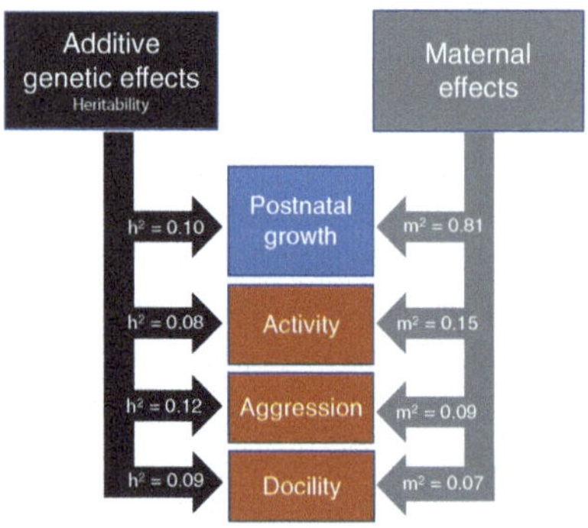

Fig. 2 Overview of heritabilities and maternal effects on postnatal growth rates and behavioral traits (docility, activity, aggression) using a quantitative genetics approach. Heritabilities (h^2) and maternal effects (m^2) were estimated using a multigenerational pedigree-based approach or cross-fostering (McAdam et al. 2002; Taylor et al. 2012). Maternal effects represent both the maternal environmental and maternal genetic effects. Highlighted here are our observation that heritabilities and maternal effects are nearly equal for behavioral traits, but offspring growth experiences very large maternal effects relative to its heritability, reflecting the potential for the mother to heavily influence offspring growth rates

influencing the offspring phenotype (McAdam et al. 2014). We have, therefore, also tested if the following specific features of the maternal phenotype or environment affect offspring characteristics.

Maternal Attentiveness

Laboratory studies in rodents often use pup retrieval latencies to quantify maternal behavior (Numan and Insel 2006; Champagne et al. 2007). In these situations, pups are removed from the nest (such as moved to the other side of the housing container), and the amount of time it takes for the pups to be retrieved and put back into the nest is used to estimate parental attentiveness (e.g., Seip and Morrell 2008). We have adopted this approach in red squirrels where we quantify the amount of time it takes for females to return to their nest to retrieve pups following our two nest entries (described above). Red squirrels often have more than one inhabitable nest on their territory, and, following a nest intrusion (by human observers), they will return to their nest and move pups individually to another nest. We believe this is an antipredator behavioral response to nest predators such as mustelids that can climb trees and access squirrel nests (Studd et al. 2015), although we have never directly observed mustelids killing squirrels or visiting their nests at our study sites. Maternal attentiveness characterizes the latency from after the pups were returned to the nest until the mother returns to move the pups to a new nest (Westrick et al. 2020). We have previously found that females exhibit repeatable differences in maternal attentiveness where some females exhibit consistently shorter latencies to retrieve pups following the nest disturbance than others (Westrick et al. 2020). On average, females took about 394 s to return to their nest containing their pups after an observer put the pups back into the nest. There was variability among females, but significant within-female repeatability ($R = 0.25$, 95% confidence interval = 0.13–0.36; Westrick et al. 2020).

Maternal Physiology

Changes in maternal hormone levels can influence offspring characteristics either by direct transmission of the hormone, changes to maternal behavior, or inducing changes in offspring behavior (Moore and Power 1986; Weinstock et al. 1992; Barbazanges et al. 1996; Champagne and Meane 2006; Wilcoxon and Redei 2007). These effects on offspring can be profound in laboratory studies, exerting potentially long-lasting changes in offspring morphology, physiology, and behavior (Weinstock 2008; Harris and Seckl 2011). For the past 15 years, we have focused on examining the impacts of variation in maternal glucocorticoid levels (GCs) on offspring characteristics in wild red squirrels. This is because maternal GCs respond to changes in the ecological agents of natural selection in this population, such as squirrel density (Dantzer et al. 2013; Guindre-Parker et al. 2019). We measure maternal GCs by first acquiring fecal samples from underneath traps when females are captured and then using an enzyme immunoassay to quantify fecal glucocorticoid metabolite levels (Dantzer et al. 2010). These measures of fecal glucocorticoid metabolites are unaffected by temporary capture and restraint. This is because the time it takes for plasma GCs to be excreted in the feces in red squirrels is 10.9 ± 2.3 h, whereas squirrels are only in traps for a maximum of 1–2 h, and we do not measure fecal glucocorticoid metabolites in samples obtained within 72 h of live trapping (Dantzer et al. 2010). We have also quantified fecal androgen metabolite levels in breeding females and found that they peak around mid-lactation, which occurs at the same time when nest use (where pups are located) is lowest and territory defense behavior is highest (Dantzer et al. 2011). This suggests that increased androgens in female red squirrels mediate a trade-off between investment in reproduction and self-maintenance.

5 Impact of Mothers on Offspring in Red Squirrels

We have focused intensively on offspring growth rates as they are often closely related to the ability of offspring to acquire a territory and survive their first winter (McAdam and Boutin 2003; Fisher et al. 2017; Hendrix et al. 2020) and accordingly sometimes experience strong selection (McAdam and Boutin 2003; Dantzer et al. 2013; Fisher et al. 2017). We also summarize recent work describing the association between maternal attentiveness and maternal reproductive success, which is itself the outcome of offspring survival.

Offspring Postnatal Growth Rates

We have been able to quantify offspring postnatal growth rates of thousands of individual red squirrels using the method described above where we estimate the change in offspring body mass from soon after birth until the offspring are approximately 25 days of age. Although offspring postnatal growth rates may serve as a proxy of parental investment (especially in studies in wild birds), this ignores the contributions to growth from offspring themselves (such as additive genetic effects of parents on growth rates). Consequently, we have instead considered how mothers or their characteristics impact offspring growth rates, which may or may not accurately estimate maternal investment in this species.

Offspring postnatal growth rates in red squirrels exhibit relatively low heritability ($h^2 = 0.10 \pm 0.001$; McAdam et al. 2002) and are highly responsive to changes in food availability (McAdam and Boutin 2003; Dantzer et al. 2013, 2020a, b). Of particular note is that maternal effects (the genetic and environmental impacts of mothers on offspring) explain a much higher proportion of the variance in offspring growth rates than direct genetic effects of mothers (McAdam et al. 2002; McAdam and Boutin 2003). Specifically, the identity of the mother explained up to 81% of the variation in offspring growth rate (McAdam et al. 2002). This indicates that mothers, whether it be specific characteristics of their phenotype or their environment, play an important role in shaping offspring growth rates (Fig. 2).

A previous study highlighted that the vast majority of variation in offspring growth rates explained by fixed effects in statistical models (69%) was explained by two characteristics of the mother: litter size and the birthdates of the offspring (McAdam et al. 2002). Here, we discuss three additional possible mechanisms by which mothers influence offspring growth rates, two involving maternal behavior (nest site selection and attentiveness) and one caused by variation in maternal hormone levels. First, female red squirrels can influence offspring growth rates through nest site selection or attentiveness toward offspring. As mentioned above, lactating female red squirrels often have multiple nests on or near their territory. Females seem to place offspring in nests that provide the thermal environment that may optimize offspring growth and development by reducing their exposure to heat stress (Guillemette et al. 2009). Specifically, older pups with fur are located in nests with less insulation than those of younger pups without fur, and lactating females with dependent pups have them in less insulated nests when it is warmer outside of the nest (Guillemette et al. 2009). Second, females that are more attentive toward their offspring (i.e., exhibit a shorter latency to retrieve pups following nest disturbance) also produce offspring that grow quicker, though this depends upon litter size (Westrick et al. 2020). Red squirrels exhibit the basic life history trade-off between offspring size and number where individual offspring from larger litters grow more slowly than those from smaller litters (McAdam et al. 2002; Dantzer et al. 2013). However, we found that mothers that are more attentive can lessen this trade-off where pups from larger litters exhibited significantly faster growth rates if their mother was more attentive ($n = 524$ pups from 167 mothers, interaction between attentiveness and litter size for pups at the first nest entry: $t_{196.22} = -2.22$, $P = 0.028$; Westrick et al. 2020).

Finally, we have previously shown that mothers with elevated GCs during pregnancy, likely due to experiencing increased conspecific densities (Dantzer et al. 2013; Guindre-Parker et al. 2019), produce faster growing offspring. This is evident when GCs during pregnancy are naturally elevated due to females experiencing increased conspecific densities (Dantzer et al. 2013) or when GCs were experimentally increased during pregnancy (Dantzer et al. 2013, 2020a, b). For example, offspring from mothers treated with GCs during pregnancy grew 17.0% faster than those produced by control mothers, and there were no sex-specific treatment effects (Dantzer et al. 2020a, b). Interestingly, females with experimentally elevated GCs during lactation produced offspring that grew 34.8% slower than those produced by control mothers (Dantzer et al. 2020a, b). There were no sex-specific treatment effects, and the maternal treatments did not affect the structural size of offspring (Dantzer et al. 2020a, b). These results indicate the importance of the timing of exposure to elevated maternal GCs on offspring characteristics: elevated GCs during pregnancy promoted offspring growth, whereas elevated GCs during lactation decreased it (Fig. 3).

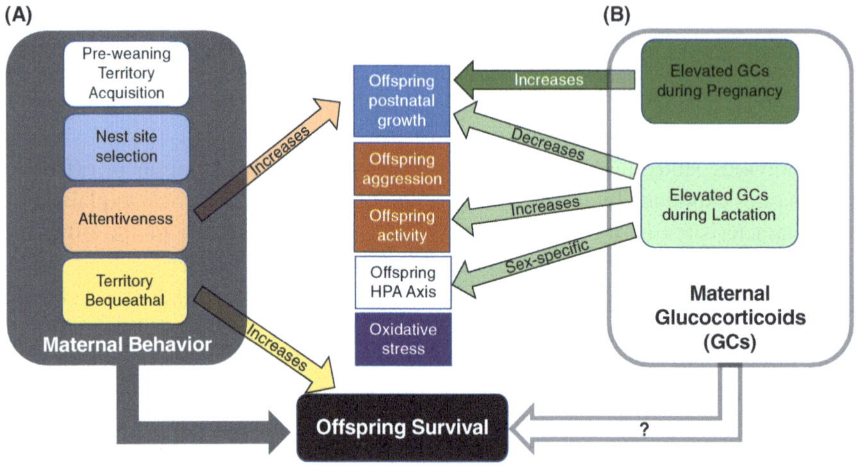

Fig. 3 Overview of effects of maternal behavior and glucocorticoids on offspring characteristics and survival in North American red squirrels. (**a**) We have quantified four maternal behaviors: where mothers acquire additional territories that they subsequently bequeath to offspring, nest site selection (thermal properties of nests where females raise pups), attentiveness (latency to return to the nest after disturbance), and territory bequeathal (where females give offspring their territory). It is likely that all of these behaviors influence offspring survival, such as females selecting the correct thermal environment for pups that optimizes the thermal environment for postnatal growth. Offspring survival is increased if their mother bequeaths their territory to them. Females generally exhibit a negative correlation between litter size and offspring growth rate, but those with higher attentiveness produce faster growing offspring. (**b**) We have also investigated how experimental elevations in maternal glucocorticoids (GCs) influence several traits in offspring. Elevated GCs during pregnancy increase offspring growth, but do not influence offspring behavior, hypothalamic-pituitary-adrenal (HPA) axis responsiveness to a stressor or ability to exert negative feedback following a stressor. Elevated GCs during lactation reduce offspring growth, but increase offspring activity and the ability of males (but not females) to exert negative feedback on the HPA axis following a stressor. There were no effects of elevated GCs during pregnancy or lactation on oxidative stress levels in offspring, which we quantified as antioxidants and oxidative damage in liver, heart, and blood as well as telomere lengths in DNA from liver tissue. We have yet to identify how GCs influence maternal behavior or if they affect offspring survival. The lack of arrows connecting maternal behavior or GCs and offspring traits indicates a lack of significant association. This summary is generated from previous studies (Boutin et al. 2000; Guillemette et al. 2009; Lane et al. 2015; Dantzer et al. 2020a, b; Westrick et al. 2020, 2021)

Offspring Behavior

Studies of laboratory rodents have shown extensive impacts of mothers on offspring behavior. The effects of perinatal stress experienced by the mother on offspring behavior have been well documented (Weinstock 2008). For example, female rats and mice that experience elevated gestational stress can produce offspring that exhibit higher anxiety-like behavior (Schulz et al. 2011; Salari et al. 2016). Additionally, behavioral traits in many animals appear to be linked with fitness

(Dingemanse and Réale 2005; Cote et al. 2008; Ballew et al. 2017; Lapiedra et al. 2018), and this is true in red squirrels as well (see below). Consequently, we have focused much attention on documenting the influence of mothers on offspring behavior through the following approaches: (1) using quantitative genetics techniques to partition the variance in offspring behavioral traits into heritable genetic effects, maternal effects, and permanent environmental effects (Fig. 2; Taylor et al. 2012) and (2) the impacts of increases in maternal GCs on offspring behavior (Fig. 3; Westrick et al. 2021). Before discussing these results, we first describe how we measure behavioral traits in red squirrels and their potential importance.

We have used struggle tests during handling (time spent struggling after removal from the trap) to quantify docility (Taylor et al. 2012), standardized open-field trials to quantify activity (Walsh and Cummins 1976; Martin and Réale 2008), and mirror image stimulation tests to quantify aggression (Svendsen and Armitage 1973). By testing the same individuals multiple times, we have shown that these three behavioral traits are significantly repeatable within individual squirrels (Taylor et al. 2012; Martinig et al. 2021). Specifically, in a study of 366 adult squirrels that were tested one to five times in our open-field trials and mirror image simulation tests to measure activity and aggression (n = 556 trials in total), respectively, the repeatability of activity was 0.51 (95% credible intervals (CI) = 0.40–0.63) and the repeatability of aggression was 0.44 (95% CI = 0.33–0.56; Taylor et al. 2012). Docility was also significantly repeatable (n = 3122 tests on 592 squirrels, R = 0.41, 95% CI = 0.36–0.49; Taylor et al. 2012). Although these results indicate within-individual consistency in these behaviors, we have subsequently showed age-related changes in these behavioral traits by measuring activity and aggression of juvenile red squirrels, either from initial emergence from the nest to weaning or from the weaning to yearling stage (Kelley et al. 2015; Martinig et al. 2021). For example, the repeatability of activity (R = 0.29, 95% CI = 0.23–0.33, n = 102 individuals) and aggression (R = 0.30, 95% CI = 0.24–0.34, n = 102 individuals) of juveniles early in life (from emergence to weaning) was lower than what we documented for adults (Martinig et al. 2021). One explanation for the observation that repeatabilities increase from the early developmental stage to adulthood is that those juveniles with inconsistent behavior have higher mortality (i.e., selectively disappear). However, we do not find support for this hypothesis, and these results instead suggest some within-individual developmental process (Martinig et al. 2021).

Are these behavioral traits that are measured under highly artificial (but standardized) conditions meaningful or important for red squirrels? We have shown that they do correspond to natural behavior as squirrels that were more active in the open-field trials also exhibited higher risk-taking behavior where they were trapped more often and at a larger number of locations on our study areas (Boon et al. 2008). Interestingly, females that were more active were less likely to survive to the following year (Boon et al. 2008), but they produced faster growing offspring at least in some years (Boon et al. 2007), and there is also some evidence (though nonsignificant) that females that were more active produced pups that were more likely to survive to the following year when conspecific densities were elevated (Taylor et al. 2014). Females that were more aggressive were more likely to produce offspring that survived their first winter (Boon et al. 2007) especially when conspecific density was elevated (Taylor et al. 2014). These studies suggested that females that were more active or aggressive had higher annual reproductive success, especially in some years or when density was high, but when assessing the association between female activity and aggression with lifespan and lifetime reproductive success (LRS; total number of pups produced), we found no significant association between female activity/aggression and lifespan or between activity and LRS but more aggressive females had higher LRS (Haines et al. 2020). Although we have not yet done similarly in-depth research on the effects of offspring activity/aggression on their survival as we have done for adult females, in a small sample size of juveniles (relative to our studies of adults), we found that more aggressive offspring were less to survive their first winter in one of two years (Martinig et al. 2021). We note that a consistent theme of this work is that the fitness consequences of variation in these behavioral traits seem to vary upon the study year (Boon et al. 2007; Taylor et al. 2014; Haines et al. 2020; Martinig et al. 2021), likely due to the pronounced fluctuations in food and density that squirrels experience.

Given that these behavioral traits seem to be linked to real-world behavior and fitness in our studies of adult red squirrels, we have focused on assessing the influence of mothers on offspring behavioral traits. First, we used a quantitative genetics approach to show that activity ($h^2 = 0.08$, 95% CI = 0.03–0.19, $n = 366$ individuals), aggression ($h^2 = 0.12$, 95% CI = 0.03–0.22, $n = 366$ individuals), and docility ($h^2 = 0.09$, 95% CI = 0.05–0.19, $n = 592$ individuals) are slightly but significantly heritable (Taylor et al. 2012). Given that selection on these traits seems to fluctuate by year (see above), these low heritabilities relative to other studies on the heritability of personality traits are not surprising (Taylor et al. 2012). Interestingly, we have also documented that these three behavioral traits experience significant maternal effects ($m^2 = 0.07–0.15$) that are as large as the heritability estimates (Fig. 2; Taylor et al. 2012). This indicates that the mother can have as large of an impact on these behavioral traits as additive genetic effects. In particular, offspring activity and aggression experienced a positive maternal effects correlation (0.58), suggesting that some specific feature about the mother generated the positive association between these two behavioral traits in offspring (Taylor et al. 2012). Once again, this approach does not allow us to assess the mechanism by which the maternal phenotype or environment influences offspring, but it does indicate variation among mothers in their effects on offspring behavioral traits and the ability for the maternal phenotype to generate both more aggressive and more active offspring.

We next focused on how changes in maternal hormones impacted offspring behavior. There is a large body of literature in laboratory rodents illustrating the profound influence of maternal stress or increases in maternal GCs on offspring behavior (Weinstock 2008). We focused on how elevated maternal GCs impact offspring behavior given that we had previously shown that they respond to conspecific density and affect offspring growth (see above). To do so, we experimentally elevated maternal GCs either during pregnancy or lactation. This caused a significant, but biologically realistic, elevation in circulating GCs over a 24 h period (van Kesteren et al. 2019). For example, in blood samples obtained within 3 min after a squirrel entered a trap (true baseline samples), plasma cortisol levels in cortisol-treated squirrels (treated with 8 mg hydrocortisone per day: 611:45 ± 104:4 ng/mL) were 185% higher than control squirrels (treated with 0 mg hydrocortisone per day: 214.5 ± 41.3 ng/mL; van Kesteren et al. 2019). These increases in circulating cortisol levels in red squirrels are similar to those obtained from control squirrels 60 min after we challenge them with adrenocorticotropin hormone that stimulates adrenal cortisol production (604.9 ± 93.6 ng/mL), indicating that our cortisol treatments simulated one stressful event. However, the true baseline cortisol levels of cortisol-treated squirrels drop back to being no different from those of controls about 24 h after treating them, so we provision females with these treatments once per day (van Kesteren et al. 2019).

Using this experimental paradigm, we found that weaned offspring from mothers treated with GCs during pregnancy did not differ in their activity (effect of treatment: $b = 0.03$, 95% confidence interval = −0.78–0.84, $P = 0.95$) or aggression ($b = 0.40$, 95% confidence interval = −1.09–0.30, $P = 0.28$) compared to controls. However, weaned offspring produced by mothers with elevated GCs during lactation exhibited significantly higher activity levels compared to controls ($b = 2.49$, 95% confidence interval = 1.04–3.94, $P = 0.01$), but there were no changes in aggression ($b = -0.81$, 95% confidence interval = −2.39–0.76, $P = 0.33$; Westrick et al. 2021). As in our studies of the effects of these maternal treatments on offspring growth, we found no sex-specific effects on offspring behavior (Fig. 3). These results indicate that changes in maternal GCs during pregnancy have no impacts on offspring activity and aggression, unlike many studies in laboratory rodents. Instead, some offspring behavior is affected by maternal GCs during lactation, and the significant positive association between offspring activity and aggression that we have previously documented (Taylor et al. 2012) does not appear to be influenced by maternal GCs as there was no evidence of a significant interaction between one of the behavioral variables and maternal treatment on the variation in the other behavioral variable (Westrick et al. 2021).

Offspring Physiology

Laboratory studies in rodents have illustrated how changes in the maternal phenotype (behavior, physiology, energetic state, etc.) or environment (exposure to stressors, food availability, etc.) can exert substantial changes in offspring physiology. As noted above, much of this work has focused on the impacts of perinatal stress or changes in maternal GCs on offspring physiology (Harris and Seckl 2011), and we have developed a parallel research program that investigates these questions in wild red squirrels. We next focus on studies about how elevated maternal GCs impact offspring HPA axis physiology, markers of oxidative stress (protein damage and telomere lengths), and antioxidant levels (Dantzer et al. 2020b; Westrick et al. 2021). We have focused on these responses because studies in laboratory rodents, humans, and other species indicate that offspring produced by mothers with elevated GCs (or exposed to perinatal stress) can exhibit hyper-responsiveness to stressors (Harris and Seckl 2011) or oxidative imbalance characterized by reduced levels of antioxidants and/or increased oxidative damage to proteins or telomeres (Haussmann and Heidinger 2015). The latter was particularly important to us as we had been interested in understanding if there were costs to offspring for being produced by a mother experiencing perinatal stress or with elevated GCs (Haussmann et al. 2012; Entringer et al. 2013; Blaze et al. 2017; Send et al. 2017) or for exhibiting faster somatic growth rates (Nussey et al. 2009; McLennan et al. 2016; Smith et al. 2016), all of which may carry some oxidative cost (e.g., reduced antioxidants, increased oxidative damage, shortened telomeres).

The HPA axis responsiveness of weaned offspring produced by female red squirrels treated with GCs either during pregnancy or lactation was first tested using a standardized HPA axis challenge procedure. An initial blood sample was obtained following capture of offspring, and then a second blood sample was obtained ~60 min following an injection with dexamethasone, which binds to GC receptors in the brain and allows us to assess the ability to terminate the production of GCs (van Kesteren et al. 2019). Indeed, plasma cortisol levels in both juvenile and adult red squirrels are substantially lower than cortisol levels measured in this initial blood sample, which is elevated due to the effects of trapping and handling stress (van Kesteren et al. 2019; Westrick et al. 2021). This provided us with our first response variable to assess, the *response to dexamethasone*, which was defined as the percentage reduction in plasma cortisol levels from the initial blood sample to cortisol levels measured in the sample obtained 60 min after dexamethasone injection. We then obtained two more blood samples 30 and 60 min after injection of the same squirrels with adrenocorticotropin hormone (ACTH), which stimulates adrenal production of GCs (van Kesteren et al. 2019). To characterize *HPA axis responsiveness*, we estimated the area under the curve from cortisol levels measured in the blood sampled following dexamethasone injection to cortisol levels measured in the sample obtained 60 min following ACTH injection (Westrick et al. 2021). There was some evidence of sex-specific effects of maternal GC treatments, but overall, there were minimal effects on the HPA axis in offspring (Westrick et al. 2021). Both females and males from mothers treated with GCs during pregnancy (effect of treatment: $b = 5.34$, 95% confidence interval = −12.48–23.16, $P = 0.56$) or lactation ($b = -15.48$, 95% confidence interval = −33.23–2.27, $P = 0.11$) had a similar HPA axis responsiveness to ACTH (area under the curve). For offspring from mothers treated during pregnancy, there was a significant interaction between treatment and sex on the response to dexamethasone ($b = 30.79$, 95% confidence interval = 9.01–52.57, $P = 0.01$). Specifically, males but not females that were from mothers treated with GCs during pregnancy exhibited significantly lower plasma cortisol levels after the dexamethasone administration compared with males from control mothers, indicative of their ability to exert greater negative feedback on the HPA axis ($b = -29.75$, $P = 0.002$). In females from mothers treated during pregnancy, there was no difference in plasma cortisol levels following dexamethasone administration between those from mothers treated with GCs or controls ($b = 1.04$, $P = 0.99$). Similarly, there were no treatment effects on the ability of offspring to exert negative feedback on the HPA axis in those from mothers treated with GCs during lactation ($b = -3.44$, 95% confidence interval = −27.40–20.51, $P = 0.78$), and there was no evidence of an interaction between sex of the offspring and maternal treatment ($b = 6.98$, 95% confidence interval = −25.42–39.39, $P = 0.68$).

Similar to the relative lack of effects on experimental elevation of maternal GCs on offspring HPA axis dynamics, we have also found few impacts of maternal GCs on offspring oxidative damage and antioxidant levels. Offspring from mothers treated with GCs during pregnancy (P-values for effect of treatment were > 0.32) or lactation (P-values for effect of treatment were > 0.11) did not differ from controls in total antioxidant capacity or superoxide dismutase levels (antioxidants levels) or protein carbonyls (oxidative protein damage) in blood samples, liver, or heart tissue (Dantzer et al. 2020b). There was also no impact of our maternal GC treatments on telomere lengths in DNA obtained from liver tissue where offspring from mothers treated with GCs during pregnancy (effect of treatment: $b = 0.42$, $t_{15.5}$, $P = 0.65$) or lactation (effect of treatment: $b = 0.52$, $t_{11.3}$, $P = 0.39$) exhibited similar telomere lengths to those from control mothers (Dantzer et al. 2020a, b). There was also no association between offspring postnatal growth rates and oxidative damage in any of the three tissues or telomere length (P-values for effect of treatment were > 0.22). Together, these results suggest that while maternal GCs can modulate offspring growth rates, it seems to do so without carrying any oxidative cost. This was not due to selective disappearance of low-quality offspring (Dantzer et al. 2020b) but could be because increased maternal GCs (or altered maternal behavior caused by the elevation in maternal GCs) promoted the production of enzymes (such as telomerase) that protect offspring telomeres from attrition (Beery et al. 2012; Blount et al. 2016; Noguera et al. 2020). Interestingly, these results also provide little evidence of an oxidative cost of growing quickly early in life (Fig. 3).

Offspring Survival

Most of the work described above has been focused on understanding how mothers influence offspring traits in such a way to make them better suited for the environment that offspring will encounter (i.e., adaptive plasticity induced by parental effects; Mousseau and Fox 1998; Badyaev and Uller 2009). However, we have also investigated more directly how specific attributes of maternal behavior may affect offspring survival. As indicated above, in most cases, offspring must acquire a territory prior to experiencing their first winter if they are to survive and "recruit" into the population (Larsen and Boutin 1994; Hendrix et al. 2020). We have previously found that mothers can increase the likelihood of offspring acquiring a territory and surviving their first winter through bequeathal, where mothers give a territory to their offspring (Price and Boutin 1993; Larsen and Boutin 1994; Berteaux and Boutin 2000; Lane et al. 2015). For most litters, offspring disperse away from their natal territory, but some mothers (on average ~19% of them per year; Lane et al. 2015) leave the territory where they produced a litter of offspring (while offspring are still present; Boutin et al. 1993) to acquire another territory that the mother then defends. Not all mothers exhibit bequeathal, but on average about 30.3% of females do so at least once in their lifetime (Lane et al. 2015). Mothers are more likely to bequeath their territory in anticipation of high food availability, usually reflecting an upcoming mast cone crop (effect of current year spruce cone abundance: $b = 0.54$, 95% credible interval = 0.25–0.76; Lane et al. 2015). Older mothers are also slightly more likely to bequeath their territory (effect of age: $b = 0.30$, 95% credible interval = 0.03–0.54; Lane et al. 2015). Interestingly, female red squirrels seem to acquire additional territories in anticipation of this possibility of bequeathal where they acquire vacant territories and then defend them up to 4 months before they have mated and up to 10 months prior to their offspring being weaned (Boutin et al. 2000). In an experimental study where the owners of territories adjacent to breeding females were permanently removed, Boutin et al. (2000) showed that females started to produce territorial vocalizations on or near the suddenly vacant territories a few hours after the owner was removed. These same females that took over those territories also owned them in the following spring, but this was not by leaving their previous territory and moving to the one that was newly vacated. Instead, some of the females expanded their territories and now defended their original one along with the one that had been vacated through experimental removal of the previous owner. Interestingly, among removed females (all between 2 and 4 years old), only those with prior breeding experience (had previously produced a litter of offspring) exhibited this behavior, whereas inexperienced females (which had never produced offspring) did not (Boutin et al. 2000). Among the females that acquired one of these additional territories, in most cases, an offspring settled on one of these acquired territories, suggesting that this served as type of anticipatory parental care where mothers preemptively accrued a resource for offspring up to 10 months before they would need it.

Territory bequeathal is beneficial to both offspring and mothers in terms of enhancing the number of offspring that survive their first winter (Berteaux and Boutin 2000) and enhances maternal reproductive success (Lane et al. 2015). Specifically, overwinter survival of juveniles (from soon after birth to the following spring) is around 26%, whereas the overwinter survival of juveniles that were bequeathed a territory from their mother soon after they were born increased to 79% (Fig. 3; Lane et al. 2015). Female red squirrels that exhibited bequeathal also experienced appreciable increases in their reproductive success as the number of offspring they produced that survived to the following spring relative to others in the study population was 1.42 pups, whereas it was only 0.90 pups in females that did not exhibit bequeathal, resulting in a significant increase in their relative reproductive success (Lane et al. 2015). Interestingly, females that bequeathed their territory to their offspring did not exhibit a difference in their own overwinter survival compared to females that did not bequeath, indicating that bequeathal increases female reproductive success without a reduction in their survival (Lane et al. 2015). Taken together, bequeathal behavior in red squirrels reflects postweaning parental care that is beneficial for both offspring (increases their survival over their first winter) and their mother (increases their relative reproductive success without a survival cost).

Finally, we have also discovered that female red squirrels that are more attentive to their offspring (quantified as the latency to retrieve pups following nest disturbances) have higher lifetime reproductive success (Westrick et al. 2020). Here, lifetime reproductive success is estimated as the total number of offspring surviving their first winter (i.e., recruiting into the population). Mothers that exhibited more attentiveness toward their offspring (returned to their nest more quickly after nest disturbance) had significantly higher lifetime reproductive success ($z = -2.09$, $P = 0.037$; Westrick et al. 2021). Based upon the estimated lifetime reproductive success of females that survive to breeding age (1.54 pups, McAdam et al. 2007), the effect size we found here shows that females that returned right after the pups were replaced into the nest (very high levels of attentiveness) had about a 1 more pup added to their lifetime reproductive success, which is a ~165% increase (Westrick et al. 2021). This study added a new dimension to the maternal styles literature from primates, laboratory rodents, and domestic animals (described above) highlighting that not only is there substantial individual variation among mothers in how attentive they are toward offspring but that this has important fitness consequences.

6 Future Directions

Red squirrels have provided an opportunity to examine the influences of mothers on offspring in nature. There remain a number of unanswered questions engendered by our research outcomes, and we discuss only a few of these below. First, why is there individual variation in maternal attentiveness? Females that are more attentive have higher lifetime reproductive success, so why do some females exhibit consistently lower attentiveness? This could reflect maternal state or their resources on hand whereby females in a better state or with greater levels of resources exhibit higher levels of attentiveness and have higher fitness (*sensu* van Noordwijk and de Jong 1986). Alternatively, high levels of attentiveness may be very costly for females if there is a nest predator (costly in that it may cause death), and so we might predict that younger females or those with higher residual reproductive value would exhibit lower attentiveness. Given that the number of nest predators at our study areas seems to vary by year (Studd et al. 2015), the costs and benefits of attentiveness may also vary by year, such as being more costly when nest predators are common. This would suggest the possibility of fluctuating selection acting upon maternal attentiveness, which we have found evidence for acting on other behavioral traits in females (Boon et al. 2007; Taylor et al. 2014). To date, we have not studied the costs of high levels of maternal attentiveness or maternal investment in offspring in general.

Second, we have focused extensively on the impacts of maternal GCs on offspring growth, behavior, and physiology. This was purposeful given the profound impacts that changes in maternal GCs have on offspring in the laboratory (Weinstock 2008; Harris and Seckl 2011; Haussmann and Heidinger 2015). Bringing it all together, we showed that offspring from mothers treated with GCs during pregnancy exhibited significantly faster postnatal growth, but there was no change in their HPA axis or behavior and no evidence for some type of oxidative cost. By contrast, offspring from mothers treated with GCs during lactation grew significantly slower and were more active, the HPA axis of males (but not females) exhibited stronger negative feedback, and again there was no oxidative cost of being produced by a female with elevated GCs (Fig. 3).

At present, we do not know if changes in maternal GCs impact offspring traits (such as growth) through direct transmission of the GCs to offspring, changes in maternal behavior, or changes in offspring behavior (or some combination). We also have a narrow understanding of how other features of maternal physiology change in response to varying environmental conditions and how they impact offspring characteristics. Red squirrels in the Yukon experience substantial variation in food abundance, conspecific density, and predators (Dantzer et al. 2020a), but we know relatively little about the physiological or behavioral responses to these environmental fluctuations besides how maternal GCs respond to changes in density. Thus, while we have documented that mothers influence offspring phenotypes (growth rates and behavior) and that maternal GCs may act as one mechanism of these changes in offspring growth, behavior, and physiology, our understanding of the mechanisms by which maternal phenotype or environment impacts offspring is still quite limited.

Finally, we have only scratched the surface for identifying how variation in nursing behavior, maternal care behaviors (e.g., offspring licking and grooming), or postweaning maternal care (bequeathal) impacts offspring characteristics and survival. This is largely because red squirrels, like most mammals, exhibit parental care mostly in secret while they are inside the nest. Whether or not our measure of maternal attentiveness correlates with other measures of maternal care (e.g., nursing duration) is not known at this time. Future studies, perhaps using accelerometers (Studd et al. 2019; Siracusa et al. 2019) or more sophisticated bio-loggers (Gaidica and Dantzer 2020), may reveal some of the ways in which maternal behavior shapes offspring.

7 Conclusions

We have adopted an integrative framework using concepts and methods from physiological ecology, behavioral neuroendocrinology, animal behavior, and evolutionary ecology to understand the causes and consequences of variation in maternal care in wild North American red squirrels. This has provided a few specific insights that can be compared to laboratory studies in rodents. First, aligned with studies in laboratory rodents, mothers do exert a significant impact on the growth and behavior of their offspring. This is of course not surprising, but it is noticeable that these effects are significant in light of all of the other sources of variation that can impact offspring phenotypes in nature. Second, in contrast to laboratory studies in rodents, we see few impacts of elevated maternal GCs on offspring behavior and HPA axis physiology or on oxidative damage and antioxidants. These effects may be visible if we are able to sample offspring across a greater number of developmental stages or use more biomarkers of HPA axis responsiveness and oxidative damage or antioxidants. Third, our research program shows how mothers may induce adaptive plasticity in offspring traits that could prepare them for future environments. Specifically, mothers with elevated GCs during pregnancy in response to increased conspecific densities produce faster growing offspring, which should increase offspring survival under high density conditions (Dantzer et al. 2013, 2020b). Additionally, mothers that are more attentive can lessen the trade-off between litter size and offspring growth rates, and those that are more attentive also exhibit higher lifetime reproductive success (Westrick et al. 2020). These studies provide a rare example from wild animals quantifying the fitness consequences of variation in maternal behavior and emphasize the need for laboratory studies to examine the ecological context in which offspring phenotypes are expressed if they are to suggest the valence of these impacts (i.e., adaptive or maladaptive; Sheriff et al. 2017). Doing so will require more work in species that experience fluctuating natural selection on offspring characteristics where plasticity in maternal behavior or other traits may be beneficial. More importantly, our studies as a whole reflect the benefits of a research program that synergizes results from laboratory studies and tests them in wild animals where fitness is quantifiable.

Acknowledgments We thank Agnes MacDonald and her family for long-term access to her trapline and Champagne and Aishihik First Nations for allowing us to conduct our work within their traditional territory. This project is the result of the hard work and creativity of numerous volunteers, field technicians, graduate students, and postdoctoral researchers, and we thank them for their efforts. We especially thank Sarah E. Westrick and Ryan W. Taylor for their contributions to understanding the impacts of mothers on offspring. This study was funded mostly by the Natural Sciences and Engineering Research Council (to SB, AGM, and JEL), National Science Foundation (IOS-1749627 to BD, DEB-0515849 to AGM), and University of Michigan (BD). This is publication #119 of the Kluane Red Squirrel Project.

References

Albers PCH, Timmermans PJA, Vossen JMH. Evidence for the existence of mothering styles in guinea pigs (*Cavia apererea f. porcellus*). Behaviour. 1999;136:469–79.

Badyaev AV, Ghalambor CK. Evolution of life histories along elevational gradients: trade-off between parental care and fecundity. Ecology. 2001;82:2948–60.

Badyaev AV, Uller T. Parental effects in ecology and evolution: mechanisms, processes and implications. Philos Trans R Soc Lond Ser B Biol Sci. 2009;364:1169–77.

Bales KL, Kramer KM, Lewis-Reese AD, Carter CS. Effects of stress on parental care are sexually dimorphic in prairie voles. Physiol Behav. 2006;87:424–9.

Ballew NG, Mittelbach GG, Scribner KT. Fitness consequences of boldness in juvenile and adult largemouth bass. Am Nat. 2017;189:396–406.

Balshine S. Patterns of parental care in vertebrates. In: Royle NJ, Smiseth PT, Kölliker M, editors. The evolution of parental care. Oxford: Oxford University Press; 2012. p. 62–80.

Barbazanges A, Piazza PV, Le Moal M, Maccari S. Maternal glucocorticoid secretion mediates long-term effects of prenatal stress. J Neurosci. 1996;16:3943–9.
Beery AK, Lin J, Biddle JS, Francis DD, Blackburn EH, Epel ES. Chronic stress elevates telomerase activity in rats. Biol Lett. 2012;8:1063–6.
Berman CM. Consistency in maternal behavior within families of free-ranging rhesus monkeys: an extension of the concept of maternal style. Am J Primatol. 1990;22:159–69.
Berteaux D, Boutin S. Breeding dispersal in female North American red squirrels. Ecology. 2000;81(5):1311–26. https://doi.org/10.1890/0012-9658(200)081[1331:BDIFNA]2.0.CO;2.
Blaze J, Asok A, Borrelli K, Tulbert C, Bollinger J, Ronca AE, Roth TL. Intrauterine exposure to maternal stress alters *Bdnf IV* DNA methylation and telomere length in the brain of adult rat offspring. Int J Dev Neurosci. 2017;62:56–62.
Blount JD, Vitikainen EIK, Stott I, Cant MA. Oxidative shielding and the cost of reproduction. Biol Rev. 2016;91:483–97.
Boon AK, Réale D, Boutin S. The interaction between personality, offspring fitness and food abundance in North American red squirrels. Ecol Lett. 2007;10:1094–104.
Boon AK, Réale D, Boutin S. Personality, habitat use, and their consequences for survival in North Americanred squirrels Tamiasciurus hudsonicus Oikos 2008;117:1321–1328.
Boonstra R, Krebs CJ, Gilbert B, Schweiger S. Voles and mice. In: Krebs CJ, Boutin S, Boonstra R, editors. Ecosystem dynamics of the boreal forest: the Kluane project. New York: Oxford University Press; 2001. p. 215–39.
Boutin S, Larsen KW. Does food availability affect growth and survival of males and females differently in a promiscuous small mammal, *Tamiasciurus hudsonicus*? J Anim Ecol. 1993;62:364–70. https://doi.org/10.2307/5367.
Boutin S, Tooze Z, Price K. Post-breeding dispersal by female red squirrels (Tamiasciurus hudsonicus): the effect of local vacancies. Behav Ecol. 1993;4(2):151–5. https://doi.org/10.1093/beheco/4.2.151.
Boutin S, Larsen KW, Berteaux D. Anticipatory parental care: acquiring resources for offspring prior to conception. Proc R Soc Lond Ser B Biol Sci. 2000;267(1457):2081–5. https://doi.org/10.1098/rspb.200.1252.
Boutin S, Wauters LA, McAdam AG, Humphries MM, Tosi G, Dhondt AA. Anticipatory reproduction and population growth in seed predators. Science. 2006;314:1928–30. https://doi.org/10.1126/science.1135520.
Brown JL, Morales V, Summers K. A key ecological trait drove the evolution of biparental care and monogamy in an amphibian. Am Nat. 2010;175:436–46.
Brummelte S, Galea LAM. Chronic corticosterone during pregnancy and postpartum affects maternal care, cell proliferation and depressive-like behavior in the dam. Horm Behav. 2010;58:769–79.
Champagne FA, Meane MJ. Stress during gestation alters postpartum maternal care and the development of the offspring in a rodent model. Biol Psychiatry. 2006;59:1227–35.
Champagne FA, Francis DD, Mar A, Meaney MJ. Variations in maternal care in the rat as a mediating influence for the effects of environment on development. Physiol Behav. 2003;79:359–71.
Champagne FA, Curley JP, Keverne EG, Bateson P. Natural variations in postpartum maternal care in inbred and outbred mice. Physiol Behav. 2007;91:325–34.
Charmantier A, Garant D, Kruuk LEB, editors. Quantitative genetics in the wild. Oxford: Oxford University Press; 2014.
Clark MM, Galef BG Jr. A testosterone-mediated trade-off between parental and sexual effort in male Mongolian gerbils (*Meriones unguiculatus*). J Comp Psychol. 1999;113:388–95.
Clutton-Brock TH. The evolution of parental care. Princeton: Princeton University Press; 1991.
Clutton-Brock T. Mammal societies. Wiley; 2016.
Cooper EB, Taylor RW, Kelley AD, Martinig AR, Boutin S, Humphries MM, et al. Personality is correlated with natal dispersal in North American red squirrels (*Tamiasciurus hudsonicus*). Behaviour. 2017;154:939–61. https://doi.org/10.1163/1568539X-00003450.

Cote J, Dreiss A, Clobert J. Social personality trait and fitness. Proc R Soc B Biol Sci. 2008;275:2851–8.

Dantzer B, McAdam AG, Palme R, Fletcher QE, Boutin S, Humphries MM, Boonstra R. Fecal cortisol metabolite levels in free-ranging North American red squirrels: assay validation and the effects of reproductive condition. Gen Comp Endocrinol. 2010;167(2):279–86. https://doi.org/10.1016/j.ygcen.2010.03.024.

Dantzer B, McAdam AG, Palme R, Boutin S, Boonstra R. How does diet affect fecal steroid hormone metabolite concentrations? An experimental examination in red squirrels. Gen Comp Endocrinol. 2011;174(2):124–31. https://doi.org/10.1016/j.ygcen.2011.08.010.

Dantzer B, Boutin S, Humphries MM, McAdam AG. Behavioral responses of territorial red squirrels to natural and experimental variation in population density. Behav Ecol Sociobiol. 2012;66(6):865–78. https://doi.org/10.1007/S00265-012-1335-2.

Dantzer B, Newman AE, Boonstra R, Palme R, Boutin S, Humphries MM, McAdam AG. Density triggers maternal hormones that increase adaptive offspring growth in a wild mammal. Science. 2013;340(6137):1215–7. https://doi.org/10.1126/science.1235765.

Dantzer B, McAdam AG, Humphries MM, Lane JE, Boutin S. Decoupling the effects of food and density on life-history plasticity of wild animals using field experiments: insights from the steward who sits in the shadow of its tail, the North American red squirrel. J Anim Ecol. 2020a;89:2397–414.

Dantzer B, van Kesteren F, Westrick SE, Boutin S, McAdam AG, Lane JE, et al. Maternal glucocorticoids promote offspring growth without inducing oxidative stress or shortening telomeres in wild red squirrels. J Exp Biol. 2020b;223(1) https://doi.org/10.1242/jeb.212373.

Descamps S, Boutin S, McAdam AG, Bearteux D, Gaillard JM. Survival costs of reproduction vary with age in North American red squirrels. Proc R Soc B Biol Sci. 2009;276:1129–35. https://doi.org/10.1098/rspb.2008.1401.

Dingemanse NJ, Réale D. Natural selection and animal personality. Behaviour. 2005;142:1159–84.

Donald JL, Boutin S. Intraspecific cache pilferage by larder-hoarding red squirrels (*Tamiasciurus hudsonicus*). J Mammal. 2011;92(5):1013–20. https://doi.org/10.1644/10-MAMM-A-340.1.

Dwyer CM, Lawrence AB. Maternal behaviour in domestic sheep (*Ovis aries*): constancy and change with maternal experience. Behaviour. 2000;137:1391–413.

Entringer S, Epel ES, Lin J, Buss C, Shahbaba B, Blackburn EH, Simhan HN, Wadhwa PD. Maternal psychosocial stress during pregnancy is associated with newborn leukocyte telomere length. Am J Obstet Gynecol. 2013;208:134–e1.

Fairbanks LA. Early experience and cross-generational continuity of mother-infant contact in vervet monkeys. Dev Psychobiol. 1989;22:669–81.

Fairbanks LA. Individual differences in maternal style: causes and consequences for mothers and offspring. Adv Study Behav. 1996;25:579–611.

Falconer DS, Mackay TFC. Introduction to quantitative genetics. 4th ed. London: Prentice Hall; 1996.

Fisher DN, Boutin S, Dantzer B, Humphries MM, Lane JE, McAdam AG. Multilevel and sex-specific selection on competitive traits in North American red squirrels. Evolution. 2017;71(7):1841–54. https://doi.org/10.1111/evo.13270.

Fisher DN, Haines JA, Boutin S, Dantzer B, Lane JE, Coltman DW, McAdam AG. Indirect effects on fitness between individuals that have never met via an extended phenotype. Ecol Lett. 2019;22(4):697–706. https://doi.org/10.1111/ele.13230.

Fletcher QE, Boutin S, Lane JE, LaMontagne JM, McAdam AG, Krebs CJ, Humphries MM. The functional response of a hoarding seed predator to mast seeding. Ecology. 2010;91(9):2673–83. https://doi.org/10.1890/09-1816.1.

Fletcher QE, Landry-Cuerrier M, Boutin S, McAdam AG, Speakman JR, Humphries MM. Reproductive timing and reliance on hoarded capital resources by lactating red squirrels. Oecologia. 2013;173(4):1203–15. https://doi.org/10.1007/s00442-013-2699-3.

Furness AI, Capellini I. The evolution of parental care diversity in amphibians. Nat Commun. 2019;10:1–12.

Gaidica M, Dantzer B. Quantifying the autonomic response to stressors – one way to expand the definition of "stress" in animals. Integr Comp Biol. 2020;60:113–25.
Gonzalez-Voyer A, Fitzpatrick JL, Kolm N. Sexual selection determines parental care patterns in cichlid fishes. Evolution. 2008;62:2016–26.
Gorrell JC, McAdam AG, Coltman DW, Humphries MM, Boutin S. Adopting kin enhances inclusive fitness in asocial red squirrels. Nat Commun. 2010;1:22.
Gross MR. The evolution of parental care. Q Rev Biol. 2005;80:37–46.
Guillemette CU, Fletcher QE, Boutin S, Hodges RM, McAdam AG, Humphries MM. Lactating red squirrels experiencing high heat load occupy less insulated nests. Biol Lett. 2009;5:166–8.
Guindre-Parker S, McAdam AG, van Kesteren F, Palme R, Boonstra R, Boutin S, Lane JE, Dantzer B. Individual variation in phenotypic plasticity of the stress axis. Biol Lett. 2019;15:20190260.
Gunn MR, Dawson DA, Leviston A, Hartnup K, Davis CS, Strobeck C, Slate JON, Coltman DW. Isolation of 18 polymorphic microsatellite loci from the North American red squirrel, *Tamiasciurus hudsonicus* (Sciuridae, Rodentia), and their cross-utility in other species. Mol Ecol Notes. 2005;5:650–3.
Haines JA, Nason SE, Skurdal AMM, Bourchier T, Boutin S, Taylor RW, McAdam AG, Lane JE, Kelley AD, Humphries MM, Gorrell JC. Sex- and context-specific associations between personality and a measure of fitness but no link with life history traits. Anim Behav. 2020;167:23–39.
Hämäläinen A, McAdam AG, Dantzer B, Lane JE, Haines JA, Humphries MM, Boutin S. Fitness consequences of peak reproductive effort in a resource pulse system. Sci Rep. 2019;7:1–10.
Harris A, Seckl J. Glucocorticoids, prenatal stress and the programming of disease. Horm Behav. 2011;59:279–89.
Harris BN, De Jong TR, Yang V, Saltzman W. Chronic variable stress in fathers alters paternal and social behavior but not pup development in the biparental California mouse (*Peromyscus californicus*). Horm Behav. 2013;64:799–811.
Haussmann MF, Heidinger BJ. Telomere dynamics may link stress exposure and ageing across generations. Biol Lett. 2015;11:20150396.
Haussmann MF, Longenecker AS, Marchetto NM, Juliano SA, Bowden RM. Embryonic exposure to corticosterone modifies the juvenile stress response, oxidative stress and telomere length. Proc R Soc B Biol Sci. 2012;279:1447–56.
Hendrix JG, Fisher DN, Martinig AR, Boutin S, Dantzer B, Lane JE, McAdam AG. Territory acquisition mediates the influence of predators and climate on juvenile red squirrel survival. J Anim Ecol. 2020;89:1408–18.
Hill HM, Greer T, Solangi M, Kuczaj SA II. All mothers are not the same: maternal styles in bottlenose dolphins (*Tursiops truncates*). Int J Comp Psychol. 2007;20:35–54.
Keebaugh AC, Barrett CE, Laprairie JL, Jenkins JJ, Young LJ. RNAi knockdown of oxytocin receptor in the nucleus accumbens inhabits social attachment and parental care in monogamous female prairie voles. Soc Neurosci. 2015;10:561–70.
Kelley AD, Humphries MM, McAdam AG, Boutin S. Changes in wild red squirrel personality across ontogeny: activity and aggression regress toward the mean. Behaviour. 2015;152:1291–306.
Klug H, Bonsall MB. Life history and the evolution of parental care. Evolution. 2010;64:823–35.
Klug H, Alonzo SH, Bonsall MB. Theoretical foundations of parental care. In: Royle NJ, Smiseth PT, Kölliker M, editors. The evolution of parental care. Oxford: Oxford University Press; 2012. p. 21–39.
Kruuk LE. Estimating genetic parameters in natural populations using the 'animal model'. Philosophical Transactions ofthe Royal Society of London. Series B: Biological Sciences, 2004;359(1446):873–890.
Kruuk LEB, Hadfield JD. How to separate genetic and environmental causes of similarity between relatives. J Evol Biol. 2007;20:1890–903.
Lane JE, Boutin S, Gunn MR, Slate J, Coltman DW. Female multiple mating and paternity in free-ranging North American red squirrels. Anim Behav. 2008;75(6):1927–37. https://doi.org/10.1016/j.anbehav.2007.10.038.

Lane JE, Boutin S, Gunn MR, Coltman DW. Sexually selected behaviour: red squirrel males search for reproductive success. J Anim Ecol. 2009;78(2):296–304. https://doi.org/10.1111/j.1365-2656.2008.01502.x.

Lane JE, McAdam AG, Charmantier A, Humphries MM, Coltman DW, Fletcher Q, et al. Post-weaning parental care increases fitness but is not heritable in North American red squirrels. J Evol Biol. 2015;28(6):1203–12. https://doi.org/10.1111/jeb.12633.

Lapiedra O, Schoener TW, Leal M, Losos JB, Kolbe JJ. Predator-driven natural selection on risk-taking behavior in anole lizards. Science. 2018;360:1017–20.

Larivée ML, Boutin S, Speakman JR, McAdam AG, Humphries MM. Associations between over-winter survival and resting metabolic rate in juvenile North American red squirrels. Funct Ecol. 2010;24(3):597–607. https://doi.org/10.1111/j.1365-2435.2009.01680.x.

Larsen KW, Boutin S. Movements, survival, and settlement of red squirrel (*Tamiasciurus hudsonicus*) offspring. Ecology. 1994;75(1):214–23. https://doi.org/10.2307/1939395.

Larsen KW, Boutin S. Exploring territory quality in the North American red squirrel through removal experiments. Can J Zool. 1995;73(6):1115–22. https://doi.org/10.1139/z95-133.

Levin SA. The problem of pattern and scale in ecology. Ecology. 1992;73:1943–67.

Li Y, Lian Z, Wang B, Tai F, Wu R, Hao P, Qiao X. Natural variation in paternal behavior is associated with central estrogen receptor alpha and oxytocin levels. J Comp Physiol A. 2015;201:285–93.

Lamontagne JM, Boutin S. Local-scale synchrony and variability in mast seed production patterns of Picea glauca. Journal of Ecology 2007;95:991–1000.

Maestripieri D. Social and demographic influences on mothering style in pigtail macaques. Ethology. 1998;104:379–85.

Maestripieri D, Mateo JM, editors. Maternal effects in mammals. Chicago: University of Chicago Press; 2009.

Mank JE, Promislow DEL, Avise JC. Phylogenetic perspectives in the evolution of parental care in ray-finned fishes. Evolution. 2005;59:1570–8.

Martin JG, Réale D. Temperament, risk assessment and habituation to novelty in eastern chipmunks, *Tamias striatus*. Anim Behav. 2008;75:309–18.

Martinig AR, McAdam AG, Dantzer B, Lane JE, Coltman DW, Boutin S. The new kid on the block: immigrant males win big whereas females pay fitness cost after dispersal. Ecol Lett. 2020;23:430–8.

Martinig AR, Mathot KJ, Lane JE, Dantzer B, Boutin S. Selective disappearance does not underlie age-related changes in trait repeatability in red squirrels. Behav Ecol. 2021;32:306–15.

McAdam AG, Boutin S. Effects of food abundance on genetic and maternal variation in the growth rate of juvenile red squirrels. J Evol Biol. 2003;16(6):1249–56. https://doi.org/10.1046/j.1420-9101.2003.00630.x.

McAdam AG, Boutin S, Réale D, Berteaux D. Maternal effects and the potential for evolution in a natural population of animals. Evolution. 2002;56:846–51. https://doi.org/10.1111/j.0014-3820.2002.tb01396.x.

McAdam AG, Boutin S, Sykes AK, Humphries MM. Life histories of female red squirrels and their contributions to population growth and lifetime fitness. Ecoscience. 2007;14(3):362. https://doi.org/10.2980/1195-6860(2007)14[362:LHOFRS]2.0.CO;2.

McAdam AG, Garant D, Wilson AJ. The effects of others' genes: maternal and other indirect genetic effects. In: Charmantier A, Garant D, Kruuk LEB, editors. Quantitative genetics in the wild. Oxford: Oxford University Press; 2014. p. 84–103.

McAdam AG, Boutin S, Dantzer B, Lane JE. Seed masting causes fluctuations in optimum litter size and lag load in a seed predator. Am Nat. 2019;194(4):574–89. https://doi.org/10.1086/703743.

McFarlane SE, Lane JE, Taylor RW, Gorrell JC, Coltman DW, Humphries MM, Boutin S, McAdam AG. The heritability of multiple male mating in a promiscuous mammal. Biol Lett. 2011;7:368–71.

McFarlane SE, Gorrell JC, Coltman DW, Humphries MM, Boutin S, McAdam AG. Very low levels of direct additive genetic variance in fitness and fitness components in a red squirrel population. Ecol Evol. 2014;4(10):1729–38. https://doi.org/10.1002/ece3.982.

McLennan D, Armstrong JD, Stewart DC, Mckelvey S, Boner W, Monaghan P, Monaghan P, Metcalfe NB. Interactions between parental traits, environmental harshness and growth rate in determining telomere length in wild juvenile salmon. Mol Ecol. 2016;25:5425–38.

Moore CL, Power KL. Prenatal stress affects mother-infant interaction in Norway rats. Dev Psychobiol. 1986;19:235–45.

Mousseau TA, Fox CW. The adaptive significance of maternal effects. Trends Ecol Evol. 1998;13:403–7.

Noguera JC, da Silva A, Velando A. Egg corticosterone can stimulate telomerase activity and promote longer telomeres during embryo development. Mol Ecol. 2020; https://doi.org/10.1111/mec.15694.

Numan M. Motivational systems and the neural circuitry of maternal behavior in the rat. Dev Psychobiol. 2007;49:12–21.

Numan M, Insel TR. The neurobiology of parental behavior. Springer; 2003.

Numan M, Insel TR. The Neurobiology of Parental Behavior (Vol. 1). Springer Science & Business Media; 2006.

Nussey DH, Pemberton JM, Pilkington JG, Blount JD. Life history correlates of oxidative damage in a free-living mammal population. Funct Ecol. 2009;23:809–17.

Olazábal DE, Young LJ. Oxytocin receptors in the nucleus accumbens facilitate "spontaneous" maternal behavior in adult female prairie voles. Neuroscience. 2006;14:559–68.

Pedersen CA, Boccia ML. Oxytocin antagonism alters rat dams' oral grooming and upright posturing over pups. Physiol Behav. 2003;80:233–41.

Pedersen CA, Caldwell JD, Walker C, Ayers G, Mason GA. Oxytocin activates the postpartum onset of rat maternal behavior in the ventral tegmental and medial preoptic areas. Behav Neurosci. 1994;108:1163.

Price K, Boutin S. Territorial bequeathal by red squirrel mothers. Behavioral Ecology, 1993;4:144–150.

Réale D, McAdam AG, Boutin S, Berteaux D. Genetic and plastic responses of a northern mammal to climate change. Proc R Soc Lond Ser B Biol Sci. 2003;270(1515):591–6. https://doi.org/10.1098/rspb.2002.2224.

Réale D, Reader SM, Sol D, McDougall PT, Dingemanse NJ. Integrating animal temperament within ecology and evolution. Biol Rev. 2007;82:291–318.

Ren T, Boutin S, Humphries MM, Dantzer B, Gorrell JC, Coltman DW, et al. Seasonal, spatial, and maternal effects on gut microbiome in wild red squirrels. Microbiome. 2017;5(1):163. https://doi.org/10.1186/S40168-017-0382-3.

Royle NJ, Smiseth PT, Kölliker M. The evolution of parental care. Oxford: Oxford University Press; 2012.

Schino G, D'Amato FR, Troisi A. Mother-infant relationships in Japanese macaques: sources of inter-individual variation. Anim Behav. 1995;49:151–8.

Schulz KM, Pearson JN, Neeley EW, Berger R, Leonard S, Adams CE, Stevens KE. Maternal stressduring pregnancy causes sex-specific alterations in offspring memory performance, social interactions, indices of anxiety, andbody mass. Physiology & Behavior, 2011;104(2):340–347.

Schulz TM, Bowen WD. The evolution of lactation strategies in pinnipeds: a phylogenetic analysis. Ecol Monogr. 2005;75:159–77.

Seip KM, Morrell JI. Exposure to pups influences the strength of maternal motivation in virgin female rats. Physiol Behav. 2008;95:599–608.

Send TS, Gilles M, Codd V, Wolf I, Bardtke S, Streit F, Strohmaier J, Frank J, Schendel D, Sütterlin MW, Denniff M. Telomere length in newborns is related to maternal stress during pregnancy. Neuropsychopharmacology. 2017;42:2407–13.

Sheriff MJ, Bell A, Boonstra R, Dantzer B, Lavergne SG, McGhee KE, MacLeod KJ, Winandy L, Zimmer C, Love OP. Integrating ecological and evolutionary context in the study of maternal stress. Integr Comp Biol. 2017;57:437–49.

Siracusa E, Morandini M, Boutin S, Humphries MM, Dantzer B, Lane JE, McAdam AG. Red squirrel territorial vocalizations deter intrusions by conspecific rivals. Behaviour. 2017a;154(13–15):1259–73.

Siracusa E, Boutin S, Humphries MM, Gorrell JC, Coltman DW, Dantzer B, Lane JE, McAdam AG. Familiarity with neighbors affects intrusion risk in territorial red squirrels. Anim Behav. 2017b;133:11–20.

Siracusa ER, Wilson DR, Studd EK, Boutin S, Humphries MM, Dantzer B, Lane JE, McAdam AG. North American red squirrels mitigate costs of territory defence through social plasticity. Anim Behav. 2019;151:29–42.

Siracusa ER, Boutin S, Dantzer B, Lane JE, Coltman DW, McAdam AG. Familiar neighbors, but not relatives, enhance fitness in a territorial mammal. Curr Biol. 2021;31:1–8.

Skibiel AL, Downing LM, Orr TJ, Hood WR. The evolution of the nutrient composition of mammalian milks. J Anim Ecol. 2013;82:1254–64.

Smith CC. The adaptive nature of social organization in the genus of tree squirrels *Tamiasciurus*. Ecol Monogr. 1968;38:31–64.

Smith CC. Structure and function of the vocalizations of tree squirrels (*Tamiasciurus*). J Mammal. 1978;59:793–808.

Smith SM, Nager RG, Costantini D. Meta-analysis indicates that oxidative stress is both a constraint on and a cost of growth. Ecol Evol. 2016;6:2833–42.

Spinka M, Illmann G, de Jonge F, Andersson M, Schuurmann T, Jensen P. Dimensions of maternal behaviour characteristics in domestic and wild X domestic crossbred sows. Appl Anim Behav Sci. 2000;70:99–114.

Stewart FE, McAdam AG. Seasonal plasticity of maternal behaviour in *Peromyscus maniculatus*. Behaviour. 2014;151:1641–62.

Stuart-Smith AK, Boutin S. Behavioural differences between surviving and depredated juvenile red squirrels. Ecoscience. 1995;2(1):34–40. https://doi.org/10.1080/11956860.1995.11682266.

Studd EK, Boutin S, McAdam AG, Krebs CJ, Humphries MM. Predators, energetics and fitness drive neonatal reproductive failure in red squirrels. J Anim Ecol. 2015;84:249–59.

Studd EK, Landry-Cuerrier M, Menzies AK, Boutin S, McAdam AG, Lane JE, Humphries MM. Behavioral classification of low-frequency acceleration and temperature data from a free-ranging small mammal. Ecol Evol. 2019;9:619–30.

Svendsen GE, Armitage KB. Mirror-image stimulation applied to field behavioral studies. Ecology. 1973;54:623–7.

Salari AA, Fatehi-Gharehlar L, Motayagheni N, Homberg JR. Fluoxetine normalizes the effects of prenatalmaternal stress on depression-and anxiety-like behaviors in mouse dams and male offspring. Behavioural Brain Research, 2016;311:354–367.

Tanaka I. Variability in the development of mother-infant relationships among free-ranging Japanese macaques. Primates. 1989;30:477–91.

Taylor RW, Boon AK, Dantzer B, Réale D, Humphries MM, Boutin S, Gorrell JC, Coltman DW, McAdam AG. Low heritabilities, but genetic and maternal correlations between red squirrel behaviours. J Evol Biol. 2012;25:614–24.

Taylor RW, Boutin S, Humphries MM, McAdam AG. Selection on female behaviour fluctuates with offspring environment. J Evol Biol. 2014;27:2308–21.

Trainor BC, Marler CA. Testosterone promotes paternal behaviour in a monogamous mammal via conversion to oestrogen. Proc R Soc Lond Ser B Biol Sci. 2002;269:823–9.

van Kesteren F, Delehanty B, Westrick SE, Palme R, Boonstra R, Lane JE, et al. Experimental increases in glucocorticoids alter function of the HPA axis in wild red squirrels without negatively impacting survival and reproduction. Physiol Biochem Zool. 2019;92:445–58.

van Noordwijk AJ, de Jong G. Acquisition and allocation of resources: their influence on variation in life history tactics. Am Nat. 1986;128:137–42.

Walsh RN, Cummins RA. The open-field test: a critical review. Psychol Bull. 1976;83:482.

Weinstock M. The long-term behavioural consequences of prenatal stress. Neurosci Biobehav Rev. 2008;32:1073–86.

Weinstock M, Matilina E, Maor GI, Rosen H, McEwen BS. Prenatal stress selectively alters the reactivity of the hypothalamic-pituitary-adrenal system in the female rat. Brain Res. 1992;595:195–200.

Westrick SE, Taylor RW, Boutin S, Lane JE, McAdam AG, Dantzer B. Attentive red squirrel mothers have faster growing pups and higher lifetime reproductive success. Behav Ecol Sociobiol. 2020;74:1–13.

Westrick SE, van Kesteren F, Boutin S, Lane JE, McAdam AG, Dantzer B. Maternal glucocorticoids have minimal effects on HPA axis activity and behavior of juvenile wild North American red squirrels. J Exp Biol. 2021;224:jeb236620.

Wilcoxon JS, Redei EE. Maternal glucocorticoid deficit affects hypothalamic-pituitary-adrenal function and behavior of rate offspring. Horm Behav. 2007;51:321–7.

Williams CT, Wilsterman K, Kelley AD, Breton AR, Stark H, Humphries MM, McAdam AG, Barnes BM, Boutin S, Buck CL. Light loggers reveal weather-driven changes in the daily activity patterns of arboreal and semifossorial rodents. J Mammal. 2014a;95:1230–9.

Williams CT, Lane JE, Humphries MM, McAdam AG, Boutin S. Reproductive phenology of a food-hoarding mast-seed consumer: resource-and density-dependent benefits of early breeding in red squirrels. Oecologia. 2014b;174(3):777–88. https://doi.org/10.1007/S00442-013-2826-1.

Wolf JB, Wade MJ. What are maternal effects (and what are they not)? Philos Trans R Soc Lond Ser B Biol Sci. 2009;364:1107–15.

Index

G. González-Mariscal (ed.), *Patterns of Parental Behavior*, Advances in Neurobiology 27, https://doi.org/10.1007/978-3-030-97762-7

D

E

F

S

The manufacturer's authorised representative in the EU is Springer Nature Customer Service Centre GmbH, Europaplatz 3, 69115 Heidelberg, Germany. If you have any concerns regarding our products, please contact ProductSafety@springernature.com

Printed and bound by CPI Group (UK) Ltd, Croydon, CR0 4YY
28/07/2026
02179206-0004